ATOMIC PHYSICS AT ACCELERATORS: STORED PARTICLES AND FUNDAMENTAL PHYSICS

ATOMIC PHYSICS AT ACCELERATORS: STORED PARTICLES AND FUNDAMENTAL PHYSICS

Proceedings of the APAC 2001,
held in Aarhus, Denmark,
8–13 September 2001

Edited by

HELGE KNUDSEN

Department of Physics and Astronomy, Aarhus University, Denmark

JENS ULRIK ANDERSEN

Department of Physics and Astronomy, Aarhus University, Denmark

and

HEINZ-JÜRGEN KLUGE

GSI, Darmstadt, Germany

Reprinted from *Hyperfine Interactions*
Volume 146, Nos. 1–4 (2003)
Volume 147, Nos. 1–4 (2003)

SPRINGER SCIENCE+BUSINESS MEDIA, B.V.

A C.I.P. Catalogue record for this book is available from the Library of Congress.

ISBN 978-94-010-3749-5 ISBN 978-94-007-0946-1 (eBook)
DOI 10.1007/978-94-007-0946-1

Printed on acid-free paper

Table of Contents

4. Atomic lifetimes and spectra

5. Ionization, fragmentation and electron capture

6. Cooling of beams in traps and storage rings

7. New developments in charged particle traps and storage rings

8. Investigations relevant for Astrophysics

9. Clusters, biomolecules, aerosols and nanoparticles

10. Atomic physics with antiparticles

Hyperfine Interactions **146/147:** 1–3, 2003.

Preface

This volume contains the proceedings of the third Euroconference on *Atomic Physics at Accelerators* (APAC 2001), with the title *Stored Particles and Fundamental Physics*. It was held in Aarhus, Denmark, from September 8 to 13 at the Marselis Hotel located near the beach and the Marselis Woods outside Aarhus, but some of the activities took place at the Department of Physics, University of Aarhus. The conference was sponsored by the Commission of the European Union (Contract No. ERBFMMACT980469) and also by the Danish Research Foundation through ACAP (**A**arhus **C**enter for **A**tomic **P**hysics).

The meeting was focused on the application of storage rings for atomic physics, and there are two fairly small rings in Aarhus, ASTRID (**A**arhus **ST**orage **R**ing for **I**ons, **D**enmark) and ELISA (**EL**ectrostatic **I**on **S**torage ring, **A**arhus). The research at these rings has contributed to the strong position of European Science in this field. Both rings are designed according to unique concepts. ASTRID is a dual-purpose ring, which half the time stores electrons for the generation of low-energy synchrotron radiation. The storage of negative particles has also been a unique feature for the application of ASTRID as an ion storage ring.

The end of 2001 marked the end of the careers of two central figures in the field of atomic physics at accelerators, Prof. Torkild Andersen and Prof. Erik Uggerhøj, who were both very active in arranging the APAC 2001 conference. Torkild Andersen has a long record of research using accelerators to study basic atomic physics, going back to beam-foil spectroscopy in the 60's. He has been a central figure in ACAP with research projects on mainly laser spectroscopy. His very accurate results for the binding energy of negative atomic ions have become benchmarks for the treatment of electron–electron correlation in the theoretical description of atoms. The photo shows Torkild Andersen in front of the undulator beam line for ionization of ions with synchrotron radiation, another example of his very successful projects. At the Department of Physics his great concern for the students will be remembered, and he became a father figure for a generation of atomic physicists graduated from the University of Aarhus.

The other photo shows Erik Uggerhøj in a characteristic situation, celebrating a new advance in the development of the underground laboratory for storage ring physics. He was the initiator of the ASTRID project and the dynamic director of the organization ISA (**I**nstitute for **S**torage Ring Facilities, **A**arhus), which develops and operates the rings. He took a keen interest in developing new cross-disciplinary

research with synchrotron radiation at ASTRID, for example, the X-ray microscope which is one of the few facilities of its kind in the world. As a physicist he will mainly be remembered for his pioneering experimental research on particle channeling in crystals. For many years he used high-energy particle beams at CERN for channeling experiments, in particular on radiation and pair production by electrons and positrons. His close connection to CERN became a great asset when ASTRID was designed and built.

The APAC conference in Aarhus was very lively, with presentation and discussion of many interesting new results. However, the scientific program was overshadowed by the momentous events on September 11. Many participants had relatives or friends in New York and were greatly concerned about their fate, and others suffered great difficulties in arranging their flights home.

On a smaller scale, we have experienced many difficulties in producing the proceedings from the conference. The conference chairman, Jeffrey Hangst, went on a leave of absence at CERN the following year to work on the formation of anti-hydrogen. This project has had great success since then but unfortunately the work on the proceedings suffered. In the end we had to transfer the responsibility to Aarhus and an unreasonable delay of the publication resulted, for which we apologize. The main burden of work has been carried by the conference secretary, Kate Andersen, assisted by Prof. Helge Knudsen. We thank the contributors for their cooperation and hope that the end result will be appreciated.

JENS ULRIK ANDERSEN
Physics Professor at the University of Aarhus and Director of ACAP

Hyperfine Interactions **146/147:** 5–12, 2003.

Recombination Measurements at Ion Storage Rings

ANDREAS WOLF and GERALD GWINNER
Max-Planck-Institut für Kernphysik, Heidelberg, Germany

Abstract. An overview of recombination measurements is given, focusing on experiments with multicharged ions and electrons at relative energies in the millielectronvolt range. High-resolution studies of near-threshold dielectronic resonances and effects of the long-range dynamics of slow electrons near multicharged ions, as revealed by merged beams experiments at very well matched velocities, are discussed.

1. Introduction

The approach of an electron to a positive ion, even at very low velocity, can initiate a large variety of processes. The simple emission of a photon, allowing the incident electron to find itself in a bound level, is known as (non-resonant) radiative recombination (RR). The impact excitation of an electron already bound to the ion by the incident electron can leave both these particles in bound orbitals, producing an autoionizing resonance embedded in the continuum; if radiative emission follows such a capture process, dielectronic recombination (DR) occurs, for which the cross section at the resonance can be orders of magnitude larger than the non-resonant RR cross section. Finally, on molecular ions, the capture of an electron under simultaneous impact excitation of the ionic core produces similar resonances, generally embedded in the ionization and the dissociation continua at the same time. In many cases the neutralized molecular system finds itself on a highly excited, dissociating electronic potential surface and readily decays into neutral fragments. This 'dissociative recombination' can proceed without any radiative emission and mostly leads to much larger recombination cross sections for molecular than for atomic ions.

The pathways to the formation of neutral matter hence reflect many aspects of the inner dynamics of highly excited atomic and molecular systems, and recombination experiments are becoming increasingly sensitive probes for such dynamics. Clean access to recombination reactions on a single-event basis is provided by collinear, merged electron and ion beams at matched or slightly detuned average velocities. With particular efficiency this technique can be used within storage rings where an ion beam is kept under well controlled conditions for extended storage times (seconds up to hours) and is recycled through the electron interaction region at a high rate (typically 10^5–10^6 s^{-1}). The main advantages of ion storage rings for recombination measurements are a good definition of the interaction conditions, in particular with regard to the collision energy and the internal state of the positive ion, an efficient detection of the recombination products, and large count rates at low background.

Recombination of atomic ions has been studied at ion storage rings for a variety of systems at a wide range of interaction energies from the meV to the keV range [1]. In the present short account we will outline the progress of studies in the milli-electronvolt energy range. Measured recombination rates at these energies can reflect in much detail the interaction of an outer electron with the atomic ion core in the presence of near-threshold autoionizing configurations. In addition, new binding mechanisms, apparently connected with the long-range dynamics of slow electrons in the attractive Coulomb potential under the possible influence of small external perturbations, are found to have effect on the recombination rates of multicharged ions at very low relative energies. Aspects of molecular recombination studies are covered elsewhere at this conference [2].

2. Outline of experimental procedures for low-energy recombination measurements

Collision energies down to a few meV can be realized in a controlled way in present storage-ring merged beams experiments, chosing small velocity detunings between the ion beam and a superimposed magnetically guided electron beam. The velocity of the stored ions is well defined in such experiments by electron cooling, which in fact uses the same merged electron beam that is also applied for the recombination measurements. Electron cooling brings the ions to exactly the same average velocity as the electrons by a friction force which tends to minimize the velocity differences between these particles on a time scale of typically 0.1 s (for ions with a charge-to-mass ratio $q/A \sim 0.5$). For recombination measurements, the relative electron–ion velocity is varied by changing the acceleration voltage of the electron beam, making sure that data are taken before the friction force between the two beams changes the ion velocity by any relevant amount. The electron energy corresponding to the longitudinal beam-velocity detuning defines the 'detuning energy' E; for typical ion and electron energies of ~5 MeV/u and 3000 eV, respectively, detuning energies of $E < 0.1$ meV can be realized in a controlled way using laboratory electron energy changes of a few eV. The collision energies are directly given by the detuning energy plus contributions from the velocity spreads of the beams, as defined by 'thermal' energies kT in co-moving reference frames. The collision energy spread due to these additional contributions varies from ~3–30 meV (given mainly by the transverse electron temperature $kT_\perp$) for small detuning energies ($E \lesssim 0.3$ eV), up to a few eV at $E \gtrsim 500$ eV. For zero detuning, the collision velocity becomes much smaller in longitudinal than in transverse direction; the corresponding longitudinal temperature $kT_\parallel$ can be as low as 0.05 meV. This reflects the small longitudinal temperatures that can be reached in magnetically guided electron beams; typical magnitudes of the longitudinal magnetic field used to guide the electrons are ~0.03–0.1 T.

The recombination processes are observed by direct counting of the recombination products, which are separated from the stored ion beam at the next storage-ring bending magnet following the electron interaction zone. On each passage, ions are merged with electrons for ~50 ns and recombination products are detected after a flight time of ~0.3 μs. Also recombination products formed in excited states are detected, and most measurements so far have not used any identification of the final states; however, field ionization in the storage-ring magnets prevents the detection of excited product ions which are bound by less than ~0.3 eV under typical conditions. The ion storage rings which are being used for recombination experiments on atomic ions are CRYRING at the University of Stockholm (Sweden), ESR at the GSI, Darmstadt (Germany), and TSR at the Max-Planck Institute for Nuclear Physics, Heidelberg (Germany) [3].

3. Splitting patterns of near-threshold autoionizing configurations

Depending on the initial electronic structure of the positive ions, the cross section of the dielectronic recombination process $A^{q+} + e \rightarrow [A^{(q-1)+}]^{**} \rightarrow A^{(q-1)+} + h\nu$ reveals the energetic positions and intrinsic properties of doubly excited electronic states $[A^{(q-1)+}]^{**}$ of the recombined system. In particular, the high energy resolution inherent to merged beams experiments at low relative energy makes it possible to focus on the detailed splitting patterns and absolute energetic positions of doubly excited configurations. The information from such experiments has recently provided a sensitive test of relativistic many-body calculations on Be-like autoionizing levels $1s^2\,2pnl$. Levels $1s^2\,2p4l$ of C^{2+}, with the lowest resonance at ~180 meV, were studied by DR of C^{3+} ions [4, 5] and levels $1s^2\,2p6l$ of F^{5+}, reaching down to only a few meV incident electron energy, could recently be accessed by DR of F^{6+} ions [6]. For the Be-like system F^{5+} the storage-ring results stimulated a highly accurate theoretical study [6] employing relativistic many-body methods in combination with complex rotation for the description of resonances. Experimental and theoretical data for this case are shown in Figure 1. By the presence of several broad resonances (width of order 10 meV) the photorecombination cross is increased with respect to the non-resonant level (from photon emission directly by the incoming electron) by a factor of ~3–4 at all energies up to ~600 meV, with a correspondingly large influence of resonant contributions on the thermal recombination rate coefficient of F^{6+} in the temperature range of several 1000 K. The calculations include significant amounts of correlation energy between the $2p$ and $6l$ electrons (of order 10 meV for several resonances) [6] and an *ab initio* value of the $2s$–$2p_j$ splitting of the Li-like F^{6+} core accurate to ~1 meV. A $J = 1$ resonance (position 6.9 meV, width 25.8 meV) is actually overlapping the ionization threshold and causes most of the rise of the calculated rate at the lowest energies. In addition, a narrow resonance, broadened by the experimental velocity spread, dominates the

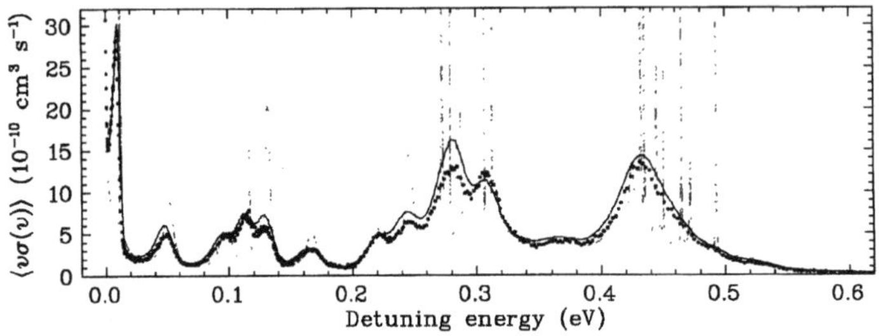

Figure 1. Near-threshold recombination spectrum of F^{6+} measured at the TSR [dots; error of the absolute rate coefficient (energy) scale $\pm 15\%$ ($\pm 3\%$)] compared to relativistic many-body calculations of the photorecombination rate including all F^{5+} $1s^2\,2p6l$ autoionizing resonances [6]; the gray and the black solid lines represent the theoretical result $v\sigma(v)$ before and after averaging over the experimental velocity distribution, respectively.

observed low-energy rate coefficient (experimental position 10 ± 0.3 meV, theoretically predicted at 11.4 meV with 0.1 meV width); the observed shape of this resonance provides a useful *in situ* diagnostics of the experimental collision energy spreads $kT_\perp$ and $kT_\parallel$ [7].

While for Be-like states an impressive level of agreement between theory and experiment is reached, theory still fails for more complex electronic structures and near-threshold dielectronic resonances observed with meV accuracies still call for improved many-body calculations. Experiments with detailed theoretical comparisons have so far been performed on a selected number of systems, aiming in particular at broad resonant structures in Ar-like ions (Ti^{4+}, Sc^{3+}) [8, 9] as well as at QED effects in Cu-like Pb ions (Pb^{53+}) [10].

4. Merged-beams recombination rates of multicharged ions at very low relative energy

For atomic ions and relative electron energies in the meV range, non-resonant radiative recombination (RR) is the usual reaction mechanism, apart from dielectronic resonances coincidentally placed near the ionization threshold. For bare ions in particular, the measured rates increase smoothly with decreasing detuning energy E and one should expect the radiative recombination rates to be safely predictable by elementary quantum mechanical theory. Indeed, at detuning energies E above a few meV the experimental results are well reproduced by calculations which apply a straightforward averaging of the elementary quantum mechanical photorecombination cross section over the distribution of the collision velocities. For C^{6+} ions, Figure 2 demonstrates such an agreement at energies in the range $E \gtrsim 3$ meV. The lower limit of this range lies already well below the transverse electron energy spread ($kT_\perp = 10$ meV in this example) so that, after averaging over the collision velocities, only a slight further increase of the recombination rate is predicted if E is lowered even further (shaded area in Figure 2). In the

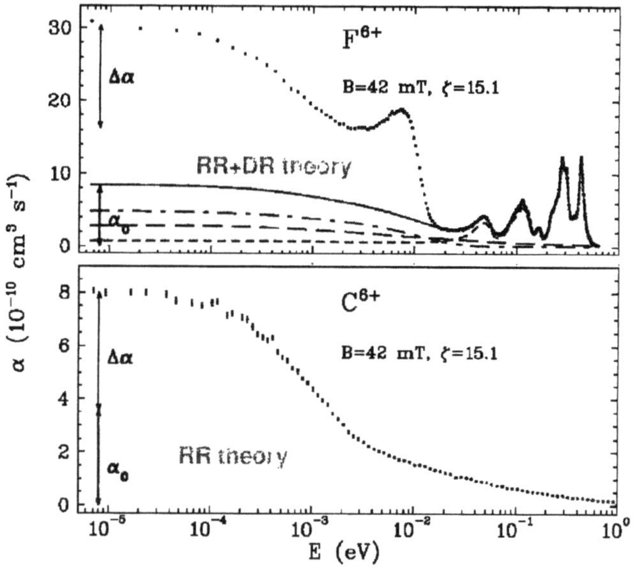

Figure 2. Recombination rate coefficient at low detuning energies E measured for F^{6+} and C^{6+} ions at the ion storage ring TSR [7]. For F^{6+} non-resonant and resonant contributions to the calculated rate are indicated separate lines (see text for definition of α_0).

experiment, however, an additional increase of the recombination rate sets in at ~3 meV and for the smallest detuning energies ($E \lesssim 0.1$ meV) the rates end up at values which are higher than the straightforward predictions (α_0) by a factor of 2.3.

Similar excess rates $\Delta\alpha$ with enhancement factors $1 + \Delta\alpha/\alpha_0$ in the typical range of 1.5–3 were clearly identified in the data from a large number experiments on multicharged ions. The relative enhancement $\Delta\alpha/\alpha_0$ at various laboratories was found to increase with the ion charge q [11], ranging from small values $\Delta\alpha/\alpha_0 \ll 1$ for $q = 1$ via $\Delta\alpha/\alpha_0 \sim 1.2$ for $q = 6$ up to $\Delta\alpha/\alpha_0 \sim 3.8$ for $q = 92$ [12], with an approximate scaling of $\Delta\alpha/\alpha_0 \propto q^{0.6}$. The onset energy of the enhancement (~3 meV for $q = 6$) also tends to increase with increasing q, reaching ~20 meV for $q = 92$ [12]. The measurements shown in Figure 2 compare the bare ion C^{6+} with Li-like F^{6+}. An enhancement with the same onset energy and the same overall trend is observed in both cases. A reference recombination rate α_0 was defined for F^{6+} by considering not only direct RR, but also all resonances that significantly contribute to the photorecombination cross section at the ionization threshold. This includes the broad $J = 1$ resonance at 6.9 meV (cf. Section 3), but not the narrow resonance at 11.4 meV, and increases α_0 by a factor of 2.4 as compared to its value for C^{6+}, while the enhancement $\Delta\alpha$ increases by a similar factor (~3) with respect to C^{6+}. In this sense, the resonant contributions seem [7] to be subject to a similar relative enhancement as non-resonant RR.

5. Long-range dynamics of slow electrons near multicharged ions

A recent study [7] of the excess recombination rate under a controlled variation of experimental parameters has given support to the interpretation that dynamics of the electron motion on a 'mesoscopic' length scale in the vicinity of an ion is at the origin of the effect. In essence, the excess rate is found to depend sensitively on the spread of relative collision velocities in *longitudinal* direction (as expressed by $T_{\|}$), which corresponds to motion along the magnetic guiding field in the interaction region. In the averaging of an isotropic (field free) recombination cross section over the distribution of collision velocities, the effect of any variations of $T_{\|}$ should be minor as $kT_{\|} \lesssim (1/10)kT_{\perp}$. However, a considerable dependence of $\Delta\alpha$ on $T_{\|}$ and also on the field strength B of the magnetic guiding field is observed (see Figure 3).

Length scales relevant for the interpretation of the experimental results are obtained by considering the distance of closest approach between an electron and an ion that must occur in order to form the final states experimentally observed. The states bound by more than ~0.3 eV which are observed on the recombination detector correspond to principal quantum numbers of roughly $n \lesssim 6.7q$ and the size of these states is roughly $45qa_0 \lesssim 2.4$ nm·q; the incident electron must approach the ion to this distance for radiative stabilization into the one of the observable final levels. In contrast, the average radius r_c of the cycloidal electron motion in the magnetic field is of the order of 5 μm (B = 40 mT, $kT_{\perp}$ = 5 meV), i.e., larger by at least 10^3. Electron motion on the length scale of the cycloidal radius r_c is expected to be influenced by both the longitudinal temperature $T_{\|}$ and the magnetic field strength, while no such dependence should exist on the length scale corresponding to efficient radiative stabilization. On the latter scale, the electron motion is expected to be governed by $T_{\perp}$ for low detuning energies $E < kT_{\perp}$. The

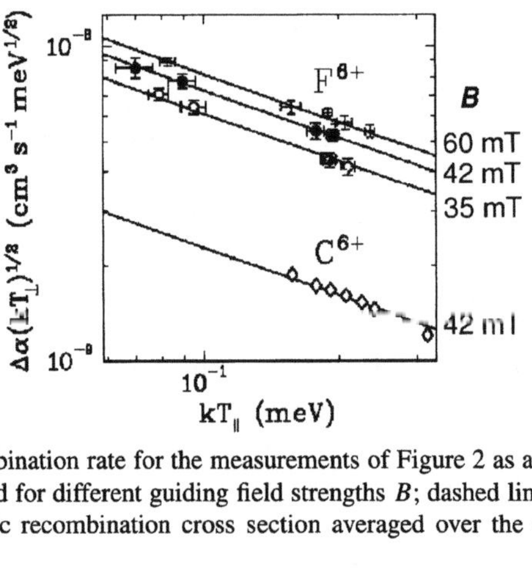

Figure 3. Excess recombination rate for the measurements of Figure 2 as a function of the longitudinal temperature [7] and for different guiding field strengths B; dashed lines indicate the expected variation for an isotropic recombination cross section averaged over the distribution of collision velocities.

observed dependence of the $\Delta\alpha$ on $T_{\parallel}$ and B therefore indicates the importance of electron dynamics on the μm length scale, and a plausible explanation for the observed excess recombination rate into bound states with ~0.3 eV could be an increase of the average electron density near an ion due to such dynamics [7].

A few simple considerations suggest that 'magnetized' electron motion at a low *longitudinal* relative velocity could indeed lead to an increased average electron density near a positive ion. Firstly, for an electron moving at a typical distance of $d \sim r_c$ from the ion, the electrostatic interaction energy $qe^2/4\pi\varepsilon_0 d$ is about q times as large as the longitudinal thermal energy $kT_{\parallel}$; hence, the field of the ion at the typical distance is strong enough to significantly alter the longitudinal electron motion. The interaction with the ion can in particular lead to an exchange of energy between the longitudinal and the transverse degrees of freedom; a small increase of the transverse and a corresponding decrease of the longitudinal energy of an electron during a distant encounter with an ion can therefore lead to a longitudinally bound motion which, although it does not represent a stable binding, tends to keep the electron in the vicinity of an ion for extended time as compared to a similar encounter without external field. Secondly, it is interesting to note that, at the given electron density $n_e \sim 0.5$–3×10^7 cm^{-3}, the average distance from any point to the next electron is ~30 μm (a value close to r_c). Hence, one or even several electrons are (on average) strongly coupled *at each time* to an ion, comparing again the electrostatic interaction energy to the longitudinal kinetic energy of $\sim kT_{\parallel}$. This illustrates that the 'turn on' of the electron–ion interaction in the time-domain, i.e. the merging between the electron and the ion beam, may also have a considerable influence on the long-range dynamics, with the possibility of forming long-range bound states.

In both cases discussed, the magnetic field present in the experimental setup appears to be important, giving a special role to the much lower longitudinal velocity spread and making it possible to consider binding with respect to the longitudinal degree of freedom, even at considerable transverse energy, in a system that might be described as a 'magnetic atom'. Several attempts to describe in more detail the long-range electron dynamics, both in a steady-state situation and after a 'turn-on' of the electron-ion interaction, are underway [13–16].

Acknowledgements

We thank our collaborators in the low-energy recombination experiments, A. Müller, S. Schippers, S. Kieslich, A. Hoffknecht, T. Bartsch, C. Brandau and W. Shi, University of Giessen, P. Glans, R. Schuch and P. Eklöw, University of Stockholm, as well as M. Tokman and E. Lindroth on the theory side. Special thanks are due to D. Schwalm for his support, advise, and collaboration. The efficient operation of the TSR by M. Grieser and M. Beutelspacher is gratefully acknowledged.

References

1. Schippers, S., *Phys. Scripta* **T80** (1999), 158.
2. Zajfman, D., Schwalm, D. and Wolf, A., this volume.
3. For a more complete overview of storage ring facilities and their experimental program see, e.g., Wolf, A., In: Beyer, H. F. and Shevelko, V. P. (eds), *Atomic Physics with Highly Charged Ions*, Springer, Berlin, 1999, chapter I.
4. Mannervik, S., deWitt, D., Engström, L., Lindroth, E., Schuch, R. and Zong, W., *Phys. Rev. Lett.* **81** (1998), 313.
5. Schippers, S., Müller, A., Gwinner, G., Linkemann, J., Saghiri, A. A. and Wolf, A., *Astrophys. J.* **555** (2001), 1027.
6. Tokman, M., Eklöw, N., Glans, P., Lindroth, E., Schuch, R., Gwinner, G., Schwalm, D., Wolf, A., Hoffknecht, A., Müller, A. and Schippers, S., *Phys. Rev. A* **66** (2002), 012703.
7. Gwinner, G., Hoffknecht, A., Bartsch, T., Beutelspacher, M., Eklöw, N., Glans, P., Grieser, M., Krohn, S., Lindroth, E., Müller, A., Saghiri, A. A., Schippers, S., Schramm, U., Schwalm, D., Tokman, M., Wissler, G. and Wolf, A., *Phys. Rev. Lett.* **84** (2000), 4822.
8. Schippers, S., Bartsch, T., Brandau, C., Linkemann, J., Gwinner, G., Müller, A., Saghiri, A. A. and Wolf, A., *J. Phys. B* **31** (1998) 4873.
9. Schippers, S. *et al.*, *Phys. Rev. A* **65** (2002), 042723.
10. Lindroth, E., Danared, H., Glans, P., Pesic, Z., Tokman, M., Vikor, G. and Schuch, R., *Phys. Rev. Lett.* **86** (2001), 5027.
11. Gao, H., DeWitt, D. R., Schuch, R., Zong, W., Asp, S. and Pajek, M., *Phys. Rev. Lett.* **75** (1995), 4381.
12. Shi, W., Bartsch, T., Böhm, S., Böhme, C., Brandau, C., Hoffknecht, A., Kieslich, S., Knopp, H., Schippers, S., Müller, A., Kozhuharov, C., Bosch, F., Franzke, B., Mokler, P. H., Steck, M., Stöhlker, Th. and Stachura, Z., In: Datz, S., Bannister, M. E., Krause, H. F., Saddiq, L. H., Schultz, D. R. and Vane, C. R. (eds), *XXII ICPEAC Abstracts of Contributed Papers*, Rinton Press, Princeton, 2001, p. 328.
13. Eichler, J., Yoshihama, Y., Toshima, N., *Phys. Rev. A* **65** (2002), 033404.
14. Hoerndl, M. *et al.*, this volume.
15. Heerlein, C. *et al.*, this volume.
16. Gwinner, G., Schwalm, D. and Wolf, A., in same Proceedings as Ref. [12], p. 326.

Hyperfine Interactions **146/147:** 13–17, 2003.

Enhancement of Low-Energy Electron Ion Recombination in a Magnetic Field

MARIA HÖRNDL[1], SHUHEI YOSHIDA[1], KAROLY TÖKÉSI[2] and JOACHIM BURGDÖRFER[1]
[1]*Institute for Theoretical Physics, Vienna University of Technology, Vienna, Austria*
[2]*Institute of Nuclear Research of the Hungarian Academy of Sciences, Debrecen, Hungary*

Abstract. Electron–ion recombination in *cold* magnetized plasmas shows a dramatic enhancement of the radiative recombination rate for bare highly charged ions relative to what standard radiative recombination rates predict. To understand the mechanism of this enhancement we investigate the classical chaotic dynamics of an electron in the combined Coulomb field of the ion and the magnetic field in the electron cooler. An increased flux of electrons *visiting* the vicinity of the target ion leads to an enhancement of the recombination process.

Key words: radiative recombination, non-linear dynamics, electron cooler.

1. Introduction

Electron–ion recombination is an important process in many areas of physics such as astrophysics [1], plasma physics [2] and accelerator physics. The recombination is a significant ion loss mechanism during electron cooling in a storage ring. Only with the recent advent of heavy-ion storage rings such as CRYRING (Sweden), ASTRID (Denmark), TSR and ESR (Germany), the study of the recombination at low temperatures under well-controlled conditions has become possible. One of the major surprises found in these experiments [4–8] is a dramatic enhancement of the radiative recombination rate for highly charged ions with low energy electrons relative to what standard theory predicts [3].

Several proposals have been put forward [8–11] to explain this surprising discrepancy with theory for one of the supposedly well understood elementary quantum-electrodynamic processes. They range from the influence of three-body recombination and density enhancement due to plasma screening effects to transient electric field induced recombination. None of these models can, so far, account quantitatively for the observed enhancement. In the following, we discuss an alternative model which focuses on the classical chaotic dynamics of the electron in the combined Coulomb field of the ion and the magnetic guiding field in the electron cooler. The phase space structure of this irregular scattering problem is such that generalized deflection functions, specifically the visit function to be discussed below, are dramatically changed compared to the pure Coulomb field. As a result, the net flux of electrons towards the immediate vicinity of the ion is enhanced. A first comparison of this theoretical approach with experimental data shows promising agreement.

2. Modification of the standard theory: Influence of chaotic dynamics

The cross section for radiative recombination (RR)

$$e^- + A^{Z+} \rightarrow A^{(Z-1)+} + h\nu, \tag{1}$$

where a free electron is captured into a bound state of an ion with simultaneous emission of a photon can be approximated for bare ions by the Kramers formula [3]

$$\sigma^{RR}(n, E_e) = \left(2.1 \times 10^{-22}\ \mathrm{cm}^2\right) G_n(E_e) \frac{Z^4 E_0^2}{n E_e (Z^2 E_0 + n^2 E_e)}. \tag{2}$$

Z is the charge of the ion, $E_0 = 13.6$ eV and the Gaunt factors $G_n(E_e)$ [12] account for deviations from the correct quantum result at low n and high E_e. The total recombination cross section is obtained by summing up the contributions from all possible final states, $\sigma^{RR}_{\mathrm{total}}(E_e) = \sum_{n=1}^{n_{\max}} \sigma^{RR}(n, E_e)$. $n_{\max}$ is determined by field ionization of states in the motional electric field of the bending magnet downstream from the electron cooler.

The measured recombination rate can be evaluated as

$$\alpha_0(v_{\mathrm{rel}}) = \left\langle v \sigma^{RR}_{\mathrm{total}} \right\rangle = \int \sigma^{RR}_{\mathrm{total}} v f(\vec{v}, v_{\mathrm{rel}})\, \mathrm{d}\vec{v}. \tag{3}$$

In the experiment, the velocity distribution $f(\vec{v}, v_{\mathrm{rel}})$ is highly anisotropic and characterized by two temperatures, $T_\perp$ and $T_\parallel$ ($T_\perp \gg T_\parallel$), which correspond to the transverse ($v_\perp$) and the longitudinal ($v_\parallel$) velocity distributions relative to the direction of the magnetic field. The most frequently used distribution is a product of two Maxwell distributions

$$f(\vec{v}, v_{\mathrm{rel}}) = \frac{1}{2\pi k_B T_\perp} \mathrm{e}^{-v_\perp^2/2k_B T_\perp} \sqrt{\frac{1}{2\pi k_B T_\parallel}} \mathrm{e}^{-(v_\parallel - v_{\mathrm{rel}})^2/2k_B T_\parallel}, \tag{4}$$

where v_{rel} is the detuning velocity between the merged ion and electron beam which defines the relative energy.

Under the influence of a magnetic field, the wavefunctions used to evaluate the recombination cross sections σ^{RR} need to be modified. The difficulty in determining the scattering wavefunction accurately comes from the fact that the system involves vastly different length scales, i.e. of the order of 10 a.u. for the Bohr radius of the final state and 10^5 a.u. for the cyclotron radius. Therefore, it is not trivial to evaluate σ^{RR} non-perturbatively. However, considering the fact that the magnetic length is much shorter than the cyclotron radius, the electron is, on the average, in high Landau quantum numbers for which a classical description should be valid. Thus, the enhancement rates may be evaluated with a classical trajectory Monte Carlo (CTMC) method whose biggest advantage is that the dynamics is easily treated non-perturbatively. The electron distribution is represented by an ensemble of classical trajectories with randomly generated initial conditions which mimic the Maxwell distribution (Equation (4)) in the asymptotic region. The flux of electrons near the ion in the presence of a magnetic field can be extracted through the probability for finding the electron in the Coulomb zone $S_C = \{\vec{r}|r < r^{RR}\}$, where radiative recombination takes place. Each visit of S_C by a given trajectory is weighted with a weighting function $w(r_{\rm min})$, where $r_{\rm min}$ corresponds to the minimum distance to the ion associated with this visit. The weighting function is obtained by breaking the Coulomb zone S_C down into different n, l subshells and by assuming each visit has a weight according to the unperturbed cross section σ_{nl}^{RR} for RR into the (n, l) state,

$$w(r) = \frac{1}{\sigma_{\rm total}^{RR}} \sum_{n=1}^{n_{\rm max}} \sum_{l=0}^{n-1} \sigma_{nl}^{RR} \Theta\left(\frac{nl}{Z} - r\right). \tag{5}$$

The visit function $V(\vec{r}, \vec{v})$ is evaluated by summing up all the weighted visits of a trajectory launched from the asymptotic coordinate $(\vec{r}, \vec{v})$, i.e. $V(\vec{r}, \vec{v}) = \sum_i w(r_{\rm min}^i)$ where the index $i = 1, 2, \ldots$ counts the number of repeated visits (if any) for a given trajectory. The visit functions (V) with and without a magnetic field (V_C) can thus be determined. Note that in the pure Coulomb case $i = 0$ or 1, that is at most one visit per trajectory takes place. Since each visit is considered to contribute to the recombination, the ratio V/V_C corresponds to the "enhancement" of the recombination. Accordingly, the recombination rate in the presence of a magnetic field becomes

$$\alpha(v_{\rm rel}) = \int \sigma_{\rm total}^{RR} v \frac{\int {\rm d}\vec{r}\, V(\vec{r}, \vec{v})}{\int {\rm d}\vec{r}\, V_C(\vec{r}, \vec{v})} f(\vec{v}, v_{\rm rel})\, {\rm d}\vec{v}. \tag{6}$$

In the pure Coulomb case, V_C is a smooth function and non-zero only in a single connected phase space region (Figure 1a). In the magnetic field a connected region at $\rho \approx 0.3$ can be found as in the pure Coulomb case, however at larger impact parameter. In the presence of a magnetic field the electron can be scattered by the ion many times allowing for multiple visits. With a longer total time propagation the visit function shows a high sensitivity to the initial coordinate ρ which is a typical signature of chaos manifesting itself in the complicated structure in Figure 1b. Irregular functions such as V demand a more general definition of integration such as the Lebesgue integral, as ordinary Riemann integration fails. In the simulation a point by point integration of V is performed over phase space coordinates (Equation (6)) requiring high levels of Monte Carlo statistics to reach converged results for measures of irregular scattering. This difficulty limits the accuracy of the present results. Note that the number of visits is counted only within a time duration corresponding to the temporal overlap between electron and ion beam in the cooler.

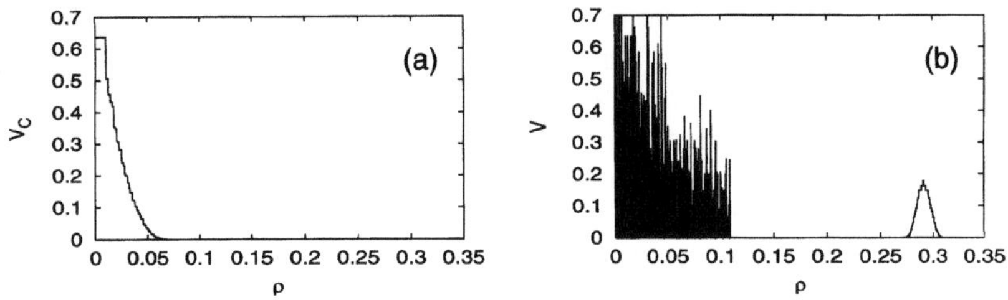

Figure 1. Visit function V as a function of the impact parameter at fixed scaled perpendicular and parallel velocities $v_\perp(BZ/c)^{-1/3} = 0.15$, $v_{||}(BZ/c)^{-1/3} = 0.3$ and $\phi = 1.3$ (angle between $\vec{\rho}$ and $\vec{v_\perp}$) without (a) and with magnetic field (b).

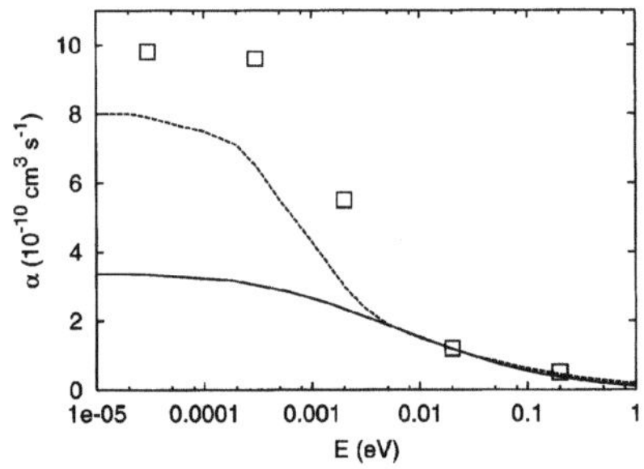

Figure 2. Radiative recombination rate for C^{6+} at $B = 42$ mT as a function of relative energy between electrons and ions. The standard RR theory with $n_{\max} = 30$ (Equations (3), (4)) (solid line), the experimental data [7] (dashed line), and CTMC results (boxes) (Equation (6)) are shown.

3. Results and conclusions

As a first application we compare our simulation with the experimental data [7] for C^{6+}. The electron temperatures in this experiment are $k_B T_{||} \approx 0.2$ meV and $k_B T_\perp \approx 10$ meV and the magnetic guiding field of the electron beam is $B = 42$ mT. Figure 2 shows the recombination rate as a function of the electron energy E_e. The modified recombination rates qualitatively agree with the measurements. In fact, the present calculation seems to overestimate the enhancement by 20 to 30%. Several effects in the experiment are not taken into account within the theory, which may contribute to this overestimate. The simulation assumes a perfectly homogeneous magnetic field neglecting any inhomogeneity. Another uncertainty in the comparison with the experiment is the exact determination of the electron temperatures influencing the recombination rate. For the transverse temperature an error of about ± 2 meV is estimated [13]. Fluctuating fields and perturbations due to neighboring ions and electrons, i.e. many-body effects in the plasma [9] are ignored. Furthermore, future investigations should include non-adiabatic switch-on and -off effects of the magnetic field at the entrance and the exit of the cooler which may have an influence on the recombination rate [10].

In summary, the present results indicate that irregular scattering in the simultaneous presence of a Coulomb field and a magnetic field can cause significant enhancement of the radiative recombination rate. First comparisons with the experimental data look promising.

Note added in proof

We have meanwhile extended our investigations to the scaling behaviour of the excess rate $\Delta\alpha = \alpha - \alpha_0$ with magnetic field B and nuclear charge Z. The magnetic field dependence of the recombination rate $\alpha \sim B^{0.5}$ obtained in the measurements is not confirmed by our simulations. By contrast, the scaling with Z exhibits a trend in agreement seen in the experiment, however, the excess recombination rate $\Delta\alpha$ scales approximately with $\sim Z^{1.6}$ while experimental data were fitted by $\Delta\alpha \sim Z^{2.6}$ [14].

Acknowledgement

This work is supported by Fonds zur Förderung der wissenschaftlichen Forschung.

References

1. Tucker, W. H., *Radiation Processes in Astrophysics*, MIT Univ. Press, Cambridge, 1975.
2. Summers, H. P. *et al.*, *Application of Recombination, NATO ASI Series B*, Plenum, New York, 1992.
3. Kramers, H. A., *Philos. Mag.* **46** (1923), 836.
4. Uwira, O. *et al.*, *Hyp. Interact.* **108** (1997), 149.
5. Baird, S. *et al.*, *Phys. Lett. B* **361** (1995), 184.
6. Hoffknecht, A. *et al.*, *J. Phys. B* **31** (1998), 2415.
7. Gwinner, G. *et al.*, *Phys. Rev. Lett.* **84** (2000), 4822.
8. Gao, H. *et al.*, *J. Phys. B* **30** (1997), L499.
9. Spreiter, Q. and Toepffer, C., *J. Phys. B* **33** (2000), 2347.
10. Gwinner, G., *et al.*, XXII-ICPEAC, Abstracts of Contr. Papers, 2001, p. 326.
11. Trost, J. and Burgdörfer, J., *Verhandl. DPG (VI)* **34** (1999), 257.
12. Zerrad, E. and Hahn, Y., *J. Quant. Spectrosc. Radiat. Transfer* **59** (1998), 637.
13. Private communication with G. Gwinner.
14. Shi, W. *et al.*, *Eur. Phys. J. D* **15** (2001), 145.

Hyperfine Interactions **146/147:** 19–22, 2003.

Recombination Enhancement in Electron Coolers

C. HEERLEIN[1] and C. TOEPFFER[2]
[1]*Institut für Theoretische Physik II, Universität Erlangen, Staudtstr. 7, 91058 Erlangen, Germany*
[2]*LPGP, Université Paris-Sud, Bât 210, F-91405 Orsay, Cedex, France*

Abstract. Recent experiments on ion–electron recombination in electron coolers show an enhancement of the recombination rate with respect to the standard theory of radiative recombination evaluated with the bulk electron density. Possible theoretical explanations of this effect have to stand a comparison with the observed scaling of the enhancement with the experimental parameters. We propose a calculation that allows to reduce the quantum mechanical problem to classical phase space trajectories. These are solved by backward characteristics and yield agreement with experiments.

1. Physical picture

A very effective method to improve beam quality in heavy ion storage rings is electron cooling [1]. A beam of cold electrons is injected along the ion beam, and consequently reduces the velocity spread of the ions. The electrons are guided by a strong magnetic field **B** parallel to the beam.

Within the electron gas ion–electron recombination occurs. We focus on radiative recombination, as the enhancement is equally observed in experiments with bare ions, where clearly no dielectronic recombination takes place. Because no density dependence is observed [2], we also exclude further tree body effects. It also can be shown numerically that self-consistent field of the electrons is small, even for the large longitudinal plasma parameters that can be achieved recently. Thus the Vlasov equation with the ionic field E_{ion} applies. It is formulated in six-dimensional phase space and reads in standard notation

$$\frac{\partial f}{\partial t} + \mathbf{v} \cdot \frac{\partial f}{\partial \mathbf{r}} + \mathbf{E}_{\text{ion}} \cdot \frac{\partial f}{\partial \mathbf{v}} = 0, \tag{1}$$

where plasma units $w_p^{-1} = \sqrt{(\varepsilon_0 m)/(e^2 n_0)}$, $\lambda_D = \sqrt{(\varepsilon_0 k_B T_\perp)/(e^2 n_0)}$, and $v_{\text{th}} = \lambda_D w_p$ have been employed.

To reduce the numerical complexity of the problem we take advantage of the axial symmetry inherent to the problem. By introducing the angular momentum along the magnetic field direction by $l = |\mathbf{r}_\perp \times \mathbf{p}_\perp|$, we obtain the cylindrical Vlasov equation

$$\frac{\partial f}{\partial t} + v_\perp \cdot \frac{\partial f}{\partial r_\perp} + \left(\frac{l^2}{r_\perp^3} + E_\perp + F_\Omega\right)\frac{\partial f}{\partial v_\perp} + v_\parallel \frac{\partial f}{\partial r_\parallel} + E_\parallel \frac{\partial f}{\partial v_\parallel} = 0 \tag{2}$$

with $\Omega := eB/m$ and $F_\Omega = \Omega(l - \Omega r_\perp^2/2)/r_\perp$ in a five-dimensional phase space with coordinates l, $r_\perp$, $r_\parallel$, $v_\perp$, $v_\parallel$. The electrical field $(E_\perp, E_\parallel)$ in (2) is given by the Coulomb field of the ion.

The cross section according to Stobbe [3] is in the usual notation

$$\sigma_{nl}\mathcal{E} = \frac{4}{3}\frac{\alpha w^3}{c^2}\frac{\pi^2\hbar^2}{2m\mathcal{E}}\left[l\left|C_{n,l}^{E,l-1}\right|^2 + (l+1)\left|C_{n,l}^{E,l+1}\right|^2\right]^2 \tag{3}$$

with

$$C_{n,l}^{E,l\pm1} = \langle n, l|\mathbf{r}|\mathcal{E}, l\pm 1\rangle. \tag{4}$$

Due to the selection rules, the cross section depends only on the energy $\mathcal{E}$ of the incident electron. This allows to write the transition rate coefficient α_{RR} with the normalized energy distribution $g(\mathcal{E})$ of the electrons as

$$\alpha_{\text{RR}} = \int \mathrm{d}\mathcal{E}\sqrt{2\mathcal{E}}\,g(\mathcal{E})\sum_{n=0}^{n_{\text{cut}}}\sum_{l=0}^{n-1}\sigma_{nl}(\mathcal{E}) \tag{5}$$

with n_{cut} as the highest electron state that is not torn apart by the bending magnet. For magnetic field strength B much smaller that the atomic field $B_c = 2.35\cdot 10^5$ T, the cross section σ_{nl} and the angular momentum distribution $g(E,l) = (2l+1)g(E)$ is not altered by B. This connotes that the discrete spectrum with positive energies (Landau levels) does not contribute in the considered energy range. Thus (5) holds within the magnetized electron plasma.

2. Method

Additionally, initial and boundary conditions are needed to turn the partial differential Equation (2) into a properly posed problem. The magnetic field suppresses the energy exchange between the parallel and perpendicular degrees of freedom. Thus a reasonable assumption for the velocity distribution of the electrons without the perturbation of the ion is an anisotropic Maxwell distribution with the perpendicular and parallel temperatures $T_\perp$ and $T_\parallel$, respectively. With the anisotropy parameter $\zeta := T_\parallel/T_\perp$, we have

$$f_0 = \frac{1}{(2\pi)^{3/2}\zeta}\exp\left(-\frac{v_\perp^2}{2}\right)\exp\left(-\frac{v_\parallel^2}{2\zeta^2}\right), \tag{6}$$

again in plasma units.

Two ways have been proposed for the formulation of the initial and boundary conditions. One can assume an initially empty space with the Maxwell distribution (6) on a large sphere. For large times the electron density propagates inward under the influence of the magnetic field. It can be shown [4] that with these boundary conditions the magnetic field does not influence the phase space density of electrons for $t \to \infty$. However, this method might be saved by introducing the 'number of visits' as the relevant observable [5]. The physical motivation of this observable is yet to be discussed.

Consequentially we investigate the alternative initial condition: we consider an initially homogeneous electron plasma with $f(t = 0) = f_0$ from (6) with infinite spacial extent. In this magnetized plasma an ion is inserted at $t = 0$. Beyond the investigation in [5] this yields a density enhancement from bound electrons. Naturally the instantaneous turning-on of the ion is an idealized model. Even so it describes the merging region of ion- and electron beam better than a large sphere. The bound electrons that our model predicts can be indeed verified by laser probe experiments [6].

By integration of its backward characteristics (i.e., particle trajectories), the Vlasov equation (2) can be solved numerically [7]. The algorithm used is based on a splitting scheme similar to [8] and an adaptive time step with a distinctive that is based on whether the magnetic of electric forces dominate the electron propagation.

3. Results and conclusion

The calculations presented here are done with the physical parameters at CRYRING [9], i.e. $T_\parallel = 1.4$ K, $T_\perp = 0.121$ K, $n_0 = 2 \cdot 10^{13}$ m^{-3}, $B = 30$ mT, $Z = 2, 10, 14$. In Figure 1 the density enhancement around a fully stripped Si-ion is shown at half passage through the electron cooler. The numerical solution of the Vlasov equation yields a density enhancement, which contains both free and bound electrons. The recombination enhancement factor may be derived as

$$\epsilon := \frac{\alpha^{\mathrm{RR}}}{\alpha^{\mathrm{RR}}_{\mathrm{free}}} \approx \frac{\langle n_{\mathrm{free}} + n_{\mathrm{bound}} \rangle}{\langle n_{\mathrm{free}} \rangle}, \tag{7}$$

where $\alpha^{\mathrm{RR}}_{\mathrm{free}}$ is the recombination rate coefficient from Stobbe [3] and α^{RR} the actual rate coefficient from our simulation. Therefore the free and bound electron

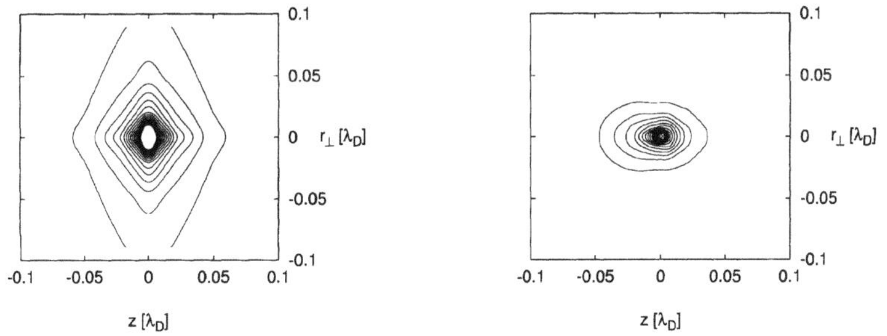

Figure 1. Density enhancement around a Si-ion at zero temperature for relative velocity zero (*left*) and thermal velocity (*right*). Plotted are counterlines for integer multiples of the bulk electron density n_0, starting from the outmost counterline at $n = 3n_0$. Physical parameters from [9].

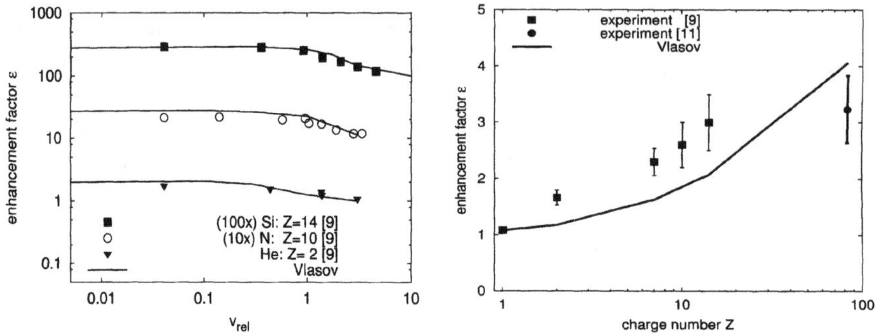

Figure 2. Left: Recombination enhancement factor ε for three ion species at varied relative ion velocity $v_{\rm rel}$ measured in units of the thermal velocity. *Right*: Scaling of the enhancement factor ε at zero relative energy ($v_{\rm rel} = 0$) at various ion charges Z.

densities, $n_{\rm free}$ and $n_{\rm bound}$ are averaged over a volume of the cyclotron radius. This reflects that only electrons from within this volume can classically reach the ion within one cyclotron period, as motivated by earlier disk models [10].

In the measurements the relative velocities $v_{\rm rel}$ of ion and electron beam are detuned. Figure 2 shows the predicted (solid curves) and measured (symbols) recombination rate coefficients for three different ion species [9]. Also shown are results for $v_{\rm rel} = 0$ as a function of ion charge. Here a result from the TSR [11] at similar parameters is included.

The above calculation agree quantitatively with the experimental findings at different electron cooler facilities. As a more realistic modeling of the beam merging seems desirable, but the instantaneous turning-on of the ionic charge explains some of the observed scaling laws with considerable precision.

References

1. Poth, H., *Physics Reports* **196** (1990), 135.
2. Gao, H. *et al.*, *Hyp. Interact.* **99** (1996), 301.
3. Stobbe, M., *Ann. Phys. (Leipzig)* **7** (1930), 661.
4. Heerlein, C., Radiative Rekombination im Elektronenkühler (Diss.), Erlangen, 2001 (in preparation).
5. Hoerndl, M., Yosida, S. and Burkdoerfer, J., this volume.
6. Borneis, S., Laserinduzierte Ein- und Zweistufen-Rekombination hochgeladener Ionen und freier Elektronen im Schwerionenspreicherring ESR (Diss.), Darmstadt, 1994.
7. Heerlein, C. and Toepffer, C., *Comp. Physics Comm.* (to be published).
8. Cheng, C. Z. and Knorr, G., *J. Comp. Phys.* **22** (1976), 330.
9. Gao, H. *et al.*, *J. Phys. B* **30** (1997), L499.
10. Spreiter, Q. and Toepffer, C., *J. Phys. B* **33** (2000), 2347.
11. Hoffknecht, A. *et al.*, Arbeitsgruppe Atomare Stöße, 20. Arbeitsbericht, 1999, p. 37.

Hyperfine Interactions **146/147:** 23–27, 2003.

Enhancement of Dielectronic Recombination by External Electromagnetic Fields

S. BÖHM[1], S. SCHIPPERS[1], W. SHI[1], A. MÜLLER[1], N. DJURIĆ[2], G. H. DUNN[2], W. ZONG[2], B. JELENKOVIĆ[3], H. DANARED[4], N. EKLÖW[5], P. GLANS[5], R. SCHUCH[5] and N. R. BADNELL[6]

[1]*Institut für Kernphysik, Universität Giessen, Germany*
[2]*JILA, University of Colorado and NIST, Boulder, CO 80309-0440, USA*
[3]*Institute of Physics, Pregrevica 118, 10800 Belgrade, Yugoslavia*
[4]*Manne Siegbahn Laboratory, 10405 Stockholm, Sweden*
[5]*Atomic Physics, Stockholm University, 10405 Stockholm, Sweden*
[6]*University of Strathclyde, Glasgow, G4 ONG, United Kingdom*

Abstract. The enhancement of the dielectronic recombination rate of lithiumlike Ne^{7+} and O^{5+} ions by external electromagnetic fields has been measured at the storage ring CRYRING. The energy range covered all $1s^2 2pnl$ dielectronic recombination resonances attached to the $2s \rightarrow 2p$ core excitation. Electric fields up to 1436 V/cm were applied in the Ne^{7+} experiment and the saturation of the enhancement with increasing electric field could clearly be seen. In the O^{5+} experiment the enhancement was studied as a function of the Rydberg quantum number n.

Key words: dielectronic recombination enhancement.

Dielectronic recombination (DR) of an ion and an electron is a resonant two-step process. In the first step, the dielectronic capture, a continuum electron is captured into a bound state with simultaneous excitation of a bound electron. In the second step the resulting doubly excited state decays by the emission of a photon to a bound state below the first ionization limit. For the present case, DR can be approximately represented as

$$e^- + A^{q+}(1s^2 2s) \rightarrow A^{(q-1)+}(1s^2 2pnl) \rightarrow A^{(q-1)+}(1s^2 2snl) + h\nu.$$

The weakly bound highly excited intermediate Rydberg states are easily perturbed by external influences, such as electromagnetic fields. The experimental investigation of dielectronic recombination in external fields (DRF) at heavy ion storage rings is an ongoing effort. Here we report on experiments with Ne^{7+} [5] and O^{5+} [6] ions, that extend a series of measurements on lithium-like ions (Si^{11+} [2], Cl^{14+} [3], Ti^{19+} [4] and Ni^{25+} [8]) towards lower values of the nuclear charge Z. High resolution zero field DR of Ne^{7+} and O^{5+} has been measured previously by Zong *et al.* [10] and Andersen *et al.* [1], respectively.

The experiments have been performed at the heavy-ion storage-ring CRYRING of the Manne Siegbahn Laboratory in Stockholm. For the measurements $^{20}Ne^{7+}$

($^{16}O^{5+}$) ions were injected into the ring, accelerated, and cooled by merging the ion beam with a beam of cold electrons. Inside the electron cooler the electron beam is guided by a magnetic field denoted by $B_{\parallel}$ in the following. For the subsequent DR measurement the electron cooler was used as an electron target. To obtain a DR spectrum the electron energy was ramped. In order to study field effects, external motional electric fields ($E_{\perp} = vB_{\perp}$) were introduced in the cooler by applying a defined transverse magnetic field $B_{\perp}$ using the steering coils of the cooler. With this technique the complete energy range of the resonances attached to the $2s \rightarrow 2p$ core excitation was scanned for different electric fields.

In storage ring experiments, the maximum Rydberg quantum number of resonant states contributing to the detected signal is limited by field ionization. As the recombined ions travel towards the detector, they encounter three regions with different transverse magnetic fields and hence different motional electric fields. These regions are (1) the toroid which guides the electron beam out of the overlap with the ion beam ($\approx 4 \times 10^4$ V/cm), (2) a dipole magnet ($\approx 1 \times 10^5$ V/cm) just behind the cooler that corrects for the ion beam displacement by the toroidal fields in the cooler, and (3) the bending dipole magnet in which the detector is placed and which is used as the charge analyzing magnet ($\approx 5 \times 10^5$ V/cm). The values of the electric fields given in brackets correspond to 11.4 MeV/u Ne^{7+} ions. These fields ionize Rydberg states with quantum numbers n higher than some cutoff value that can be expressed as $n_F = (q^3/9F)^{1/4}$ where F is the electric field in atomic units (1 au $= 5.142 \times 10^9$ V/cm). For 11.4 MeV/u Ne^{7+} ions in the CRYRING, i.e. with the electric fields given above, one obtains $n_F = 46$ in the toroid, $n_F = 35$ in the correction magnet, and $n_F = 24$ in the charge analyzing dipole. Only recombined ions with $n < n_F$ can reach the detector. Since the simple formula for n_F does not consider radiative decay of Rydberg states on the way from the cooler to the different field ionization zones, it is only indicative. In a more detailed field ionization model [9] we have considered all three field ionization regions mentioned above to obtain nl-dependent Rydberg detection probabilities.

Figure 1(a) shows DR spectra of Ne^{7+} ions for five selected electric fields. Clearly the measured DR rate coefficient in the energy region 11.65–15.9 eV increases with increasing electric field strength. For a longitudinal magnetic field $B_{\parallel}$ of 180 mT the electric field was varied between 0 and 1436 V/cm in 25 different steps. For a second set of measurements $B_{\parallel}$ was set to 30 mT and the electric field was varied between 0 and 144 V/cm in 15 different steps.

The enhancement of the DR via high Rydberg states is quantified by extracting rate coefficients I_{hi} integrated over the energy range 11.65–15.9 eV for Ne^{7+} and 9.4–12 eV for O^{5+}, respectively. For normalization purposes we also monitored the integral I_{lo} at low energies where the electric field has no significant influence on the recombination rate (4.5–8.5 eV and 1.3–7.25 eV, respectively). The high Rydberg contributions I_{hi} monotonically increase with the electric field, while the lower-n contribution I_{lo} remains constant when the electric field is

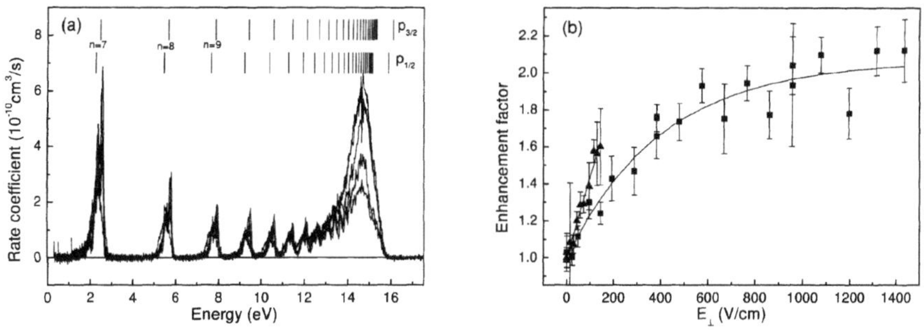

Figure 1. (a) Ne^{7+} DR, spectra for 5 different electric fields 0, 144, 479, 1077, 1316 V/cm ($B_\parallel$ = 180 mT). (b) Electric field enhancement factors for Ne^{7+}-ions for 2 different longitudinal magnetic fields, 180 mT (squares) and 30 mT (triangles).

changed. In Figure 1(b) we have plotted the electric-field enhancement factor [4, 8]

$$r(E_\perp, B_\parallel) = C(B_\parallel)\frac{I_{\text{hi}}(E_\perp, B_\parallel)}{I_{\text{lo}}(E_\perp, B_\parallel)} \tag{1}$$

with the constant $C(B_\parallel)$ chosen such that fits to the data points (see below) yield $r^{(\text{fit})}(0, B_\parallel) = 1.0$. The formula

$$r^{(\text{fit})}(E_\perp, B_\parallel) = 1 + s(B_\parallel)E_{\text{sat}}(B_\parallel)\left\{1 - \exp\left[\frac{-E_\perp}{E_{\text{sat}}(B_\parallel)}\right]\right\}, \tag{2}$$

that we have fitted to the measured enhancement factors (full lines in Figure 1(b)), provides an easy parameterization of our data. The parameter $E_{\text{sat}}(B_\parallel)$ is the saturation field and $s(B_\parallel)$ is the initial slope, that is, the tangent to $r^{(\text{fit})}(E_\perp, B_\parallel)$ at $E_\perp = 0$. The enhancement factor first grows linearly and then saturates for high electric fields. This saturation was already indicated by other measurements [3] but was not seen previously as clearly as in the experiment with Ne^{7+} ions.

The initial slope of the enhancement factor for the measurements at 30 mT is much higher than the slope at $B_\parallel$ = 180 mT, i.e. the magnetic field reduces the effect of the perpendicular electric field on DR. This B field dependence of DR rates was theoretically predicted [7] and experimentally verified [3, 4] only recently.

Compared to other measurements where the detected enhancement of the recombination rate due to an electric field increased when going from Ni^{25+} ($Z = 28$) to Si^{11+} ($Z = 14$) the enhancement of the detected recombination rate decreases again in the present experiments with lower Z ions. This decrease is caused by field ionization of high Rydberg states that for low Z sets in at relatively low n. The results of the Ne^{7+} experiment have been reported in detail previously in [5].

In the O^{5+} experiment DRF was studied as a function of the Rydberg quantum number n. This was done by utilizing the unavoidable field ionization described above which limits the maximum detected Rydberg state to a value $n_F =$

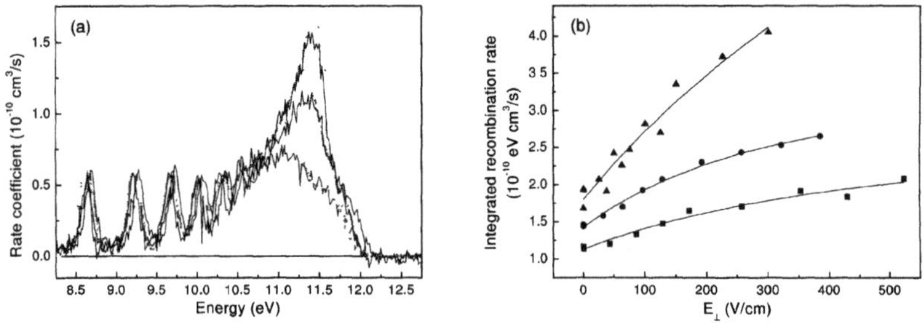

Figure 2. (a) O^{5+} DR spectra for three different ion energies with no electric fields applied to the interaction region. Included are the theoretical results obtained by AUTOSTRUCTURE to which our field ionization model has been applied (thick lines), (b) Integrals I_{hi} from 9.4–12 eV for all three ion energies (9.4 MeV/u squares, 5 MeV/u circles, and 3.26 MeV/u triangles).

$(q^3/9F)^{1/4}$. The field F depends on the ion velocity. As the ion velocity is decreased the magnetic field of the bending dipole magnets can be reduced and therewith the motional electric field $F = vB$. This shifts the cutoff quantum number n_F to higher values.

Three different ion energies (9.4, 5 and 3.26 MeV/u) were used in the O^{5+} experiment. The corresponding cutoff quantum numbers n_F are 19, 22, and 25, respectively. The effect of this increase of n_F is seen in Figure 2(a). As the ion energy is decreased and n_F grows, the recombination rate coefficient in the high energy part of the spectra increases.

The experimental spectra are compared to theoretical results from Badnell obtained by the AUTOSTRUCTURE atomic structure code to which our field ionization model has been applied. The thick lines in Figure 2(a) show the results which agree very well with the experiment. From this we conclude that field ionization is sufficiently well understood in our experiments.

The enhancement of the DR via high Rydberg states has been quantified by extracting the integrated recombination rate of the high energy part of the spectra (9.4–12 eV) as described above. These integrals I_{hi} are shown in Figure 2 for all three ion energies. The difference between these curves is due to the bandwidths of n states added by lowering the ion energy ($n = 19$–22 and $n = 22$–25). By subtracting the curves from one another one gets the integrated recombination rate only due to these two bandwiths. The increasing difference between these curves as the electric field increases shows that the higher Rydberg states ($n = 22$–25) are much more sensitive to external fields then the lower ones ($n = 19$–22). The analysis of the n differential DR data has not been completed yet but it is obvious that this technique allows us to investigate n differential DRF.

References

1. Andersen, L. H., Bolko, J. and Kvistgaard, P., *Phys. Rev. A* **41** (1990), 1293.
2. Bartsch, T., Müller, A., Spies, W., Linkemann, J. *et al.*, *Phys. Rev. Lett.* **79** (1997), 2233.

3. Bartsch, T., Schippers, S., Müller, A., Brandau, C. *et al.*, *Phys. Rev. Lett.* **82** (1999), 3779.
4. Bartsch, T., Schippers, S., Beutelspacher, M., Böhm, S. *et al.*, *J. Phys. B* **33** (2000), L453.
5. Böhm, S., Schippers, S., Shi, W., Müller, A. *et al.*, *Phys. Rev. A* **64** (2001), 032707.
6. Böhm, S., Schippers, S., Shi, W. Müller, A. *et al.*, *Phys. Rev. A* **65** (2002), 052728.
7. Robicheaux, F. and Pindzola, M. S., *Phys. Rev. Lett.* **79** (1997), 2237.
8. Schippers, S., Bartsch, T., Brandau, C., Müller, A. *et al.*, *Phys. Rev. A.* **62** (2000), 022708.
9. Schippers, S., Müller, A., Gwinner, G., Linkemann, J. *et al.*, *Astrophys. J.* **555** (2001), 1027.
10. Zong, W., Schuch, R., Gao, H., DeWitt, D. R. *et al.*, *J. Phys. B* **31** (1998), 3729.

Hyperfine Interactions **146/147:** 29–34, 2003.

Radiative Electron Capture into the K- and L-Shell of H-, He-, and Li-Like Uranium Ions at Relativistic Energies

G. BEDNARZ[1], A. WARCZAK[1], D. SIERPOWSKI[1], T. STÖHLKER[2], S. HAGMANN[2], F. BOSCH[2], A. GUMBERIDZE[2], C. KOZHUHAROV[2], D. LIESEN[2], P. H. MOKLER[2], X. MA[3] and Z. STACHURA[4]
[1]*Institute of Physics, Jagiellonian University, Cracow, Poland*
[2]*Gesellschaft für Schwerionenforschung, Darmstadt, Germany*
[3]*Institute of Modern Physics, Lanzhou, China*
[4]*Institute for Nuclear Physics, Cracow, Poland*

Abstract. We report on the experimental study of the angular distributions of photons for the time-reversed photoionization process, i.e., the radiative electron capture (REC), in relativistic atomic collisions involving high-Z projectiles. Whereas former investigations focused on initially bare projectiles, we here extend our studies to heavy few-electron ions (H-, He- and Li-like uranium) in order to elucidate the importance of electron–electron interaction for photoionization in the realm of high-Z ions.

Key words: radiative electron capture, photoionization, relativistic effects, electron correlations.

1. Introduction

Radiative electron capture (REC) is the most important channel for charge exchange in collisions of highly charged ions with light target atoms [1]. As, in addition, the process is the time reversal of photoionization, its investigation opens new possibilities for advanced studies of radiation–matter interaction [2]. In particular, the photon angular distributions of REC are a very sensitive probe of relativistic effects mediated by the strong Coulomb fields of heavy ions. Here, spin-flip transitions were observed, for the first time, in fast collisions of bare U-ions with low Z-ions [3]. Even at very low projectile velocities, corresponding to photoionization close to the threshold, the presence of these transitions was confirmed as well [4]. So far, mainly bare high-Z ions were used in the experiments. In contrast, the present study concentrates around heavy few-electron ions, i.e., H-, He-, and Li-like uranium projectiles. This enables us to investigate for the REC process the role of additional electrons present in the projectile prior to the collision.

2. Experimental method

The experiment was performed at the experimental storage ring (ESR) at GSI-Darmstadt by using H-, He-, and Li-like uranium ions at an energy of 216 MeV/u. For the experiment, the X-ray spectroscopy environment of the internal jet-target was used. Here, projectile X-rays produced in collisions of the stored ion beams with a N_2 jet-target were detected by an array of solid state detectors (for details see [5]). This made it possible to measure simultaneously at observation angles of 13°, 35°, 60°, 90°, 120° and 150° with respect to the beam axis.

The U^{91+}-, U^{90+}- and U^{89+}-ions, at an energy of 216 MeV/u, colliding with N_2-target were used. After the beam passes through the reaction target, the emerging charge states (initial charge state Q, Q−1 for capture, and Q+1 for ionization) were separated by the dipole magnet mounted next to the target. There, the down- and up-charged ions were registered (for the study of capture and ionization processes) with position-sensitive particle detectors located in the inner and outer part of the storage ring. This enabled us in particular to measure the X-ray production in the target in coincidence with the outgoing charge state.

In the following, only preliminary results, concerning electron capture, are presented and discussed.

3. Experimental results and discussion

Figure 1 shows a typical X-ray spectrum observed at 90° for U^{91+}-ions associated with capture of one electron. The broad structures arise from REC into the projectile K-, L- and M-shells. The line-widths are dominated by the Compton profile of the target electrons. Electron capture into excited states (L, M and higher shells)

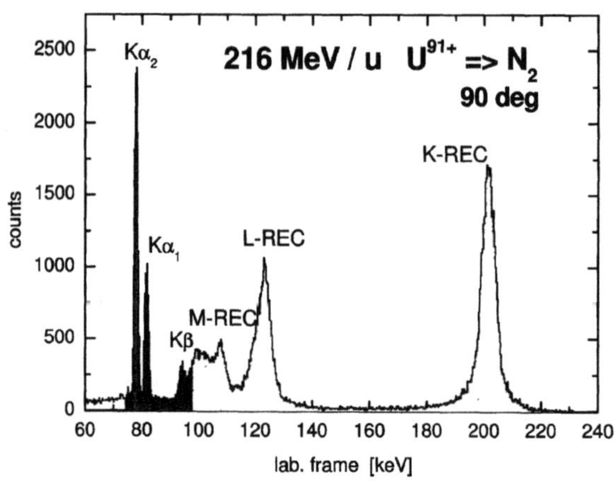

Figure 1. X-ray spectrum associated with one-electron capture (corrected for random events and detection efficiency) observed at an angle of 90°.

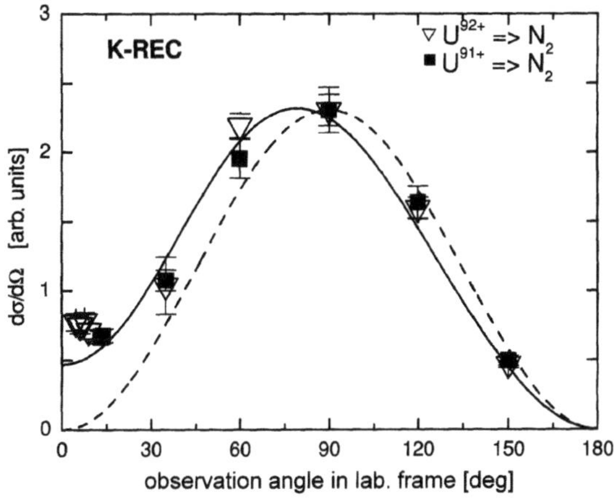

Figure 2. K-REC differential cross sections for 216 MeV/u $U^{91+} \rightarrow N_2$ (squares) and for 310 MeV/u $U^{92+} \rightarrow N_2$ (triangles); dashed line – nonrelativistic description, solid line – relativistic predictions for bare U-ions [6].

leads, via cascades, to the characteristic Kα- and Kβ-transitions, clearly seen in the spectrum.

All the X-ray spectra were first energy calibrated and corrected for random events and for detection efficiency. Then, the yields of K- and L-REC photons were determined by dedicated fitting routines (see [7, 8]). Simultaneously, the intensity of the Kα_2-line ($2^3S_1, 2^3P_1 \rightarrow 1^1S_0$) were extracted ($U^{91+} \rightarrow N_2$). Assuming, that the K$\alpha_2$-line (similarly to the Lyα_2-line) is isotropic in the emitter system [2, 3], all the angular distributions (U^{91+}, U^{90+}, $U^{89+} \rightarrow N_2$) for K- and L-REC were normalized relative to the Kα_2 intensity pattern in the laboratory frame.

In Figure 2, the measured differential cross section (photon angular distribution) for REC into the K-shell of 216 MeV/u U^{91+}-ions (present experiment) and 310 MeV/u U^{92+}-ions (taken from [3]), are plotted as a function of the laboratory observation angle and compared with predictions (for $U^{92+} \rightarrow N_2$) based on rigorous relativistic calculations, see [6]. All the experimental and theoretical cross sections were normalized to an arbitrary value at 90°. Figure 2 shows that the shape of the angular distribution is almost unchanged when comparing the results for bare (triangles) and H-like (squares) U-ions. The experimental data are in accordance with a fully relativistic theoretical description (solid line). The REC emission pattern deviates considerably from symmetry around 90° in contrast to non-relativistic predictions (see dashed line in Figure 2) [10]. Although the latter approach is a fully non-relativistic one, it includes the full retardation as well as Lorentz transformation to the laboratory frame. Note that the non-vanishing cross sections observed close to 0° point to the occurrence of magnetic (spin-flip) transitions for REC into high-Z projectiles, as discussed recently [4, 9].

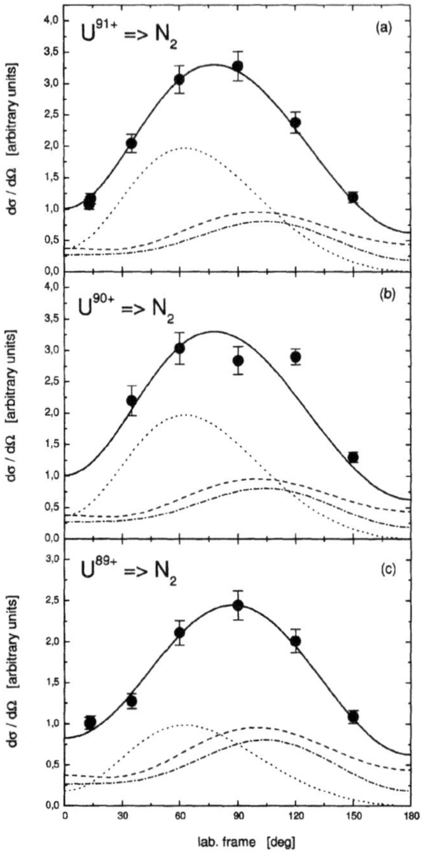

Figure 3. Differential cross sections for capture into the L-shell of U^{91+} (a), U^{90+} (b), U^{89+} (c); Dotted lines: capture into the $2s_{1/2}$-shell, dashed-lines: $2p_{1/2}$-shell, dot-dashed-line: $2p_{3/2}$-shell, solid lines: sum of all contributions from different subshells (relativistic calculations [9]).

The differential cross sections for L-REC into U^{91+}-, U^{90+}- and U^{89+}-ions are presented in Figures 3a, 3b, 3c, respectively, along with theoretical calculations for the different subshells [9]. A basic feature of all the L-REC angular distributions displayed in Figure 3 is the asymmetry between the forward and the backward photon emission. The main contribution for H-like (Figure 3a) and He-like U-ions (Figure 3b) arises from the capture into the 2s-shell (dotted-line in Figure 3a and Figure 3b) which has a pronounced maximum at the forward direction (close to 60°). Similar to K-REC, the non-vanishing cross-section at 0° for REC into the 2s-state is a clear signature of magnetic (spin-flip) transitions. Contributions from the 2p-shells to the L-REC photon angular distribution (dashed line – $2p_{1/2}$, dashed-dotted line – $2p_{3/2}$) reach their maxima at backward angles. They also show non-vanishing cross-sections at 0° and 180° as well. However, as has been discussed in [9], this is a result of the coupling of the electron spin to the angular momentum and not an indication of spin-flip, as in the $2s_{1/2}$ case.

In contrast to H- and He-like uranium (Figure 3c), the total L-REC emission pattern is for the Li-like species almost symmetric to 90° due to the partially blocked 2s-shell (compare dotted lines in Figure 3a – empty 2s-shell and Figure 3c – one electron in the 2s-shell). Here, again the experimental data agree well with theoretical predictions.

4. Summary

In summary, we measured the photon angular distributions for K- and L-REC into H-, He-, and Li-like U-ions at an energy of 216 MeV/u. The data are found to be asymmetric and more pronounced at forward directions. The emission pattern for K-REC stays unchanged for bare and H-like U-ions. Due to the dominance of capture into the 2s-state, the L-REC distribution for H- and He-like uranium is compatible with the certain forward peaking expected from calculations for the 2s-contribution. However, for Li-like uranium with only one 2s-vacancy, the photon emission pattern shows an almost symmetric distribution with respect to 90° which can be explained by the partially blocked 2s-shell. The experimental data for the photon angular distributions prove that at high energies, the REC process is not influenced by electron-electron interaction effects. This is in agreement with the commonly made assumption that for photoionization at high energies, neutral atoms can be treated even as H-like systems. This is because at high energies the photoionization cross-sections are determined by the photon–electron interaction at small distances to the nucleus, i.e., distances of the order of the Compton wavelength. At such distances, electron wave functions are point Coulombic and are not affected by screening or correlation effects. Consequently, at high energies, even neutral atoms can be treated as H-like systems. However, in the intermediate and low energy domain where the energy of the ejected electron is comparable or lower than the initial binding energy, much larger distances to the nucleus are involved. As a consequence, the electron–electron interaction may considerably alter the angular distribution of the ejected electrons. Therefore, future REC angular distribution studies will focus on decelerated heavy few-electron systems as it is now possible at the ESR storage ring [11].

Acknowledgement

This research has been supported by a Marie Curie Fellowship of the European Community Programme IHP under contract number HPMT-CT-2000-00197.

References

1. For a review of relativistic ion-atom collisions see, e.g., Eichler, J. and Meyerhof, W. E., *Relativistic Atomic Collisions*, Academic Press, San Diego, 1995.
2. Stöhlker, Th., Mokler, P. H., Kozhuharov, C. and Warczak, A., *Com. At. Mol. Phys.* **33** (1997), 271.
3. Stöhlker, Th., Ludziejewski, T., Bosch, F., Brinzanescu, O., Kozhuharov, C., Kriegal, A., Mokler, P. H., Beyer, H. F., Eichler, J., Hagmann, S., Krämer, A., Liesen, D., Lekki, J., Menzel, G., Reich, H., Rymuza, P., Stachura, Z., Swiat, P. and Warczak, A., *Phys. Rev. Lett.* **82** (1999), 3232.
4. Stöhlker, Th., Ma, X., Ludziejewski, T., Beyer, H. F., Bosch, F., Brinzanescu, O., Dunford, R. W., Eichler, J., Hagmann, S., Ichihara, A., Kozhuharov, C., Kramer, A., Liesen, D., Mokler, P. H., Stachura, Z., Swiat, P. and Warczak, A., *Phys. Rev. Lett.* **86** (2001), 983.
5. Stöhlker, Th., Brinzanescu, O., Kramer, A., Ludziejewski, T., Ma, X., Swiat, P. and Warczak, A. In: D. S. Gemmel, E. P. Kanter, L. Young (eds), *X-Ray and Inner Shell Processes, AIP Conference Proc. 506*, Chicago, Illinois, 1999, p. 389.
6. Ichihara, A., Shirai, T. and Eichler, J., *Phys. Rev. A* **49** (1994), 1875.
7. Stöhlker, Th., Ludziejewski, T., Reich, H., Bosch, F., Dunford, R. W., Eichler, J., Franzke, B., Kozhuharov, C., Menzel, G., Mokler, P. H., Nolden, F., Rymuza, P., Stachura, Z., Steck, M., Swiat, P. and Warczak, A., *Phys. Rev. A* **58** (1998), 2043.
8. Swiat, P., Warczak, A., Bosch, F., Franzke, B., Klepper, O., Kozhuharov, C., Menzel, G., Mokler, P. H., Reich, H., Stöhlker, Th., Dunford, R. W., Rymuza, P., Ludziejewski, T. and Stachura, Z., *Physica Scripta T* **80** (1999), 326.
9. Eichler, J., Ichihara, A. and Shirai, T., *Phys. Rev. A* **51** (1995), 3027.
10. Spindler, E., Betz, H.-D. and Bell, F., *Phys. Rev. Lett.* **42** (1979), 832.
11. Kramer, A., Kritzer, A., Reich, H. and Stöhlker, Th., *Nucl. Instr. Meth. B* **174** (2001), 205.

Hyperfine Interactions **146/147:** 35–40, 2003.

Polarization of the *Lyman*–α_1 Line Following the Radiative Recombination of Bare, High-Z Ions

A. SURZHYKOV[1], S. FRITZSCHE[1] and TH. STÖHLKER[2]
[1]*Fachbereich Physik, Universität Kassel, Heinrich-Plett-Str. 40, D-34132 Kassel, Germany; e-mail: surz@physik.uni-kassel.de*
[2]*Gesellschaft für Schwerionenforschung, Planckstr. 1, D-64291 Darmstadt, Germany*

Abstract. The radiative recombination of a free electron into the $2p_{3/2}$ state of bare, highly-charged ions and the subsequent *Lyman*–α_1 ($2p_{3/2} \to 1s_{1/2}$) decay are studied theoretically. Special attention is paid to the linear polarization of the characteristic X-ray radiation. In particular, it is found that the *angular distribution* of the *Lyman*–α_1 line polarization is remarkably modified by the interference between the leading electric dipole E1 decay channel and the – much weaker – magnetic quadrupole M2 term. Relativistic calculations have been carried out for the electron capture into bare uranium ion for a wide range of projectile energies.

Key words: radiative recombination, radiative decay, X-ray polarization, E1–M2 interference.

1. Introduction

With the recent experimental advances in heavy-ion accelerators and ion storage rings, a number of new possibilities arose to study ion–electron and ion–atom collisions. In these collisions, one of the most basic process is the transfer of an electron from the target atom to the fast moving projectile ion. For high-energy collisions of bare ions, this charge transfer is accompanied by the emission of a photon which carries away the excess energy and momentum. This process, which is known as the radiative recombination (RR) of ions, has been intensively studied during recent years at the GSI storage ring in Darmstadt [1]. So far, however, most of the measurements concerned the capture of an electron into the K-shell of the projectile ions. If, in contrast, the electron is captured into some excited state of the ion, its subsequent decay will lead to the emission of one (or several) photons until the ground state is reached. Such subsequent photon emission is characterized (apart from the well known energies) by its *angular distribution* and *polarization.* Both of these properties are closely related to the magnetic sublevel population of the excited ion as it arises from the electron capture [2, 3]. Several experiments have been carried out during last few years in order to study the angular distribution of the subsequent photons and, therefore, to derive the alignment of the residual ions [4]. Till today, no polarization measurement has yet been performed, mainly because of the lack of efficient X-ray polarization detectors. Owing to recent improvements in the detector techniques, however, such polarization experiments are likely to be carried out at the GSI storage ring within the next few years.

In this contribution, we study the polarization of the *Lyman*–α_1 radiation following the capture of a free electron into the $2p_{3/2}$ state of bare, highly-charged ions. The most natural framework for such polarization studies is given by the density matrix theory [5, 6]. For the sake of brevity, in the present paper, we omit the details of the density matrix formalism when applied to the radiative recombination of bare ions. Instead, we display the *final* formulas for the alignment of the excited ionic state and the polarization of the decay photons in Sections 2 and 3. In Section 4, we describe the computations for the linear polarization of the *Lyman*–α_1 radiation following an electron capture into the bare *uranium* ion. A brief summary is given finally in Section 5.

2. Ion alignment following the electron capture

The capture of an *unpolarized* electron into an ionic bound state $|n_b\ j_b\rangle$ leads to an equal population of all pairs of sublevels with the same modulus of the magnetic quantum numbers $|\mu_b|$. Moreover, an ion is said to be *aligned* if, for $j_b > 1/2$, different pairs of such magnetic substates are unequally populated. Usually, the alignment of the residual ion is described in terms of one (or several) parameters $\mathcal{A}_k$ which are related to the total cross sections $\sigma^{RR}(\mu_b)$ for the radiative recombination into the various magnetic sublevels $|n_b\ j_b\ \mu_b\rangle$. For the capture into the $2p_{3/2}$ level, for instance, only one parameter $\mathcal{A}_2$ is nonzero and can be expressed as [2, 3]:

$$\mathcal{A}_2 = \frac{\sigma^{RR}(\mu_b = 3/2) - \sigma^{RR}(\mu_b = 1/2)}{\sigma^{RR}(\mu_b = 3/2) + \sigma^{RR}(\mu_b = 1/2)}. \tag{1}$$

The calculation of the alignment parameter $\mathcal{A}_2$ in the framework of the Dirac theory has been discussed in detail by Eichler and co-workers [2]. It requires the computation of free-bound transition matrix elements for the electron–photon interaction. We calculated these matrix elements and utilized them to obtain both, the partial cross sections and the alignment parameters by using the DIRAC [7] and RACAH [8] programs. The DIRAC package has been developed for studying the properties and behaviour of the hydrogen-like ions and has been used before for investigating the polarization phenomena in the radiative electron capture [9].

3. Polarization of the subsequent photons

The alignment parameter $\mathcal{A}_2$ describes the population of the $2p_{3/2}$ ionic state as it arises from the electron capture. The subsequent decay of such – an *aligned* – state may lead both to an anisotropic angular distribution [2, 3] as well as to a *non-zero* linear polarization of the characteristic X-ray radiation, which is defined in terms of the *degree of polarization* [10]:

$$P_L = \frac{I_\parallel - I_\perp}{I_\parallel + I_\perp}. \tag{2}$$

Here, $I_\parallel$ and $I_\perp$ are the intensities of light, which is linearly polarized in parallel (perpendicular) respectively to the reaction plane, given by the directions of the projectile ion and characteristic photon. Of course, the linear polarization (2) will depend not only on the population of the excited ionic states but also on the angle θ, under which the decay photon is detected with respect to the ion beam. For instance, the linear polarization of the *Lyman–α_1* photons is given in the emitter frame by the standard expression [10, 11]:

$$P_L(\theta) = \frac{-\frac{3}{2}\frac{\mathcal{A}_2}{2}\sin^2\theta}{1 + \frac{\mathcal{A}_2}{2}P_2(\cos\theta)}, \tag{3}$$

where P_2 is the second-order Legendre polynomial.

Equation (3) includes only the dominant electric dipole (E1) decay channel while – the much weaker – magnetic quadrupole component (M2), which also contributes to the *Lyman–α_1* transition, is neglected. Although the electric dipole approximation (3) is certainly appropriate for light ions, it should be questioned for highly-charged ions since the contribution of the magnetic quadrupole term to the properties of the *Lyman–α_1* transition increases dramatically for the higher nuclear charges Z. As we showed recently, for instance, the *angular distribution* of the characteristic radiation is considerably modified by incorporating the magnetic quadrupole (M2) term [3]. Of course, the M2 term must also be taken into account for the *linear polarization* of the decay photons. By using the density matrix theory, Equation (3) can easily be extended to include both, the electric dipole and the magnetic quadrupole components. For the sake of brevity, we leave out the derivation of the following expression and just present the final result:

$$\begin{aligned} P_L(\theta) &= \frac{-\frac{3}{2}\frac{\mathcal{A}_2}{2}\sin^2\theta\left[\frac{|\langle||\mathbf{E1}||\rangle|^2 - |\langle||\mathbf{M2}||\rangle|^2 + \frac{2}{\sqrt{3}}\langle||\mathbf{E1}||\rangle\langle||\mathbf{M2}||\rangle^*}{|\langle||\mathbf{E1}||\rangle|^2 + |\langle||\mathbf{M2}||\rangle|^2}\right]}{1 + \frac{\mathcal{A}_2}{2}P_2(\cos\theta)\left[\frac{|\langle||\mathbf{E1}||\rangle|^2 - |\langle||\mathbf{M2}||\rangle|^2 + 2\sqrt{3}\langle||\mathbf{E1}||\rangle\langle||\mathbf{M2}||\rangle^*}{|\langle||\mathbf{E1}||\rangle|^2 + |\langle||\mathbf{M2}||\rangle|^2}\right]} \\ &\propto \frac{-\frac{3}{2}\frac{\mathcal{A}_2}{2}\sin^2\theta\left[1 + \frac{2}{\sqrt{3}}\frac{\langle||\mathbf{M2}||\rangle}{\langle||\mathbf{E1}||\rangle}\right]}{1 + \frac{\mathcal{A}_2}{2}P_2(\cos\theta)\left[1 + 2\sqrt{3}\frac{\langle||\mathbf{M2}||\rangle}{\langle||\mathbf{E1}||\rangle}\right]}, \end{aligned} \tag{4}$$

where the $\langle||\mathbf{E1}||\rangle \equiv \langle 2p_{3/2}||\alpha\mathbf{A}^{(\mathbf{e})}_{\mathbf{L=1}}||1s_{1/2}\rangle$ and $\langle||\mathbf{M2}||\rangle$ are the reduced matrix elements for the electric dipole and magnetic quadrupole transitions, respectively [3].

As seen from Equation (4), the main correction due to the magnetic quadrupole decay channel arises from the term which is proportional to the ratio of the transition amplitudes. For high-Z ions, this ratio is of the order $\propto 0.1$, which may lead to the 10–15% magnetic quadrupole correction over the electric dipole approximation (3). Moreover, due to the Z scaling rule for the ratio of the transition amplitudes $\langle||\mathbf{M2}||\rangle/\langle||\mathbf{E1}||\rangle \propto Z^{2.2}$, the non-negligible effect of a few percent remains even for the medium-Z ions. For instance, for the hydrogen-like xenon Xe^{53+} this contribution is still about 4%.

4. Calculations and discussion

Experimental studies on the linear polarization of the characteristic *Lyman*–α_1 radiation are currently planned to be carried at the GSI storage ring within the next few years. As seen from the Equations (3)–(4), such polarization measurements can provide detailed information on the population of the magnetic substates, which is produced by the electron capture processes. However, for a proper interpretation of the experimental data, the magnetic quadrupole correction (M2) should be taken into account since it modifies the linear polarization of the emitted photons. Figure 1, for instance, shows the degree of linear polarization of the *Lyman*–α_1 radiation as a function of the photon emission angle, drawn for the electron capture into the $2p_{3/2}$ state of bare *uranium* ion with energy $T_p = 220$ MeV/u. Two different approximations are shown: while the dashed line corresponds to the electric dipole approximation (i.e., when the magnetic quadrupole term is neglected), the solid line represents the degree of linear polarization $P_L(\theta)$ as defined by Equation (4). Since, moreover, the radiative recombination of ions is usually studied experimentally by collisions of high-Z projectiles with (low-Z) target atoms, the linear polarization in figure 1 is presented for angles as measured in the *laboratory system* (the rest frame of the target atoms). As seen from the figure, the strongest effect due to the magnetic quadrupole term arises around the angle of 54°, which corresponds to an angle of $\theta = 90°$ in the projectile ion system. At this angle, the degree of linear polarization is 33%, which is less than the 38% as obtained from the electric dipole approximation (3).

Until now, we studied the degree of linear polarization $P_L(\theta)$ for the electron capture into bare uranium projectile with energy $T_p = 220$ MeV/u. Apart from the photon emission angle, of course, the linear polarization also depends on the projectile energy. As seen from Equations (3)–(4), the energy dependence is given by the alignment parameter (1), which decreases for the energy range of $T_p = 110\ldots358$ MeV/u from $\mathcal{A}_2 = -0.655$ to $\mathcal{A}_2 = -0.528$ [2]. Of course, decrease of the alignment parameter also leads to a smaller degree of the linear polarization for high projectile energies. Figure 2, for instance, shows the degree of linear polarization (4), drawn for three typical projectile energies. As seen from the figure, the – maximal – linear polarization decreases by one sixth from 36% to 30% for the projectile energies $T_p = 110\ldots358$ MeV/u.

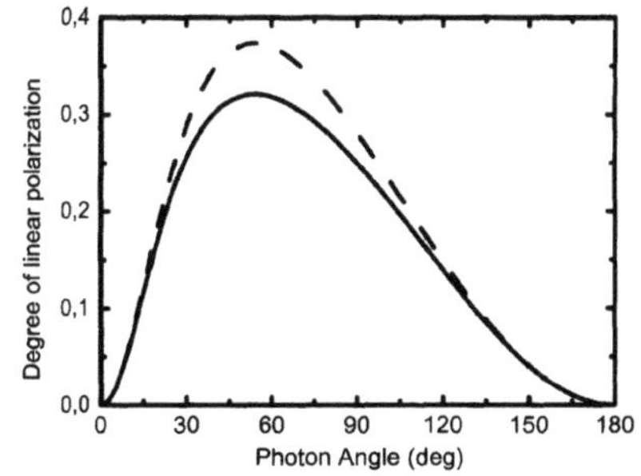

Figure 1. Degree of linear polarization $P_L(\theta)$ of the *Lyman–*α_1 radiation following the electron capture into the $2p_{3/2}$ state of bare uranium projectile with energy $T_p = 220$ MeV/u. The dashed line represents the results within the electric dipole approximation, i.e., when the magnetic quadrupole term M2 is neglected. The solid line shows the degree of linear polarization as defined by Equation (4).

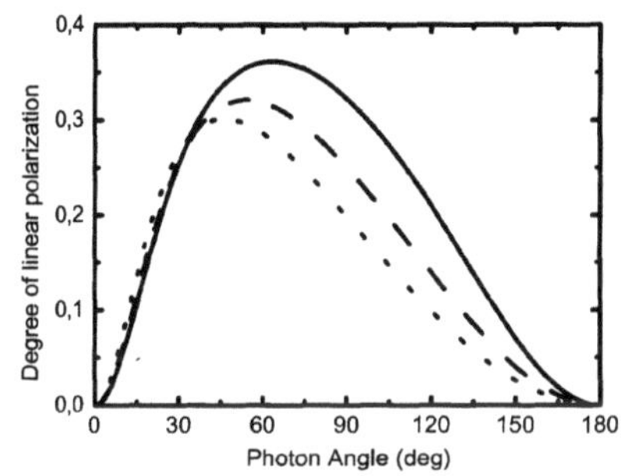

Figure 2. Degree of linear polarization $P_L(\theta)$ of the *Lyman–*α_1 radiation following the electron capture into the $2p_{3/2}$ state of bare uranium ion. Results are shown for the projectile energies: $T_p = 110$ MeV/u (—), $T_p = 220$ MeV/u (– –) and $T_p = 358$ MeV/u (- -).

5. Summary

In this contribution we studied the linear polarization of the *Lyman–*α_1 photons following the radiative recombination of a free electron into the $2p_{3/2}$ state of high-Z, bare ions. This polarization is found to be significantly affected by the interference between the leading electric dipole (E1) decay channel and the weak magnetic quadrupole branch (M2). For the hydrogen-like uranium, the E1–M2 interference leads to the 15% enhancement of the linear polarization when compared with the dipole approximation.

Similar interference effects can be expected also for the polarization of the the characteristic radiation in few-electron heavy ions. Two interest cases of this type are, for instance, the radiative decays in the He-like ions following the *KLL*-dielectronic recombination [11] or the electron-impact excitation [12, 13].

Acknowledgement

This work was supported by the GSI project KS-FRT.

References

1. Stöhlker, Th., *Physica Scripta T* **80** (1999), 165.
2. Eichler, J., Ichihara, A. and Shirai, T., *Phys. Rev. A* **58** (1998), 2128.
3. Surzhykov, A., Fritzsche, S., Stöhlker, Th. and Gumberidze, A., *Phys. Rev. Lett.* **88** (2002), 153001.
4. Stöhlker, Th., Bosch, F., Gallus, A., Kozhuharov, C., Menzel, G., Mokler, P. H., Prinz, H. T., Eichler, J., Ichihara, A., Shirai, T., Dunford, R. W., Ludziejewski, T., Rymuza, P., Stachura, Z., Swiat, P. and Warczak, A., *Phys. Rev. Lett.* **79** (1997), 3270.
5. Balashov, V. V., Grum-Grzhimailo, A. N. and Kabachnik, N. M., *Polarization and Correlation Phenomena in Atomic Collisions*, Kluwer Academic Plenum Publishers, New York, 2000.
6. Berezhko, E. G. and Kabachnik N. M., *J. Phys. B* **10** (1977), 2467.
7. Surzhykov, A. and Fritzsche, S., to be published (2003).
8. Inghoff, T., Fritzsche, S. and Fricke, B., *Comput. Phys. Commun.* **139** (2001), 297.
9. Surzhykov, A., Fritzsche, S. and Stöhlker, Th., *Phys. Lett. A* **289** (2001), 213.
10. Percival, I. C. and Seaton, M. J., *Phil. Trans. R. Soc. A* **251** (1958), 113.
11. Chen, M. H. and Scofield, J. H., *Phys. Rev. A* **52** (1995), 2057.
12. Reed, K. J. and Chen, M. H., *Phys. Rev. A* **48** (1993), 3644.
13. Inal, M. K. and Dubau, J., *J. Phys. B* **20** (1987), 4221.

Hyperfine Interactions **146/147:** 41–45, 2003.

Dielectronic Recombination of Very Heavy Lithiumlike Ions

C. BRANDAU[1], T. BARTSCH[1], S. BÖHM[1], C. BÖHME[1], A. HOFFKNECHT[1], S. KIESLICH[1], H. KNOPP[1], S. SCHIPPERS[1], W. SHI[1], A. MÜLLER[1], N. GRÜN[2], W. SCHEID[2], T. STEIH[2], F. BOSCH[3], B. FRANZKE[3], C. KOZHUHAROV[3], A. KRÄMER[3], P. H. MOKLER[3], F. NOLDEN[3], M. STECK[3], T. STÖHLKER[3] and Z. STACHURA[4]

[1]*Institut für Kernphysik, Universität Giessen, Germany*
[2]*Institut für Theoretische Physik, Universität Giessen, Germany*
[3]*Gesellschaft für Schwerionenforschung (GSI), Darmstadt, Germany*
[4]*Institute for Nuclear Physics, Kraków, Poland*

Abstract. An overview of measurements of dielectronic recombination (DR) with the heaviest lithiumlike ions is presented. The experiments have been carried out at the heavy ion storage ring ESR at GSI utilizing the electron cooler as an electron target. In particular, results for low energy dielectronic recombination of Au^{76+} and U^{89+} are discussed. In the energy range of 0 to 400 eV which can be covered with our present set-up dielectronic resonances associated with $\Delta n = 0$ core excitations from $2s_{1/2} \rightarrow 2p_{1/2,3/2}$ were found. The experimental results can only be explained within a fully relativistic treatment and are sensitive to QED contributions of the order α^2 as well as to the finite size of the atomic nucleus.

Key words: dielectronic recombination, heavy ions.

At low electron densities recombination of a free electron with an atomic ion proceeds mainly via photorecombination (PR). In PR the excess energy of the recombination process is carried away by one or more photons. One usually distinguishes between the direct process, radiative recombination (RR), and the resonant photorecombination channel, dielectronic recombination (DR). The main focus of this paper is on DR while RR is treated as a background process. DR consists of the two steps, dielectronic capture (DC) and subsequent radiative stabilization of the compound system.

$$e^- + A^{q+} \rightarrow \left(A^{(q-1)+}\right)^{**} \rightarrow A^{(q-1)+} + n \cdot h\nu. \tag{1}$$

In the first step a bound electron is excited by the simultaneous radiationless capture of a free electron (time-inverse to autoionization), and DR is completed if the doubly excited intermediate state decays to a state below the autoionization threshold.

The experimental data presented here have been obtained at the electron cooler of the storage ring ESR of GSI. Heavy lithiumlike ions have been accumulated and cooled in the ring at ion energies of typically 90–100 MeV/u. In order to introduce non-zero relative energies between ions and electrons the energy of the electrons has repeatedly been detuned for periods 30–40 ms from cooling energy with the aid of drifttubes (-5 kV to $+5$ kV) mounted around the overlap region of the beams. The recombined ions were counted with fast scintillation counters behind the next dipole bending magnet. With the knowledge of the ion and electron currents a rate coefficient $\alpha = \langle v \cdot \sigma \rangle$ can be obtained on an absolute scale. Details of the measurement procedure at the ESR can be found in [3, 6]. Due to its kinematics a merged beams arrangement provides access to high resolution and precise spectroscopic information in particular at low relative energies.

Lithiumlike ions are the simplest ions for which $\Delta n = 0$ excitations ($2s_{1/2} \rightarrow 2p_{1/2}$ and $2s_{1/2} \rightarrow 2p_{3/2}$) are possible from the ground state, thus providing DR resonances in the low-energy domain. Therefore, lithiumlike ions have been the premier choice to study DR at storage rings during the last decade. However, only a few DR experiments have been carried out for the heaviest lithiumlike ions ($Z > 70$).

For every excitation channel of the DC an infinite number of dielectronic resonances may be observed. For lithiumlike ions with $2s_{1/2} \rightarrow 2p_j$ excitations the according resonance positions $E_{\mathrm{Res}}(Z, n)$ are roughly given by

$$E_{\mathrm{Res}}(Z, n) = E(Z, 2s_{1/2} - 2p_j) - 13.6\ \mathrm{eV} \cdot \left(\frac{Z-3}{n}\right)^2. \tag{2}$$

In this equation $E(Z, 2s_{1/2} - 2p_j)$ is the $2s_{1/2} \rightarrow 2p_j$ excitation energy and $j = 1/2$ or $3/2$. With the aid of tabulated values for $E(2s_{1/2} - 2p_j)$ the lowest principal quantum number n_{min} can be estimated for which DC is energetically possible ($E_{\mathrm{Res}}(Z, n) > 0$).

Figure 1(a) shows the energy splitting $E(2s_{1/2} - 2p_j)$ for lithiumlike ions in dependence on the nuclear charge Z and Figure 1(b) the resulting n_{min}. For low-Z ions the two sublevels with $j = 1/2$ and $j = 3/2$ of the $1s^2 2p_j$ configuration are nearly degenerate, only one series of the $1s^2 2p_j$-resonances can be resolved in a DR spectrum (compare, e.g., [2]), and $n_{\mathrm{min}} = 4$ or 5 is the same for both values of j. With increasing Z both series begin to separate and for $Z \gtrsim 20$ neighbouring resonances belong to different n of the Rydberg electron. In contrast, for the heaviest ions the splitting for $j = 3/2$ is more than one order of magnitude bigger than for $j = 1/2$. As a consequence $n_{\mathrm{min}} = 20$ for $j = 1/2$ and $n_{\mathrm{min}} = 5$ or 6 for $j = 3/2$. Within the experimental energy range of 0 to 400 eV of the ESR set-up one finds a nicely resolved pattern of high n Rydberg resonances from the $2p_{1/2} n\ell_j$ series as well as $n = 5, 6$ $2p_{3/2}$-resonances the fine structure of which covers an energy range of the order of 200 eV (see Figures 2 and 3). There the combination of high Z and low n acts as a magnifying glass which facilitates a deep insight in the couplings of the core and the Rydberg electron. The individual $2p_{3/2} n\ell_j$ resonances are arranged in groups of the same angular momentum j of the outer electron. The additional structure within one group of same j stems from

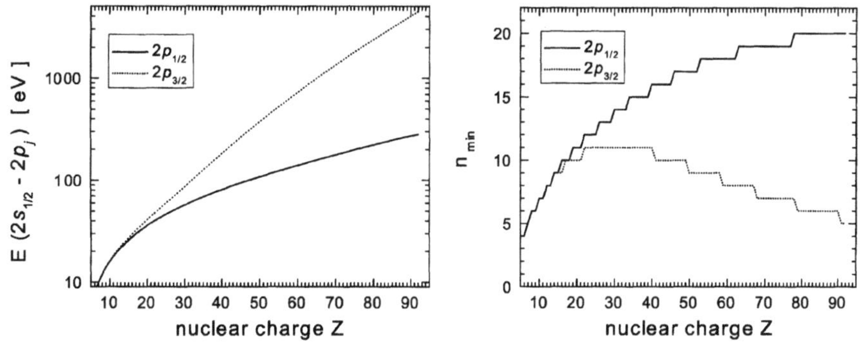

(a) (b)

Figure 1. (a) Energy splitting $E(2s_{1/2} - 2p_j)$ for lithiumlike ions for $j = 1/2$ (solid line) and $j = 3/2$ (dotted line) in dependence on the nuclear charge Z. Values have been taken from [4]. (b) Lowest principal quantum number $n_{\min}$ for which DR is energetically possible. $n_{\min}$ is estimated with equation (2) and $E(2s_{1/2} - 2p_j)$ is the same as in (a).

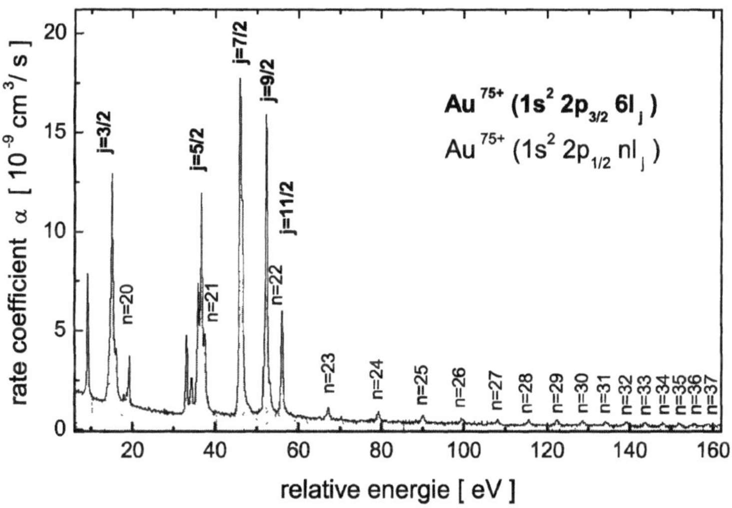

Figure 2. Experimental data for the PR of Au^{76+} (grey shaded area). The smooth background is the contribution from RR. The $2p_{1/2}n\ell_j$ resonances with $n = 21, 22$ are partially masked by the much stronger $2p_{3/2}6\ell_j$ group. Note that the $2p_{3/2}6\ell_j$ resonances with $j = 1/2$ are energetically not possible.

the different couplings of core and Rydberg electron to the total angular momentum J of the doubly excited system. For a comprehensive description of the heavy ions under investigation a fully relativistic treatment is needed [7]. Furthermore, radiative (QED) corrections scale approximately with Z^4 and become increasingly important as well as corrections due to the finite size of the atomic nucleus. As can be seen from Figure 3 a very good agreement between experiment and our fully

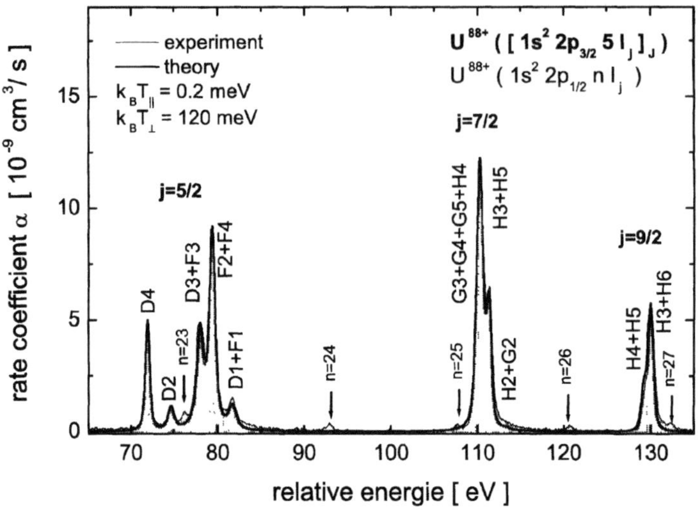

Figure 3. Measured DR spectrum of U^{89+} (thin line/grey shaded area) in comparison with our fully relativistic calculations for the $U^{88+}([1s^2 2p_{3/2} 5\ell_j]_J)$ resonances (thick line). The theory is convoluted with the experimental response function (anisotropic Maxwell–Boltzmann) and shifted by -0.6 eV. In the calculations a Fermi distribution with an rms-radius of 5.86 fm has been used for the atomic nucleus. Arrows indicate the positions of the $U^{88+}(1s^2 2p_{1/2} n\ell_j)$ resonances which are not included in the calculations. The labels are abbreviations for the spectroscopic notation of the doubly excited state: The letter denotes the angular momentum ℓ of the Rydberg electron and the number the total angular momentum J of the compound system.

relativistic DR/MCDF calculations (GRASP code) is found as long as the *shapes* and *resonance strenghts* of the DR peaks are concerned. On the other hand it is known that the GRASP code can produce uncertainties in *absolute* energies in the order of 1 eV or even more, mainly due to the approximations used to include QED in the MCDF code. Until now no strict QED calculations are available which allow for a direct comparison to the measured DR resonances of heavy lithiumlike ions and hence a testing of QED in strong fields. On the other hand the $2s$–$2p$ splitting of lithiumlike ions has been in the focus of big theoretical and experimental efforts (e.g., [5, 8]). Therefore, we have developed a novel method which utilizes the well known energy pattern of the dielectronic Rydberg resonances in order to precisely determine the energy of the underlying excitation channel itself. The main idea is to extrapolate the resonance positions of the high n Rydberg peaks to the series limit. Besides the series limit as one fitting parameter, additional fitting parameters can be introduced to improve the overall energy calibration. The work is still ongoing and not finalized yet, but an error of about 0.1–0.2 eV can be assumed. We like to note that the accuracy of this method is sufficient to test QED in strong fields on a level up to the order of α^2 and is comparable to the most precise experiments for very heavy lithiumlike ions [1, 5].

References

1. Beiersdorfer, P., Osterheld, A. L., Scofield, J. H., Crespo López-Urrutia, J. R. and Widmann, K., *Phys. Rev. Lett.* **80** (1998), 3022.
2. Böhm, S., Schippers, S., Shi, W., Müller, A., Djurić, N., Dunn, G. H., Zong, W., Jelenković, B., Danared, H., Eklöw, N., Glans, P., Schuch, R. and Badnell, N. R., this volume.
3. Hoffknecht, A., Brandau, C., Bartsch, T., Böhme, C., Knopp, H., Schippers, S., Müller, A., Kozhuharov, C., Beckert, K., Bosch, F., Franzke, B., Krämer, A., Mokler, P. H., Nolden, F., Steck, M., Stöhlker, Th. and Stachura, Z., *Phys. Rev. A* **63** (2000), 012702.
4. Kim, Y.-K., Baik, D. H., Indelicato, P. and Desclaux, J. P., *Phys. Rev. A* **44** (1991), 148.
5. Schweppe, J., Belkacem, A., Blumenfeld, L., Claytor, N., Feinberg, B., Gould, H., Kostroun, V. E., Levy, L., Misawa, S., Mowat, J. R. and Prior, M. H., *Phys. Rev. Lett.* **66** (1991), 1434.
6. Shi, W., Böhm, S., Böhme, C., Brandau, C., Hoffknecht, A., Kieslich, S., Schippers, S., Müller, A., Kozhuharov, C., Bosch, F., Franzke, B., Mokler, P. H., Steck, M., Stöhlker, Th. and Stachura, Z., *Eur. Phys. J. D* **15** (2001), 145.
7. Spies, W., Müller, A., Linkemann, J., Frank, A., Wagner, M., Kozhuharov, C., Franzke, B., Beckert, K., Bosch, F., Eickhoff, H., Jung, M., Klepper, O., König, W., Mokler, P. H., Moshammer, R., Nolden, F., Schaaf, U., Spädtke, P., Steck, M., Zimmerer, P., Grün, N., Scheid, W., Pindzola, M. S. and Badnell, N. R., *Phys. Rev. Lett.* **69** (1992), 2768.
8. Yerokhin, V. A., Artemyev, A. N., Shabaev, V. M., Sysak, M. M., Zherebtsov, O.M. and Soff, G., *Phys. Rev. A* **64** (2001), 032109.

Hyperfine Interactions **146/147:** 47–52, 2003.

Measurement of the g Factor of the Bound Electron in Hydrogen-like Oxygen $^{16}O^{7+}$

J. VERDÚ[1,2], T. BEIER[2], S. DJEKIC[1], H. HÄFFNER[2], H.-J. KLUGE[2], W. QUINT[2], T. VALENZUELA[1] and G. WERTH[1]
[1]*Institut für Physik, Universität Mainz, D-55099 Mainz, Germany*
[2]*GSI, Planckstr. 1, D-64291 Darmstadt, Germany*

Abstract. The measurement of the g factor of the electron bound in a hydrogen-like ion is a high-accuracy test of the theory of Quantum Electrodynamics (QED) in strong fields. Here we report on the measurement of the g factor of the bound electron in hydrogen-like oxygen $^{16}O^{7+}$. In our experiment a single $^{16}O^{7+}$ ion is stored in a Penning trap. Quantum jumps between the two spin states (spin *up* and spin *down)* are induced by a microwave field at the spin precession frequency of the bound electron. The g factor of the bound electron is obtained by varying the microwave frequency and counting the number of spin flips. Our experimental value for the g factor of the bound electron is $g_{\text{exp}}(^{16}O^{7+}) = 2.000\,047\,026(4)$. The theoretical prediction from non-perturbative bound-state QED calculations is $g_{\text{th}}(^{16}O^{7+}) = 2.000\,047\,0202(6)$.

Key words: magnetic moment, g factor, quantum electrodynamics, fundamental constants.

1. Introduction

A few years ago we started an experimental and theoretical program to investigate the magnetic moment anomaly of the bound electron in highly charged ions [1]. We constructed a Penning-trap quantum jump spectrometer for highly charged ions and, for the first time, applied the continuous Stern–Gerlach effect [2] to an atomic ion [3]. With a novel double-trap technique we determined the g factor of the bound electron in hydrogen-like carbon $^{12}C^{5+}$ with an accuracy on the ppb level [4]. We have performed bound-state QED calculations of the g factor of the bound electron in hydrogen-like ions with different nuclear charge Z up to uranium $^{238}U^{91+}$ in a non-perturbative treatment [5, 6]. Recent theoretical progress made it possible to determine the electron's mass in atomic units with unprecedented accuracy from the measured g factor of the electron in carbon ($^{12}C^{5+}$) [7, 8].

In this contribution we report on our measurement of the g factor of the bound electron in hydrogen-like oxygen $^{16}O^{7+}$ which is part of our efforts to extend the g-factor measurements to heavy hydrogen-like ions [9].

2. Penning-trap quantum jump spectrometer

In a Penning trap a charged particle is stored in a combination of a homogeneous magnetic field B_0 and an electrostatic quadrupole potential [10]. The magnetic field confines the particle in the plane perpendicular to the magnetic field lines, and the electrostatic potential in the direction parallel to the magnetic field lines. The three eigenmotions that result are the trap-modified cyclotron motion (frequency w_+), the magnetron motion (frequency w_-), which is a circular $E \times B$ drift motion perpendicular to the magnetic field lines, and the axial motion (parallel to the magnetic field lines, frequency w_z). The free-space cyclotron frequency $w_c = (Q/M)B_0$ of an ion with charge Q and mass M in a magnetic field B_0 can be determined from a combination of the trapped ion's three eigenfrequencies w_+, w_z, and w_- with the formula $w_c^2 = w_+^2 + w_z^2 + w_-^2$.

The principle of the continuous Stern–Gerlach effect is based on a coupling of the magnetic moment μ of the particle to its axial oscillation frequency w_z in the Penning trap. This coupling is achieved by a quadratic magnetic field component ("magnetic bottle") superimposed on the homogeneous magnetic field B_0 of the Penning trap, $B(z) = B_0 + \beta_2 z^2$ (Figure 1). Due to the interaction of the z-component μ_z of the magnetic moment with the "magnetic bottle" term the trapped ion possesses a position-dependent potential energy $V_m = -\mu_z(B_0+\beta_2 z^2)$, which adds to the potential energy V_{el} of the ion in the electrostatic well. Therefore, the effective trapping force is modified by the magnetic interaction, and the axial frequency of the trapped ion is shifted upwards or downwards, depending on the

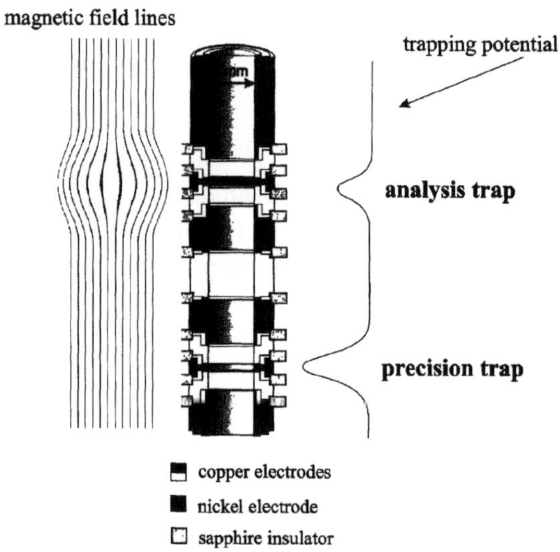

Figure 1. Sketch of the electrode structure and potential distribution of the double trap. Spin flips are induced in the precision trap by a microwave field and detected in the analysis trap via the continuous Stern–Gerlach effect.

sign of the z-component μ_z of the magnetic moment. This axial frequency shift is given by

$$\delta w_z = \frac{\beta_2 \mu_z}{M w_z}, \tag{1}$$

where w_z is the unshifted axial frequency. Due to the scaling with $1/M$, the frequency shift becomes smaller for heavier hydrogen-like ions.

In our Penning trap apparatus, which has been described in [11], there are two positions in the stack of cylindrical electrodes where ions can be trapped (Figure 1). In the precision trap, a single $^{16}O^{7+}$ ion is prepared, and the spin-flip transition of the bound electron is excited by applying a microwave field. The single $^{16}O^{7+}$ ion is then transported along the magnetic field lines to the analysis trap. The ring electrode of this trap is made out of ferromagnetic material (nickel) to produce the quadratic component of the magnetic field ($\beta_2 = 0.01$ T/mm^2) which is necessary to observe the continuous Stern–Gerlach effect. The axial oscillation frequency w_z, of the single $^{16}O^{7+}$ ion in the analysis trap is measured non-destructively with an electronic detection method through the image currents which are induced in the trap electrodes by the particle motion [12]. A *LCR* circuit resonant at $w_z, = 2\pi \times$ 369 kHz (with quality factor $Q = 2400$) is attached to one of the trap electrodes to optimize the detection sensitivity. The ion's axial frequency is determined in a fast Fourier transform (FFT) of the signal across the resonance circuit Figure 2.

3. *g*-Factor measurement

The quantum state of the $^{16}O^{7+}$ ion, i.e. the magnetic quantum number m_s of the bound electron in the $1s_{1/2}$ ground state, can be determined non-destructively in the analysis trap in measurements of the axial frequency of the trapped ion. Transitions between the two spin states $m_s = \pm 1/2$ are induced by a microwave field (at 104 GHz) resonant with the Larmor precession frequency w_L of the bound electron

$$\hbar w_L = g \frac{e\hbar}{2m_e} B = g \mu_B B. \tag{2}$$

Here, g is the g factor of the bound electron and $\mu_B = e\hbar/2m_e$ is the Bohr magneton. The spin-flip transitions are observed as discrete changes of the ion's axial frequency. Figure 2 shows such a quantum jump observed via the continuous Stern–Gerlach effect. In the case of hydrogen-like oxygen $^{16}O^{7+}$, the measured axial frequency shift for a transition between the two quantum levels is $w_z(\uparrow) - w_z(\downarrow) = 2\pi \times 0.45$ Hz. We have plotted the axial frequency versus time when we irradiate the ion with a microwave field at the Larmor precession frequency of the bound electron. Frequency jumps of 0.45 Hz can be clearly observed (Figure 3).

The observation of the continuous Stern–Gerlach effect on the hydrogen-like carbon ion $^{16}O^{7+}$ makes it possible to determine its electronic g factor to high accuracy. Using the cyclotron frequency w_c of the ion for the calibration of the

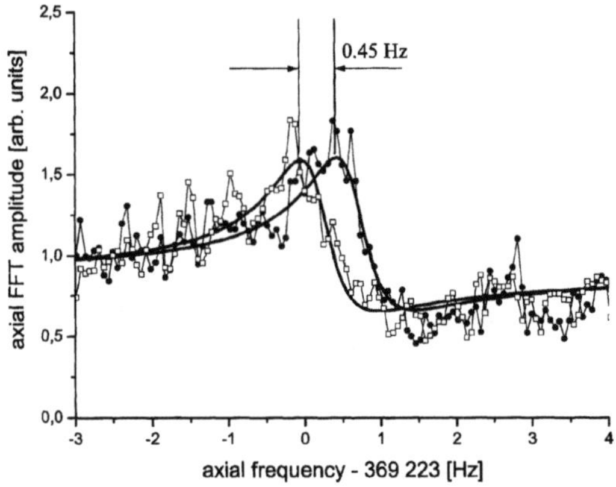

Figure 2. Axial frequencies of a single $^{16}O^{7+}$ ion for different spin directions. The averaging time for each resonance curve was 1 min (FFT: fast Fourier transform).

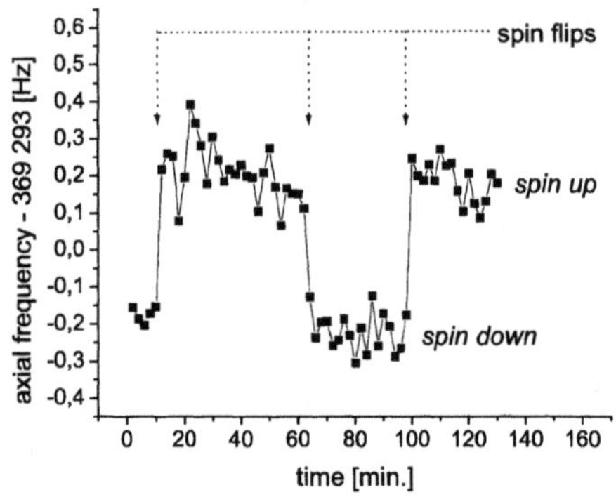

Figure 3. Quantum non-demolition measurement of the spin state: the spin-flip transitions are observed as small discrete changes of the axial frequency of the stored $^{16}O^{7+}$ ion.

magnetic field B, the g factor of the bound electron can be calculated from the ratio of the Larmor precession frequency w_L of the electron and the cyclotron frequency w_c of the $^{16}O^{7+}$ ion when the mass ratio of the ion and the electron is known

$$g = 2 \cdot \frac{w_L}{w_c} \cdot \frac{Q/M}{e/m_e}. \tag{3}$$

A resonance spectrum of the Larmor precession frequency w_L of the bound electron is obtained in the following way. First, we determine the spin direction in the analysis trap via the continuous Stern–Gerlach effect. The hydrogen-like ion is then transferred to the precision trap where spin flips are induced by a microwave field at frequency $w_{\text{mw}} \approx w_L$. Then the $^{16}O^{7+}$ ion is transferred back to the analysis trap where the spin state is analyzed again. Now the ion is moved back to the precision trap, and the measurement cycle is started again. The number of spin-flip transitions which occurred in the precision trap is counted. Then the microwave

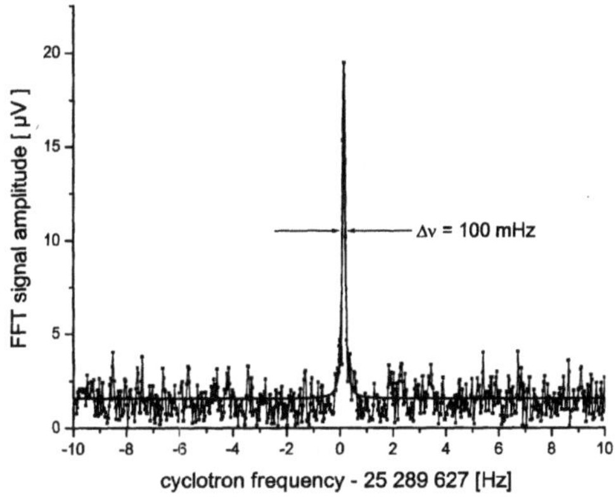

Figure 4. Fourier transform of the voltage induced in one of the electrodes from the cyclotron motion of a single $^{16}O^{7+}$ ion stored precision trap. The fractional line width of the cyclotron resonance is a few ppb.

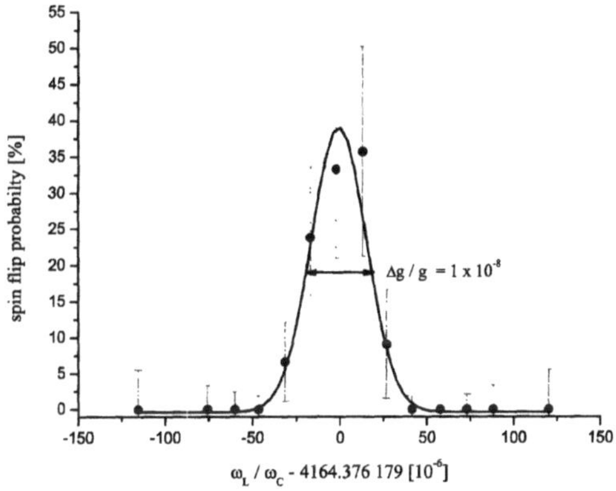

Figure 5. Larmor resonance spectrum measured in the precision trap. Spin-flip transitions are driven by a microwave field at frequency w_{mw}. The cyclotron frequency w_c^{ion} of the $^{16}O^{7+}$ ion is measured simultaneously for magnetic-field calibration. The spin-flip probability is the ratio of observed spin flips to the number of attempted excitations.

frequency w_{mw} is varied and the measurement is repeated at different excitation frequencies.

Finally, the plot of the quantum jump rate versus excitation frequency yields the resonance spectrum. For magnetic field calibration, the Larmor frequency is divided by the cyclotron frequency of the $^{16}O^{7+}$ ion which is simultaneously measured in a fast Fourier transform of the image currents induced in the trap electrodes by the ion motion (Figure 4). The Larmor resonance shown in Figure 5 was obtained from raw measurement data taken during our measurement campaign. A number of small corrections of the order of a few ppb arising mainly from the finite cyclotron energy of the ion have to be applied in order to obtain the final result.

The g factor of the bound electron in $^{16}O^{7+}$ is determined from the measured Larmor resonance by inserting the mass ratio of the ion and the electron into Equation (3) [13]. Our value for the g factor of the bound electron in $^{16}O^{7+}$ is $g_{exp}(^{16}O^{7+}) = 2.000\,047\,026(4)$. The most recent theoretical prediction from non-perturbative bound-state QED calculations is $g_{th}(^{16}O^{7+}) = 2.000\,047\,0202(6)$ [8]. Within the error bars the measurement confirms the validity of the theory on the ppb-level.

Future measurements of the g factor of the bound electron in heavier hydrogen-like ions will provide even more stringent tests of bound-state quantum electrodynamics. If QED is proven to predict the measured g-factor values, such measurements enable, in addition, a number of quite exciting novel applications [7].

Note added in proof

Additional measurements after submission of the manuscript led to a final value for the g-factor of $g_{exp}(^{16}O^{7+}) = 2.000\,047\,024\,6(46)$.

Acknowledgements

We are grateful to N. Hermanspahn, S. G. Karshenboim, V. M. Shabaev, S. Stahl, and A. S. Yelkhovsky for enlightening and stimulating discussions. Financial support was obtained from the European Union under the contract numbers ERB FMRX CT 97-0144 (EUROTRAPS network) and HPRI-CT-2001-50036 (HITRAP network).

References

1. Quint, W., *Phys. Scr.* **T59** (1995), 203.
2. Dehmelt, H., *Proc. Natl. Acad. Sci. U.S.A.* **83** (1986), 2291.
3. Hermanspahn, N., Häffner, H., Kluge, H.-J., Quint, W., Stahl, S., Verdú, J. and Werth, G., *Phys. Rev. Lett.* **84** (2000), 427.
4. Häffner, H., Beier, T., Hermanspahn, N., Kluge, H.-J., Quint, W., Stahl, S., Verdú, J. and Werth, G., *Phys. Rev. Lett.* **85** (2000), 5308.
5. Beier, T., *Phys. Rep.* **339** (2000), 79.
6. Beier, T., Lindgren, I., Persson, H., Salomonson, S., Sunnergren, P., Häffner, H. and Hermanspahn, N., *Phys. Rev. A* **62** (2000), 032510.
7. Beier, T., Häffner, H., Hermanspahn, N., Karshenboim, S. G., Kluge, H.-J., Quint, W., Stahl, S., Verdú, J. and Werth, G., *Phys. Rev. Lett.* **88** (2002), 011603.
8. Yerokhin, V. A., Indeficato, P., Shabaev, V. M., *Phys. Rev. Lett.* **89** (2002), 143001.
9. Quint, W., Dilling, J., Djekic, S., Häffner, H., Hermanspahn, N., Kluge, H.-J., Marx, G., Moore, R., Rodriguez, D., Schönfelder, J., Sikler, G., Valenzuela, T., Verdu, J., Weber, C., Werth, G., *Hyp. Interact.* **132** (2001), 453.
10. Dehmelt, H., *Rev. Mod. Phys.* **62** (1990), 525.
11. Diederich, M., Häffner, H., Hermanspahn, N., Immel, M., Quint, W., Stahl, S., Kluge, H.-J., Ley, R., Mann, R. and Werth, G., *Hyp. Interact.* **115** (1998), 185.
12. Winefand, D. J. and Dehmelt, H. G., *J. Appl. Ph.* **46** (1975), 919.
13. Mohr, P. J. and Taylor, B. N., *Rev. Mod. Phys.* **72** (2000), 351.
14. Shabaev, V. M., E-print archive, physics/0110056 (2001) (http://www.lanl.gov).

Hyperfine Interactions **146/147:** 53–57, 2003.

Mass of the Electron from the Electronic g Factor in Hydrogenlike Carbon – the Influence of Other Fundamental Parameters

THOMAS BEIER[1], H.-JÜRGEN KLUGE[1], WOLFGANG QUINT[1], HARTMUT HÄFFNER[2] and GÜNTHER WERTH[2]
[1] *GSI, Atomic Physics, Planckstr. 1, 64291 Darmstadt, Germany; e-mail: t.beier@gsi.de*
[2] *Universität Mainz, Institut für Physik, 55099 Mainz, Germany*

Abstract. The electron's mass determined from the electronic g factor of $^{12}C^{5+}$ is given by $m_e = 0.000\,548\,579\,909\,2\,(4)$ u and has a three times lower error margin than the 1998 CODATA value. We elucidate the influence of the fine-structure constant and the binding energies of the electrons in ^{12}C to this value.

Key words: electron's mass, g factor, fine structure constant, fundamental constants, QED.

Recently, we reported a new determination of the electron's mass [2] based on a high-precision measurement of the electronic g factor in hydrogenlike carbon [8]. We obtained a value of

$$m_e(g) = 0.000\,548\,579\,909\,2\,(4)\ \text{u} \tag{1}$$

which agrees with the 1998 CODATA value from [12] (based on [5]),

$$m_e(\text{CODATA}) = 0.000\,548\,579\,911\,0\,(12)\ \text{u} \tag{2}$$

within 1.5 standard deviations but is three times more precise. In this contribution we are going to elucidate the influence of other parameters from the recent listing of fundamental constants [12] on the value (1). Such a compilation also helps to clarify the relations between the different fundamental constants.

In [2], the mass of the electron is obtained from

$$m_e = \frac{g}{2}\frac{e}{q}\frac{\omega_c}{\omega_L} m_{\text{ion}}, \tag{3}$$

where e is the (positive) fundamental charge unit, q is the charge of the ion, ω_c/ω_L is the measured frequency ratio of the ion's cyclotron frequency and the Larmor precession frequency of the electron (for details cf. [8]), g is the theoretical value for the g factor ([1, 2] and references therein) and m_{ion} is the ion's mass. For $^{12}C^{5+}$,

$e/q = 1/5$, $\omega_L/\omega_c = 4\,376.210\,498\,9\,(19)(13)$ (cf. [2], statistical and systematical uncertainties are separately indicated), and

$$m\left(^{12}\mathrm{C}^{5+}\right) = m\left(^{12}\mathrm{C}\right) - 5m_e + E_B/c^2. \tag{4}$$

When inserting (4) into (3), m_e is present on both sides. The final formula thus is given by

$$m_e = \frac{1}{5}\left[m\left(^{12}\mathrm{C}\right) + E_B/c^2\right] \Big/ \left(\frac{2}{g}\frac{\omega_L}{\omega_c} + 1\right), \tag{5}$$

where in both the numerator and the denominator the second term is small compared with the first one. In particular, the influence of the cumulative binding energies E_B becomes clear. They were taken from [12] to be $E_B = 5.79835(1) \times 10^{-7}\,\mathrm{u}c^2$ where the error comes from the missing last digit for the 5th electron in [12]. As $m(^{12}\mathrm{C}) \equiv 12\,\mathrm{u}$, the binding-energy uncertainty affects the mass of the electron only on the level of 10^{-13} and is therefore negligible at present.

The major influence from other constants results from the theoretical prediction for g, given in Table I where all contributions are listed. The numerical entries in the table are not always obtained from the series expansion in the fine-structure constant α. For technical reasons, the bound-state QED terms of order (α/π) (row 5) were calculated together with the corresponding term for the g factor of the free electron, (α/π) (cf. [3]) to all orders in $Z\alpha$. The numerical result in row 5 was obtained after subtracting the free-electron g factor part (row 4) which is exactly (α/π) [13]. The indicated series expansion was not employed, and in fact only its leading term is known [7]. Only recently, a complete analytical formula for some vacuum polarization contributions [10] was presented for a point nucleus, but it accounts only for a minor proportion of the bound-state effects in $^{12}\mathrm{C}^{5+}$. The numbers $(-0.328\ldots)$ and $(1.181\ldots)$ indicated in rows 6, 7, and 8 are analytically known expansion coefficients for the g factor of the free electron. The corresponding (lengthy) formulae can be found in [9] and references therein.

The finite-nuclear-size correction (row 2) can be obtained both from the detailed derivation of [6] and also from a comparison of the results employing electron wave functions for an extended nuclear charge distribution and comparing the result with that for the point-nucleus value given in row 1,

$$g_{\text{bound}} = \frac{2}{3}\left[1 + 2\sqrt{1 - (Z\alpha)^2}\right]. \tag{6}$$

For low-Z systems like C^{5+} and the required precision, both methods yield identical results.

The recoil correction accounts for the movement of the nucleus when surrounded by an electron because it is not infinitely heavy. It has to be considered fully relativistically and in particular for heavier systems to all orders in $Z\alpha$. Only recently this problem was completely solved for spinless nuclei [14, 15] at least to order (m_e/M_N) whereas for arbitrary nuclear spin still only expansions in $Z\alpha$ exist

Table I. Theoretical contributions to $g(^{12}C^{5+})$. 1 = Dirac theory (incl. binding), cf. Equation (6), 2 = correction for nuclear size, 3 = correction for nuclear mass, 4 = QED, free, order (α/π), 5 = QED, bound, order (α/π), 6 = QED, free, orders $(\alpha/\pi)^2$ to $(\alpha/\pi)^4$, 7 = QED, bound, order $(\alpha/\pi)^2$, 8 = QED, bound, order $(\alpha/\pi)^3$ and higher. The leading terms of an expansion in α are indicated. In a few cases the terms of the next order (indicated by $\mathcal{O}(\ldots)$) contain also a factor $\log(Z\alpha)$ multiplied to the corresponding power. In rows 5, 7 and 8, the specific term of order $(Z\alpha)^4$ is still unknown. The entry in row 5 however has been calculated non-perturbatively to all orders in $(Z\alpha)$. Where no error is given, it is less than one unit of the last digit. The error for the bound-state QED of order (α/π) is purely numerical, the error for bound-state QED, order $(\alpha/\pi)^2$, results from employing a perturbation series in $Z\alpha$. M_N indicates the nuclear mass, $\lambda\!\!\!^{-} = \hbar/(m_e c)$ the Compton wavelength of the electron, and $\langle R^2\rangle^{(1/2)}$ the nuclear rms radius. The numerical results sometimes cover more terms of the analytical evaluation than are given here

1	1.998 721 354 4	$2-(2/3)(Z\alpha)^2+\mathcal{O}(Z\alpha)^4$	[4]
2	+0.000 000 000 4	$(8/3)(Z\alpha)^4\,\lambda\!\!\!^{-\,-2}\langle R^2\rangle+\mathcal{O}(Z\alpha)^6$	[6]
3	+0.000 000 087 6	$(m_e/M_N)((Z\alpha)^2-(1/12)(Z\alpha)^4+\mathcal{O}(Z\alpha)^5)$	[15]
4	+0.002 322 819 5	α/π	[13]
5	+0.000 000 844 2 (9)	$\alpha/\pi((Z\alpha)^2/6+\mathcal{O}(Z\alpha)^4)$	[3, 10]
6	−0.000 003 515 1	$2[(\alpha/\pi)^2\times(-0.328\ldots)+(\alpha/\pi)^3\times(1.181\ldots)]+\mathcal{O}(\alpha/\pi)^4$	[9]
7	−0.000 000 001 1 (4)	$2(\alpha/\pi)^2((Z\alpha)^2/6\times(-0.328\ldots)+\mathcal{O}(Z\alpha)^4)$	[7]
8	0.000 000 000 01	$2(\alpha/\pi)^3((Z\alpha)^2/6\times(1.181\ldots)+\mathcal{O}(Z\alpha)^4)$	[7]
Tot.:	2.001 041 589 9 (10)		

([11] and references therein). However, for carbon even the first term in such an expansion would be sufficient to obtain the value presented in the table.

The expansions in Table I indicate also the dependence on other parameters, i.e., the fine-structure constant α, the nuclear size, via the Compton wavelength of the electron also Planck's constant $\hbar$ and the speed of light c, and also the masses of electron and nucleus. When determining the electron's mass according to Equation (5), the dependence of g_{theo} on m_e is so weak, however, that inserting the value from (2) causes no difference on the current level of precision.

The most important dependence is that from the fine-structure constant α. The leading influence up to order α^3 is given by

$$g(\alpha) = 2+\alpha\left(\frac{1}{\pi}\right)+\alpha^2\left[\left(\frac{m_e}{M_N}-\frac{2}{3}\right)Z^2+\frac{2}{\pi^2}(-0.328\ldots)\right] + \alpha^3\left[\frac{Z^2}{6\pi}+\frac{2}{\pi^3}(1.181\ldots)\right]+\cdots. \tag{7}$$

Together with Equation (5) we obtain the leading dependence of m_e from α up to order α^3 ($A\equiv m(^{12}C)+E_B/c^2$, $B\equiv\omega_L/\omega_c$),

$$m_e(\alpha) = \frac{1}{5}\frac{A}{B+1}+\alpha\left[\frac{1}{10}\frac{1}{\pi}\frac{AB}{(B+1)^2}\right]+\frac{\alpha^2}{2}\frac{1}{10}\frac{AB}{(B+1)^2}$$

$$\times \left\{ -\frac{1}{\pi^2(B+1)} + 2\left[\left(\frac{m_e}{M_N} - \frac{2}{3} \right) Z^2 + \frac{2}{\pi^2}(-0.328\ldots) \right] \right\}$$
$$+ \frac{\alpha^3}{6} \frac{1}{10} \frac{AB}{(B+1)^2} \left\{ \frac{3}{2} \frac{1}{\pi^3} \frac{1}{(B+1)^2} - 6 \frac{1}{\pi} \frac{1}{(B+1)} \right.$$
$$\times \left[\left(\frac{m_e}{M_N} - \frac{2}{3} \right) Z^2 + (-0.328\ldots) \frac{2}{\pi^2} \right]$$
$$\left. + 6 \left[\frac{Z^2}{6\pi} + (1.181\ldots) \frac{2}{\pi^3} \right] \right\} + \cdots . \tag{8}$$

In our calculations, we employ the value for α from [12],

$$\alpha = 1/137.035\,999\,76(50). \tag{9}$$

It causes an uncertainty of 2×10^{-11} in the prediction for g and is therefore at the current level of precision negligible compared to the other sources of error [2]. It should be noted, however, that due to the Z^2 terms in (7) and (8), the sensitivity on α is higher for heavier ions. A g-factor experiment in Ca^{19+} with the same precision as in carbon would allow a determination of α with a relative precision of 1.5×10^{-8}.

The discussion presented here is an important addition to our determination of the electron's mass because they indicate the relations between the fundamental constants employed, and the correlation coefficients between them form an important topic in [12].

Note added in proof

During the course of publication of the current work, new theoretical results have been published [16] which led to a new value for the mass of the electron of $m_e = 0.000\,548\,579\,909\,3(3)$ u. This result does not affect the above considerations.

Acknowledgements

We are grateful to S. G. Karshenboim, V. M. Shabaev, and A. S. Yelkhovsky for enlightening and stimulating discussions. Financial support was obtained from the European Union under the contract number ERB FMRX CT 97-0144 (EURO-TRAPS network).

References

1. Beier, T., *Physics Reports* **339**(2–3) (2000), 79.
2. Beier, T., Häffner, H., Hermanspahn, N., Karshenboim, S. G., Kluge, H.-J., Quint, W., Stahl, S., Verdú, J. and Werth, G., 2002, *Phys. Rev. Lett.* **88**(1) (2002), 011603-1.

3. Beier, T., Lindgren, I., Persson, H., Salomonson, S., Sunnergren, P., Häffner, H. and Hermanspahn, N., *Phys. Rev. A* **62** (2000), 032510-1.
4. Breit, G., *Nature (London)* **122**(3078) (1928), 649.
5. Farnham, D. L., Van Dyck, Jr., R. S. and Schwinberg, P. B., *Phys. Rev. Lett.* **75**(20) (1995), 3598.
6. Glazov, D. A. and Shabaev, V. M., *Phys. Lett. A* **297** (2002), 408.
7. Grotch, H., *Phys. Rev. Lett.* **24**(2) (1970), 39.
8. Häffner, H., Beier, T., Hermanspahn, N., Kluge, H.-J., Quint, W., Stahl, S., Verdú, J. and Werth, G., *Phys. Rev. Lett.* **85**(25) (2000), 5308.
9. Hughes, V. W. and Kinoshita, T., *Rev. Mod. Phys.* **71**(2) (1999), 133.
10. Karshenboim, S. G., Ivanov, V. G. and Shabaev, V. M., *Zh. Eksp. Teor. Fiz.* **120** (2001), 546. [*JETP* **93** (2001), 477.]
11. Martynenko, A. P. and Faustov, R. N., *Zh. Eksp. Teor. Fiz.* **120** (2001), 539. [*JETP* **93** (2001), 471.]
12. Mohr, P. J. and Taylor, B. N., *Rev. Mod. Phys.* **72**(2) (2000), 351.
13. Schwinger, J., *Phys. Rev.* **73** (1948), 416.
14. Shabaev, V. M., *Phys. Rev. A* **64** (2001), 052104-1.
15. Shabaev, V. M. and Yerokhin, V. A., *Phys. Rev. Lett.* **88**(9) (2002), 091801-1.
16. Yerokhin, V. A., Indelicato, P. and Shabaev, V. M., *Phys. Rev. Lett.* **89**(14) (2002), 143001-1.

Hyperfine Interactions **146/147:** 59–65, 2003.

Towards a Measurement of the $n = 2$ Lamb Shift in Hydrogen-like Nitrogen Using an Electron Beam Ion Trap

K. HOSAKA[1], D. N. CROSBY[1], K. GAARDE-WIDDOWSON[1], C. J. SMITH[1], J. D. SILVER[1], E. G. MYERS[2], T. KINUGAWA[3] and S. OHTANI[3]
[1]*Department of Physics, University of Oxford, Oxford, OX1 3PU, UK*
[2]*Department of Physics, Florida State University, Tallahassee, FL 32306, USA*
[3]*Cold Trapped Ions Project, JST, University of Electro-Communications, Chofu, Tokyo 182-0024, Japan*

Abstract. Using a $^{14}C^{16}O_2$ laser the $2s_{1/2}$–$2p_{3/2}$ (fine structure – Lamb shift) transition has been induced in $^{14}N^{6+}$ ions trapped in an electron beam ion trap. Prospects for a measurement of the Lamb shift in hydrogen-like nitrogen are discussed.

Key words: electron-beam ion trap, laser spectroscopy, hydrogen-like nitrogen, Lamb shift.

1. Introduction

The electron beam ion trap (EBIT), e.g., [1, 2] is a device for producing trapped highly charged ions and has been widely used for X-ray as well as UV and visible spectroscopy in emission. Although there have been previous attempts to utilise laser spectroscopy on highly-charged ions in an EBIT, one example being the investigation into the fine-structure transitions in oxygen-like argon [3], the relatively low number of trapped ions, small transition probabilities, difficulty of discriminating against background signals, and large Doppler widths, have limited the attainability of a useful signal. Here, we have employed a continuous-wave carbon dioxide laser, tuned to the $2s_{1/2}$–$2p_{3/2}$ transition in hydrogen-like nitrogen, and optimised running conditions in order to obtain a clear laser induced signal.

Comparison of theory with experiment for transitions in hydrogen and hydrogen-like ions continues to be an important test of fundamental physics. Transitions in hydrogen-like ions can also be used to provide calculable secondary frequency standards [4]. Current interest in the theory of hydrogen and hydrogen-like ions has been directed towards the calculation of binding corrections to the two-loop self-energy [5–19]. These binding corrections scale more rapidly with atomic number, Z, than the total Lamb shift splitting. Hence useful tests may be provided by Lamb shift measurements of higher Z hydrogen-like ions even though they may be of lower precision than those in hydrogen.

Significant progress in accurate measurements of the $n = 2$ Lamb shift in moderate-Z hydrogen-like ions [20–31] has been difficult to achieve. The most sensitive measurement to date in this range of Z is that for P^{14+} [27], which has a fractional precision of 0.14% of the $n = 2$ Lamb shift. This is of similar magnitude to the two-loop binding corrections, which are of order $\alpha^2(Z\alpha)^5$ atomic units. Hence improved precision is required in order to obtain a critical test of theory. In moderate-Z hydrogen-like ions, the $2s_{1/2}$–$2p_{1/2}$ and $2s_{1/2}$–$2p_{3/2}$ transitions lie in the infrared and visible regions of the electromagnetic spectrum and thus, in principle, are amenable to laser spectroscopy. The laser resonance technique relies on the metastable $2s_{1/2}$ state. Laser radiation is used to transfer ions from the $2s_{1/2}$ state to either the $2p_{1/2}$ or the $2p_{3/2}$ state, from which they rapidly decay to the ground state. The resonance is monitored by observing the rate of emission of Lyman-α photons as a function of laser frequency. All previous $n = 2$ Lamb shift measurements for moderate-Z hydrogen-like ions have been carried out using fast ion beams. The precision of these measurements has in general been limited by poor statistics, and by systematic effects such as properly normalising the signal as the laser is tuned across the relatively wide resonance (due to the short lifetime of the $2p$ states). However, for more precise measurements, uncertainties in the Doppler shift correction for the motion of the ions can form a significant source of error. Spectroscopy of trapped ions, with no net centre of mass motion, avoids this problem.

Both the laser induced transition probability per unit intensity and the lifetime of the $2s_{1/2}$ metastable state scale as Z^{-6}. Hence, as regards choice of ion, there is a trade-off between increased sensitivity to the corrections of interest and the feasibility of obtaining a useful signal. The choice of hydrogen-like nitrogen was motivated by the ready availability of high laser power at the required wavelength and the improved signal-to-background ratio obtainable at a moderate Z. For hydrogen-like nitrogen ion, Ivanov and Karshenboim [19] have determined a theoretical value for the $2s_{1/2}$–$2p_{3/2}$ transition $^{14}N^{6+}$ of 834.931(7) cm^{-1}, where the dominant error is the uncertainty in the two-loop binding corrections. Other QED corrections, and the nuclear size correction (based on the stated error of the nuclear charge radius measurements), contribute only about 0.001 cm^{-1}. Hence an experimental precision of 0.002 cm^{-1} or better would provide an interesting test of the two loop corrections.

2. Experimental

The technique for studying the $2s_{1/2}$–$2p_{3/2}$ transition in hydrogen-like nitrogen by laser spectroscopy is illustrated in Figure 1. In hydrogen-like nitrogen the $2s_{1/2}$–$2p_{3/2}$ (fine structure – Lamb shift) transition lies at around 12 μm. This wavelength is accessible to a radioactive isotope $^{14}C^{16}O_2$ laser (wavelength range 9.9–12.1 μm). The experiment uses the laser to induce ions from the metastable $2s_{1/2}$ state, mean lifetime 1.0×10^{-6} s, which is excited by electron impact, to the $2p_{3/2}$

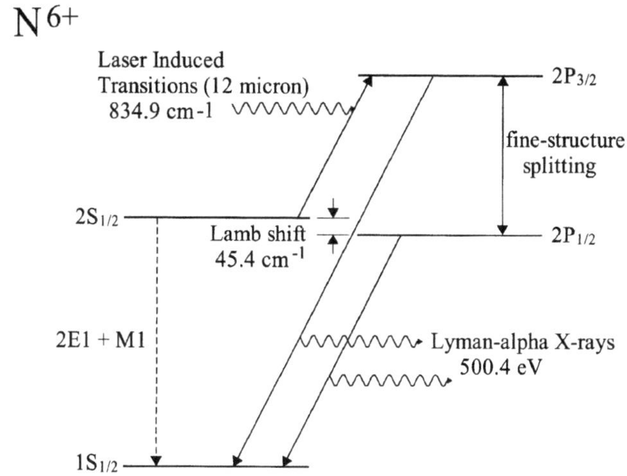

Figure 1. Partial term diagram of a hydrogen-like nitrogen ion, showing the laser induced transition. (Not to scale.)

state, mean lifetime 6.6×10^{-13} s. The $2p_{3/2}$ state decays to the ground state via electric dipole radiation, producing a Lyman-α X-ray at 500.4 eV. In order to detect these low energy X-rays and determine the Lyman-α emission rate as a function of laser wavelength, a lithium-drifted silicon (Si(Li)) detector with a quantum efficiency of 45% at 500 eV is employed. Subsequently, the $n = 2$ Lamb shift may be deduced by correcting for the $2p_{3/2}$–$2p_{1/2}$ $n = 2$ fine structure splitting, which is accurately known from theory.

The experimental arrangement is shown in Figure 2. A 1.5 keV, 30 mA electron beam was compressed to about 70 μm in diameter by the 3 tesla axial field of a pair of super-conducting Helmholtz magnets. Nitrogen gas was injected perpendicular to the electron beam wherein sequential electron impact ionisation produces highly charged ions. The resulting ions are trapped radially by a combination of the space charge of the electron beam and the magnetic field. Axial confinement is provided by voltage applied to three collinear drift tubes. The ions are thus confined to a cylindrical volume of about 2.5 cm in length. Spectroscopic measurements are made via slotted apertures in the centre drift tube. The ionisation balance obtained within the trap depends on parameters such as the electron beam energy and background gas pressure. These were optimised to maximise the observed yield of nitrogen K X-rays.

As shown in Figure 2, the laser beam was directed into the EBIT through a ZnSe viewport and intersects the electron beam at right angles. An off-axis concave mirror was used to focus the laser beam to a spotsize of approximately 1 mm in width and 4 mm in height onto the central trap region. The Si(Li) detector observed this region through another slot in the drift tube. In order to minimise the collection of X-rays from regions of the ion cloud not irradiated by the laser an additional aper-

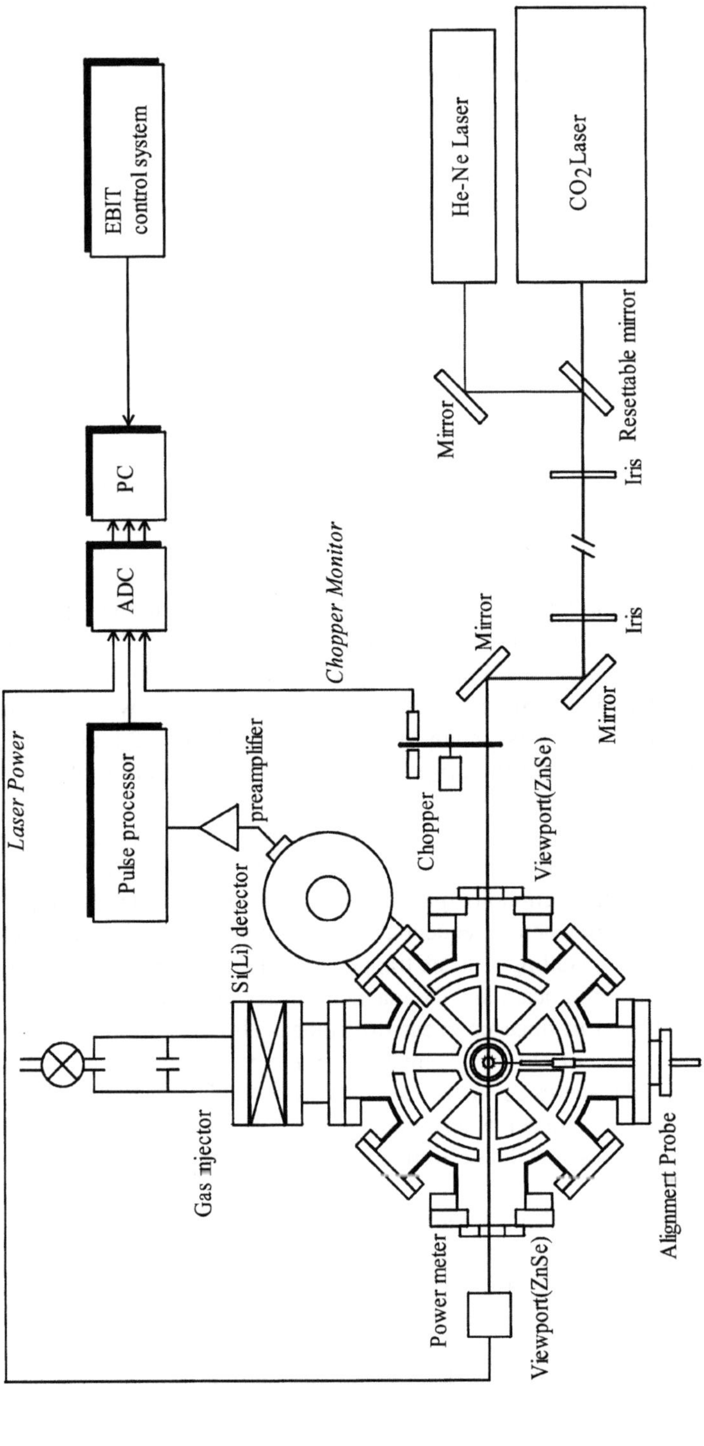

Figure 2. Schematic diagram of the experimental set up.

ture was mounted to the front of the detector. This aperture, and the geometrical constraints inside EBIT limited the detection solid angle to only 4.8 msterad. The laser beam was chopped at 125 Hz by a chopper wheel and the output power was monitored by a power meter placed on the far side of the EBIT. The pulse height spectrum of the X-rays, a chopper monitor signal, and the output of the power meter, are simultaneously fed into a personal computer allowing X-ray detection events to be recorded in coincidence with the chopper monitor signal. A helium–neon laser and a retractable probe inside the EBIT were used for alignment of the CO_2 laser beam to the electron beam.

3. Results and discussion

Figure 3 shows the enhancement of X-rays from hydrogen-like nitrogen ions induced by laser interaction as a function of channel number in the digitised output of the Si(Li) detector. For this measurement, the laser wavelength was fixed on the P36 laser line of $^{14}C^{16}O_2$ (835.537 cm^{-1}) with a power of 15.1 W. In this figure, (a) is total X-rays count and (b) is the enhancement obtained by subtracting the X-ray counts with the laser blocked by the chopper wheel (the OFF signal) from those with the laser entering the EBIT (ON). The peak at around channel 70 in Figure 3(b) indicates the laser radiation enhancement of excitation from $2s_{1/2}$ to $2p_{3/2}$ states. In the region from channel 50 to 90, which corresponds to the $1s$–$2p$ transition in hydrogen-like nitrogen, the ratio between the laser enhanced signal and the total counts is approximately 0.9%. This low signal is consistent with the estimated laser induced transition probability, and the fact that the total X-ray background includes contributions from helium-like nitrogen and contaminant ions in the trap, which are not resolved by the Si(Li) detector [32]. Furthermore, the collision strength of electron impact excitation of the $2p$ states in hydrogen-like nitrogen is about 8 times larger than that of $2s$ state at 1.5 keV [33], so a large component of the background is electron excited Lyman-α. Further work showing the wavelength dependence of the laser-induced signal, will be published elsewhere.

4. Concluding remarks

We have observed an enhancement of Lyman-α radiation from hydrogen-like nitrogen ions trapped in an EBIT due to laser induced excitation of the $2s_{1/2} - 2p_{3/2}$ transition. Hence the possibility of a measurement of the $n = 2$ Lamb shift in hydrogen-like nitrogen ions trapped in an EBIT, has been demonstrated. However, background emission from transitions in helium-like nitrogen ions, as well as from the $2p$ state in hydrogen-like nitrogen, greatly reduce the signal-to-background ratio and hence the statistical precision to which the measurement can be obtained. Improvements, such as switching off the electron beam on a sub-microsecond time-scale and increasing the detection solid angle of the Si(Li) detector, are required if a precision measurement is to be achieved. Other challenges to achieving high

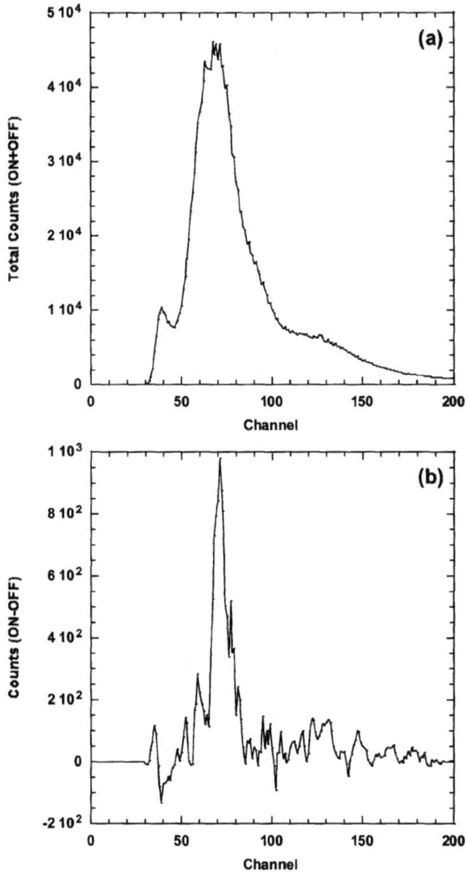

Figure 3. Enhancement of X-ray emission from hydrogen-like nitrogen ions induced by laser interaction as a function of X-ray energy (channel number of the Si(Li) spectrum). (a) is the total X-rays count and (b) corresponds to the laser induced signal, namely, X-ray counts (laser on)–X-ray counts (laser-off). (The small oscillations, which can be seen in the spectrum, are artefacts of the analogue-to-digital converter and are not intrinsic to the spectrum.)

precision are the control of the laser and ion cloud overlap and the levelling of the laser power as the laser wavelength is scanned across the resonance.

Acknowledgements

We gratefully acknowledge the useful advice of Professor Patrick Baird. We thank Peter Hirst and Tony Handford for technical support. This work is partially supported by the Japan Science and Technology Corporation as part of the "Cold Trapped Ions" International Collaborative Research Project.

References

1. Levine, M. A., Marrs, R. E., Henderson, J. R., Knapp, D. A. and Schneider, M. B., *Physica Scripta* **T22** (1988), 157.
2. Silver, J. D. *et al.*, *Rev. Sci. Instrum.* **65** (1994), 1072.
3. Back, T. V., Margolis, H. S., Oxley, P. K., Silver, J. D. and Myers, E. G., *Hyp. Interact.* **114** (1998), 203.
4. Flowers, J. L. *et al.*, In: *Proceedings of the 14th European Frequency and Time Forum*, Turin, Italy, 2000.
5. Johnson W. R. and Soff, G., *At. Dat. Nucl. Dat. Tab.* **33** (1985), 405.
6. Boshier, M. G. In: H. B. van Linden van den Heuvell, J. T. M. Walraven and M. W. Reynolds (eds), *Atomic Physics 15, Proceedings of the 15th International Conference on Atomic Physics, Amsterdam, The Netherlands, 1996*, World Scientific, 1997, p. 328.
7. Pachucki, K., Leibfried, D., Weitz, M., Huber, A., König, W. and Hänsch, T. W., *J. Phys. B* **29** (1996), 177.
8. Boshier, M. G., *Physica Scripta* **T86** (2000), 21.
9. Mohr, P. J., Plunien, G. and Soff, G., *Phys. Rep.* **293** (1998), 227.
10. Niering, M., Holzwarth, R., Reichert, J., Pokasov, P., Udem, Th., Weitz, M., Hänsch, T. W., Lemonde, P., Santarelli, G., Abgrall, M., Laurent, P., Salomon, C. and Clairon, A., *Phys. Rev. Lett.* **84** (2000), 5496.
11. Jentschura, U. D., Mohr, P. J. and Soff, G., *Phys. Rev. Lett.* **82** (1999), 53.
12. Appelquist, T. and Brodsky, S. J., *Phys. Rev. A* **2** (1970), 2293.
13. Pachucki, K., *Phys. Rev. Lett.* **72** (1994), 3154.
14. Eides, M. I. and Shelyuto, V. A., *Phys. Rev. A* **52** (1995), 954.
15. Mallampalli, S. and Sapirstein, J., *Phys. Rev. Lett.* **80** (1998), 5297.
16. Karshenboim, S. G., *JETP* **76** (1993), 541.
17. Goidenko, I., Labzowsky, L., Nefiodov, A., Plunien, G. and Soff, G., *Phys. Rev. Lett.* **83** (1999), 2312.
18. Yerokhin, V. A., *Phys. Rev. A* **62** (2000), 012508.
19. Ivanov, V. G. and Karshenboim, S. G., *The Hydrogen Atom, Precision Physics of Simple Atomic Systems*, Springer, Berlin/Heidelberg, 2001.
20. Kugel, H. W., Leventhal, M. and Murnick, D. E., *Phys. Rev. A* **6** (1972), 1306.
21. Myers, E. G. *et al.*, *The Hydrogen Atom, Precision Physics of Simple Atomic Systems*, Springer, Berlin/Heidelberg, 2001.
22. Leventhal, M., Murnick, D. E. and Kugel, H. W., *Phys. Rev. Lett.* **28** (1972), 1609.
23. Lawrence, G. P., Fan, C. Y. and Bashkin, S., *Phys. Rev. Lett.* **28** (1972), 1612.
24. Curnutte, B., Cocke, C. L. and Dubois, R. D., *Nucl. Instr. Meth.* **202** (1982), 119.
25. Kugel, H. W., Leventhal, M., Murnick, D. E., Patel, C. K. N. and Wood, O. R. II, *Phys. Rev. Lett.* **35** (1975), 647.
26. Pellegrin, P., El Masri, Y. and Palffy, L., *Phys. Rev. A* **31** (1985), 5.
27. Pross, H.-J., Budelsky, D., Kremer, L., Platte, D., von Brentano, P., Gassen, J., Müller, D., Scheuer, F., Pape, A. and Sens, J. C., *Phys. Rev. A* **48** (1993), 1875.
28. Zacek, V., Bohn, H., Brum, H., Faestermann, T., von Feilitzsch, F., Giorginis, G., Kienle, P. and Schuhbeck, S., *Z. Phys. A* **318** (1984), 7.
29. Georgiadis, A. P., Müller, D., Sträter, H.-D., Gassen, J., von Brentano, P., Sens, J. C. and Pape, A., *Phys. Lett. A* **115** (1986), 108.
30. Wood, O. R. II, Patel, C. K. N., Murnick, D. E., Nelson, E. T., Leventhal, M., Kugel, H. W. and Niv, Y., *Phys. Rev. Lett.* **48** (1982), 398.
31. Gould, H. and Marrus, R., *Phys. Rev. A* **28** (1983), 2001.
32. Bashkin, S. and Stoner, Jr. J. O., *Atomic Energy Levels and Grotrian Diagrams, Vol. I: Hydrogen I–Phosphorus XV*, North-Holland Publishing Company, Amsterdam, 1975.
33. Callaway, J., *Phys. Lett. A* **96** (1983), 83.

Hyperfine Interactions **146/147:** 67–70, 2003.

Vacuum-Polarization Corrections to Parity-Nonconserving Effects in Atomic Systems

I. BEDNYAKOV[1], W. R. JOHNSON[1,2], G. PLUNIEN[1] and G. SOFF[1]
[1] *Institut für Theoretische Physik, Technische Universität Dresden, Mommenstrasse 13, D-01062 Dresden, Germany*
[2] *Department of Physics, 225 Nieuwland Science Hall, University of Notre Dame, Notre Dame, IN 46556, USA*

Abstract. The dominant one-loop vacuum-polarization correction to atomic wave functions are evaluated for the $6s$–$7s$ parity nonconserving (PNC) E1 transition amplitude in cesium. This correction increases the size of the PNC amplitude by 0.4% and thus increases the difference between the experimental value of the weak charge Q_W and the one predicted by the standard model.

Key words: parity non-conservation, atomic systems.

1. Introduction

Parity non-conservation (PNC) in atomic systems as described within the standard model of the electroweak interaction via exchange of Z bosons between bound electrons and nuclear quarks leads to non-vanishing electric-dipole matrix elements between atomic states with the same parity. The nuclear spin-independent part of PNC matrix elements arising from the nuclear vector current is proportional to the conserved weak charge Q_W and thus is sensitive to possible "new physics" beyond the standard model such as the existence of additional neutral Z' bosons or new heavy fermions.

Measurements of the $6s$–$7s$ PNC amplitude in ^{133}Cs were carried out at the level of accuracy of about 2% by Gilbert and Wieman [1], and at about 0.3% accuracy by Wood *et al.* [2], respectively. Bennett and Wieman [4] have analyzed the differences between experimental and theoretical values deduced for transition amplitudes in ^{133}Cs and concluded that the error in the atomic-structure calculations which enter in the PNC amplitude should be reduced from 1% (this value given in [3]) to 0.4%. As a result they found that the experimentally deduced value for the weak charge in ^{133}Cs deviates from the standard-model value [5]

$$Q_W(^{133}\mathrm{Cs}) = -73.09 \pm 0.03 \tag{1}$$

by 2.3σ. Such a difference between experiment and theory suggests, e.g., the existence of a second neutral Z' boson [6].

It has been shown by Derevianko [7] that Breit corrections to the PNC amplitude decrease the magnitude of the calculated PNC amplitude by about 0.6%. As a consequence the Breit corrections reduce the deviation from the standard model to 1.6σ when assigning an error of about 0.4% to the calculated matrix elements as it has been assumed in [4]. Taking a more conservative value of a 1% error as given in [3] leads to a deviation of only about 0.9σ.

Radiative corrections to PNC matrix elements $\langle np_{1/2}|H_{\rm PNC}|n's_{1/2}\rangle$ in the strong Coulomb field of a heavy one-electron ion were considered recently by Bednyakov *et al.* [10]. These corrections were decomposed into two parts, radiative corrections to the operator $H_{\rm PNC}$ and radiative corrections to the wave functions $|n's_{1/2}\rangle$ and $|np_{1/2}\rangle$. We have found that these corrections increase the size of the PNC amplitude in ^{133}Cs by about 0.4% and, correspondingly, increase the difference between the theoretical and experimental weak charge. Moreover we found that these corrections are insensitive to electron–electron correlation effects.

2. One-loop radiative corrections to PNC effects

One-loop vacuum polarization corrections in an external Coulomb field modify the Coulomb interaction at short distances. To lowest order in $Z\alpha$ this modification is described by the Uehling potential for an extended nuclear charge distribution $\rho(r)$ [14]:

$$\delta V(r) = -\frac{2\alpha^2}{3r}\int_0^\infty \mathrm{d}x\, x\,\rho(x)\int_1^\infty \mathrm{d}t\sqrt{t^2-1} \\ \times\left(\frac{1}{t^3}+\frac{1}{2t^5}\right)\left(\mathrm{e}^{-2ct|r-x|}-\mathrm{e}^{-2ct(r+x)}\right). \tag{2}$$

Here we employ atomic units, where $\alpha = 1/137.036\ldots$ is the fine-structure constant and $c \equiv \alpha^{-1}$ is the speed of light. We have carried out calculations of the $6s$–$7s$ PNC amplitude in the presence of the additional Uehling potential. The first calculation is performed at the level of the "weak" Dirac–Hartree–Fock (DHF) approximation. The perturbation $\delta\psi_v^{\rm DHF}$ to a valence electron wave function $\psi_v^{\rm DHF}$ induced by the *weak* interaction $h_{\rm PNC}$ satisfies the inhomogeneous DHF equation

$$\left(h_0 + V^{\rm DHF} - \epsilon_v^{\rm DHF}\right)\delta\psi_v^{\rm DHF} = -h_{\rm PNC}\psi_v^{\rm DHF}. \tag{3}$$

In this equation, $V_{\rm HF}$ is the HF potential of the closed xenon-like core and $\epsilon_v^{\rm HF}$ is the eigenvalue of the unperturbed DHF equation. The PNC amplitude is then given by the sum of two terms:

$$E_{\rm PNC} = \left\langle\psi_{7s}^{\rm HF}\middle|D\middle|\delta\psi_{6s}^{\rm HF}\right\rangle + \left\langle\delta\psi_{7s}^{\rm HF}\middle|D\middle|\psi_{6s}^{\rm HF}\right\rangle, \tag{4}$$

where D denotes the dipole operator. In Table I, we provide numerical results for the individual terms of Equation (4) calculated with and without the additional Uehling potential δV of Equation (2), respectively. This one-loop radiative correction increases the magnitude of the sum of both contributions by 0.41%.

A second calculation has been performed in the "weak" random-phase approximation (RPA) which accounts for the class of correlation corrections associated

Table I. One-loop corrections to the weak PNC amplitude for the $6s$–$7s$ E1 transition in ^{133}Cs. Units: $iea_0 \times 10^{-12}(-Q_W/N)$

Type	$\langle\phi_{7s}\vert D\vert\delta\phi_{6s}\rangle$	$\langle\delta\phi_{7s}\vert D\vert\phi_{6s}\rangle$	E_{PNC}
Weak HF approximation:			
DHF	2.749183	−10.14387	−7.394685
DHF + δV	2.760414	−10.18581	−7.425391
Δ (%)	0.41	0.41	0.41
Weak RPA approximation:			
RPA	3.457036	−12.72562	−9.268581
RPA + δV	3.471169	−12.77834	−9.307166
Δ (%)	0.41	0.41	0.41

with weak perturbations of the core orbitals. This class of correlation corrections is included by solving

$$\left(h_0 + V^{\mathrm{HF}} - \epsilon_v^{\mathrm{HF}}\right)\delta\psi_v^{\mathrm{RPA}} = -\left[h_{\mathrm{PNC}} + V_{\mathrm{PNC}}^{\mathrm{HF}}\right]\psi_v^{\mathrm{PRA}}, \tag{5}$$

where $V_{\mathrm{PNC}}^{\mathrm{HF}}$ is the weak correction to the HF potential. The resulting PNC amplitude is given by

$$E_{\mathrm{PNC}} = \left\langle\psi_{7s}^{\mathrm{HF}}\middle|D\middle|\delta\psi_{6s}^{\mathrm{RPA}}\right\rangle + \left\langle\delta\psi_{7s}^{\mathrm{RPA}}\middle|D\middle|\psi_{6s}^{\mathrm{HF}}\right\rangle. \tag{6}$$

In Table I we also compare the corresponding results with and without the Uehling corrections. Again, the vacuum-polarization correction is seen to increase the results by 0.41%. From these calculations we can conclude that vacuum-polarization corrections are independent from electron–electron correlation effects.

3. Conclusion

One-loop vacuum-polarization corrections to the electron wave functions have been evaluated. The resulting wave functions are used to calculate the $6s$–$7s$ PNC transition amplitude in ^{133}Cs. They lead to an increase of the magnitude of the amplitude by 0.4%. Taking into account the Breit and the one-loop vacuum-polarization corrections we deduce the following value for the weak charge

$$Q_W^{\mathrm{expt}}\left(^{133}\mathrm{Cs}\right) = -72.12 \pm (0.28)_{\mathrm{exp}} \pm (0.34)_{\mathrm{theor}}.$$

This value differs by 2.2σ from the standard model. If we assume an error of about 1% in the theoretical amplitude, then the error in the theoretical component of Q_W^{exp} increases to $\pm(0.74)_{\mathrm{theor}}$ and the deviation from the value predicted by the standard model becomes 1.2σ.

Acknowledgements

The work of W.R.J. was supported in part by an Alexander von Humboldt Award from the Federal Republic of Germany and in part by the U.S. National Science Foundation under grant No. 99-70666. W.R.J. also owes a debt of gratitude to O. Sushkov for discussions on this problem. G.S. acknowledges support from DFG, GSI and BMBF.

References

1. Gilbert, S. L. and Wieman, C. E., *Phys. Rev. A* **34** (1986), 792.
2. Wood, C. S. *et al.*, *Science* **275** (1997), 1759.
3. Blundell, S. A., Sapirstein, J. and Johnson, W. R., *Phys. Rev. D* **45** (1992), 1602.
4. Bennett, S. C. and Wieman, C. E., *Phys. Rev. Lett.* **82** (1999), 2484; **82** (1999), 4153; **83** (1999), 889.
5. Groom, D. E. *et al.*, *Euro. Phys. J. C* **15** (2000), 1.
6. Erler, J. and Langacker, P., *Phys. Rev. Lett.* **84** (2000), 212.
7. Derevianko, A., *Phys. Rev. Lett.* **85** (2000), 1618.
8. Dzuba, V. A., Harabati, C., Johnson, W. R. and Safronova, M. S., *Phys. Rev. A* **63** (2001), 044103.
9. Kozlov, M. G., Porsev, S. G. and Tupitsyn, I. I., *Phys. Rev. Lett.* **86** (2001), 3260.
10. Bednyakov, I., Labzowsky, L., Plunien, G., Soff, G. and Karasiev, V., *Phys. Rev. A* **61** (1999), 012103.
11. Lynn, B. W. and Sandars, P. G. H., *J. Phys. B* **27** (1994), 1469.
12. Marciano, W. J. and Rosner, J. L., *Phys. Rev. Lett.* **65** (1990), 2963; **68** (1992), 898.
13. Sirlin, A., *Phys. Rev. D* **22** (1980), 971.
14. Fullerton, L. W. and Rinker, G. A., Jr., *Phys. Rev. A* **13** (1976), 1283.
15. Johnson, W. R. and Soff, G., *At. Data Nucl. Data Tables* **33** (1965), 405.

Hyperfine Interactions **146/147:** 71–75, 2003.

Toward a New Test of the Relativistic Time Dilation Factor by Laser Spectroscopy of Fast Ions in a Storage Ring

G. SAATHOFF[1], U. EISENBARTH[1], S. HANNEMANN[1], I. HOOG[2], G. HUBER[2], S. KARPUK[2,3], S. KROHN[1], J. LASSEN[2], D. SCHWALM[1], M. WEIDEMÜLLER[1], A. WOLF[1] and G. GWINNER[1]
[1]*Max-Planck-Institut für Kernphysik, 69029 Heidelberg, Germany*
[2]*Institut für Physik, Universität Mainz, 55099 Mainz, Germany*
[3]*Dept. of Laser Physics and Spectroscopy, Belarussian State Univ., Minsk, Belarus*

Abstract. The frequency measurement of Doppler-shifted optical lines of ions circulating in a storage ring at high speed permits a sensitive test of the relativistic Doppler-formula and, hence, the time dilation factor γ_{SR} of special relativity. Previous measurements at the storage ring TSR with $^7Li^+$ at $v = 0.065c$ gave a new, improved limit, but were hampered by the large observed linewidth, exceeding the natural width 15-fold. Recently we have identified the broadening to be caused by velocity-changing processes in the storage ring. Saturation spectroscopy has proven to be largely immune against these effects and has yielded linewidths only a few MHz larger than the natural one. This is the major ingredient for an improved test of γ_{SR}, which is now under way.

Key words: special relativity, laser spectroscopy, storage ring.

Laser spectroscopy of fast ions moving in a heavy-ion storage ring permits a very sensitive test of the relativistic Doppler-formula and consequently of the time dilation factor γ_{SR} of special relativity. In essence, a modern version of the Ives–Stilwell experiment [1] is performed, e.g., on a Λ-type system (see Figure 1). Two lasers with laboratory frequencies ν_a and ν_p counter- and copropagate with the ions in a straight segment of the storage ring and excite both legs of the Λ at the frequencies ν_1 and ν_2 in the ion's rest frame, respectively. Only when *both* lasers are on resonance with an ion strong fluorescence is observed, as a single resonant laser quickly pumps the ion into the other, "dark" ground state. Due to the opposite direction of the laser beams they can only be simultaneously in resonance with one velocity class, yielding (to first order) a Doppler-free "Λ-resonance". From the resonance conditions $\nu_1 = \gamma(1-\beta)\nu_p$ and $\nu_2 = \gamma(1+\beta)\nu_a$ and the well-known relation $\gamma_{SR} = 1/\sqrt{(1-\beta^2)}$ we have within the theory of special relativity

$$\nu_1\nu_2 = \nu_a\nu_p. \tag{1}$$

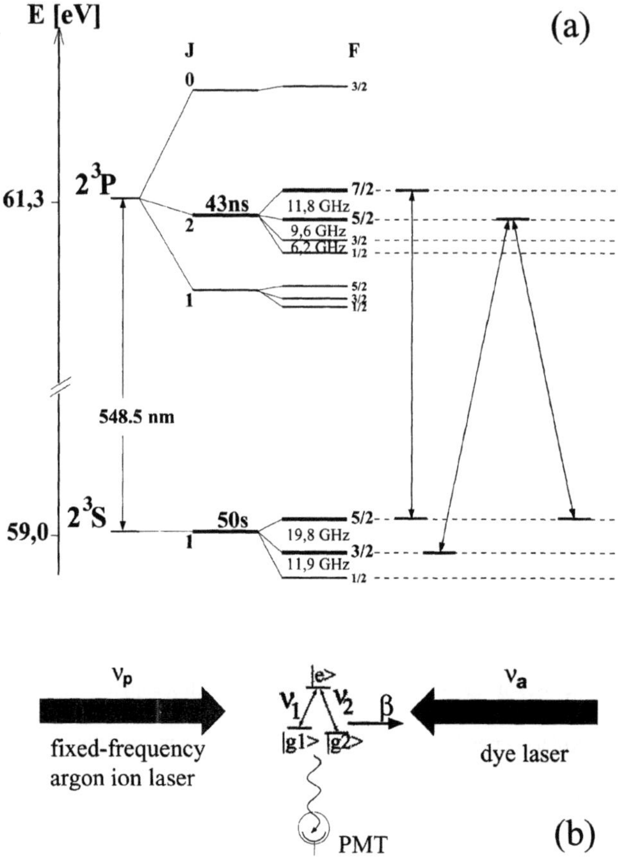

Figure 1. (a) Level diagram of heliumlike $^7Li^+$. (b) Principle of the modern Ives–Stilwell-type experiment as performed at the TSR.

Test theories of special relativity (see, e.g., [2]) parametrize possible deviations of γ from γ_{SR} by $\gamma = \gamma_{SR}(1 + \delta\alpha \cdot \beta^2 + \delta\alpha_2 \cdot \beta^4 + \cdots)$, and can be tested by comparing the transition frequencies ν_a and ν_p measured on fast moving ions to the frequencies ν_1 and ν_2 measured in ions at rest:

$$\nu_a = \frac{\nu_1\nu_2}{\nu_p}\left[1 + 2\delta\alpha\left(\beta^2 + 2\vec{\beta}\cdot\vec{\beta}' + \cdots\right)\right], \tag{2}$$

where $\vec{\beta}$ is the velocity of the ion with respect to the laboratory and $\vec{\beta}'$ is the motion of the laboratory with respect to a potential preferred reference frame (e.g., the one, in which the microwave background is isotropic). A non-zero value for $\delta\alpha$ indicates a deviaton from the theory of special relativity.

The measurements at the TSR storage ring in Heidelberg use the $2^3S_1 \to 2^3P_2$ transition in metastable $^7Li^+$ ions (Figure 1a). The ions are accelerated to 13.4 MeV

($v = 0.065c$) in a tandem Van-de-Graaff accelerator and are injected into TSR. About 20% of the ions emerge in the metastable triplet state from the gas stripper of the tandem. After five seconds of electron cooling in the ring the longitudinal momentum spread is reduced to $\Delta p/p \approx 3 \times 10^{-4}$, leading to a Doppler-width of the transition of about 2.5 GHz. Λ-spectroscopy is then performed on the 2^3S_1 ($F = 3/2$), 2^3S_1 ($F = 5/2$) and 2^3P_2 ($F = 5/2$) states.

Earlier measurements at TSR [3] have determined the Doppler-shifted resonance frequencies of this Λ-system to an accuracy of $\Delta\nu/\nu < 6 \times 10^{-9}$ ($\Delta\nu =$ 3 MHz). The comparison of these frequencies with the ions' rest-frame frequencies measured by Riis *et al.* [4] and Rong *et al.* [5] constrains deviations $\delta\alpha$ of the time dilation factor to less than 8×10^{-7} [3].

The frequency determination was mainly limited by a large broadening of the resonance to about 60 MHz (the natural linewidth is 3.8 MHz), which cannot be explained by saturation. As long as the origin of this broadening was not determined, it could not be ruled out that the same mechanism also leads to frequency *shifts*. Our recent measurements have found strong evidence that velocity-changing processes occuring typically over several to many roundtrips of the ions in the storage ring are the reason for the broadening. Λ-spectroscopy is very susceptible to such velocity changes as nothing requires the two lasers to be in resonance *simultaneously*, i.e. the back-and-forth pumping between the two ground states can occur over many roundtrips in the ring. If during that time velocity changing collisions occur, e.g., in the electron cooler, the Λ-resonance is smeared out.

To circumvent this problem we investigated saturation spectroscopy on a two-level system established by the 2^3S_1 ($F = 5/2$) and 2^3P_2 ($F = 7/2$) states. For this case one just has to replace $\nu_1\nu_2$ in Equation (2) by the square of the transition frequency of the two-level system ν_0^2. In contrast to the Λ-scheme, saturation spectroscopy is a Doppler-free method which requires simultaneity for the interaction of both lasers with an ion, i.e. saturation is only observed if both lasers are resonant during a time period corresponding to the natural lifetime of the upper state (43 ns), a time intervall much shorter than the roundtrip time in TSR of 4 μs. Under otherwise unchanged conditions, a much narrower saturation dip of only 15 MHz FWHM was observed (Figure 2b).

In order to determine the position of the saturation dip with high accuracy, one has to take into account the slope of the fluorescence background. The lifetime of the metastable ions in the storage ring is roughly 15 seconds and hence comparable to the length of a laser frequency scan, so the number of ions decreases significantly during one scan. This makes a lifetime correction necessary, which has already lead to a systematic error of 0.5 MHz in the old Λ-type experiment. In our new measurement this problem is avoided by using a scheme where both lasers are switched on and off at frequencies of up to 200 kHz (corresponding roughly to the TSR roundtrip time) with help of accousto-optic modulators (AOMs). Figure 2a shows the chopping pattern: during one time-window (c) both lasers are on and the saturation signal is recorded, in two additional windows (a, b) the lasers are

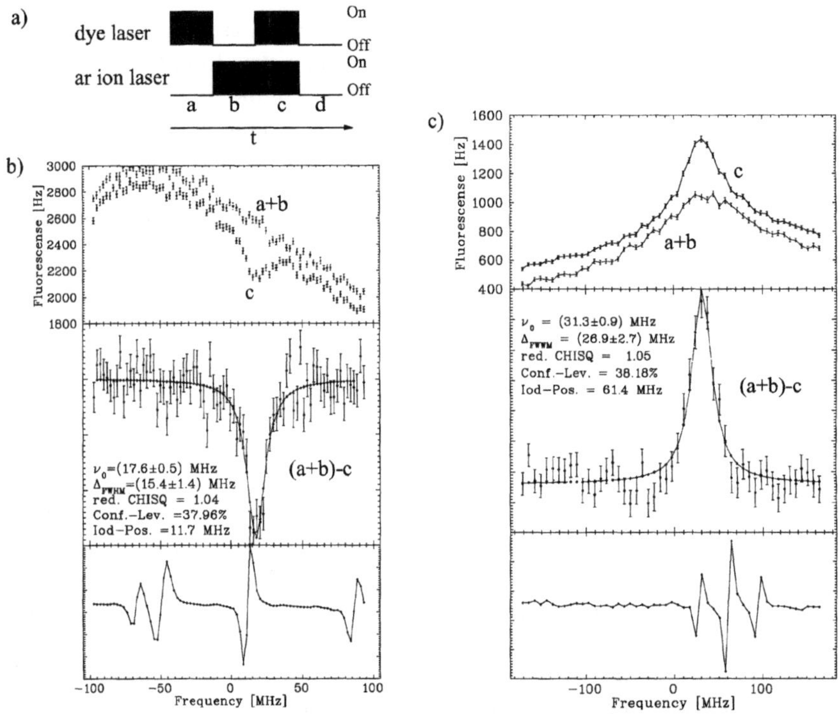

Figure 2. (a) Sequence of the laser switching scheme. (b) Saturation spectroscopy: the upper figure shows spectrum c with both lasers switched on as well as the sum of spectra a and b, which has been given an offset of 150 Hz here for better illustration. The middle graph is the difference signal reflecting the pure saturation dip. The lower figure shows hyperfine states of molecular iodine serving as a frequency standard. (c) Λ-spectroscopy: spectrum a + b again has an offset of −100 Hz.

switched on separately. The combined fluorescence from (a) and (b) can be subtracted from (c) to isolate the saturation feature from the background and is also used to normalize the Lamb dip with respect to the ion current decay in the ring.

Figure 2b shows an example of such a Lamb dip measured at laser intensities corresponding to a moderate saturation parameter. The subtraction of the fluorescence background yields a resonance with a Lorentzian shape lying on a flat background. The width of the Lamb dip of about 15 MHz is significantly smaller than the width of the Λ-resonance in the previous experiment. The situation is even more favorable, as the 15 MHz include a "trivial" broadening of roughly 2 MHz stemming from the frequency modulation used for the stabilization of the lasers, leaving at most a few MHz of unaccounted broadening. Interestingly, we have achieved linewidths narrower than those reported for slow ions [4, 5].

The laser switching method has also proven to be useful for Λ-spectroscopy (Figure 2c). A narrow component of the Λ-resonance occurs predominantly in the window where both lasers are on, and the subtracted spectrum again yields a

Lorentzian on a flat background. The statistics seem to be poorer than for saturation spectroscopy. However, this can be compensated by increasing the laser intensity – at the cost of additional power broadening. Since this is a well understood broadening mechanism, it can be tolerated.

In conclusion, we have identified the cause for the large linewidths of Λ-spectroscopy in the old TSR experiments and have implemented schemes to achieve much narrower resonances. Absolute frequency calibrations of the Lamb dip and the Λ-resonance are now under way, and new results can be expected soon.

References

1. Ives, H. E. and Stilwell, G. R., *J. Opt. Soc. Am.* **28**(7) (1938), 215.
2. Mansouri, R. and Sexl, R., *Gen. Rel. Gravit.* **8** (1977), 497.
3. Grieser, R. *et al.*, *Appl. Phys. B* **59** (1994), 127.
4. Riis, E. *et al.*, *Phys. Rev. A* **49** (1994), 207.
5. Rong, H. *et al.*, *Eur. Phys. J. D* **3** (1998), 217.

Hyperfine Interactions **146/147:** 77–82, 2003.

The Weak Interaction Studied with Trapped Ions and Atoms

H. W. WILSCHUT
Kernfysisch Versneller Instituut, Zernikelaan 25, 9747 AA Groningen, The Netherlands

Abstract. An overview of the experimental approaches to study weak interactions using traps is given. Both charged and neutral weak currents are considered.

Key words: the Standard Model, weak interaction, new physics, ion and atom traps.

1. Introduction

The study of weak interactions in nuclear and atomic physics allows one to test the Standard Model in the light quark sector. In Figure 1 the two basic interactions are shown. Either a lepton (electron or neutrino) interacts via a charged W boson to change a proton into a neutron or vice versa, or it interacts with a nucleon via the exchange of a neutral Z boson. The bosons are massive, hence the interaction is weak. The charged current interaction is easily observed from β-decay (see Figure 2) when a nucleon embedded in a nucleus "changes its nature". In contrast, the neutral interaction is not so easily detected. Fortunately, the weak interactions do not conserve parity, so that it can be observed, in principle, in parity non-conserving (PNC) interactions in atoms. We know already from the occurrence of neutrino oscillations that the Standard Model for the weak interaction is insufficient. In which way it should be extended can in part be answered by comparing the observations in β-decay of nuclei and PNC transitions in atoms to the predictions of the Standard Model. In the following we will first discuss measurements concerning β-decay (Section 2), followed by a discussion on the potential of PNC transitions (Section 3) and we summarize other approaches to find new physics using traps.

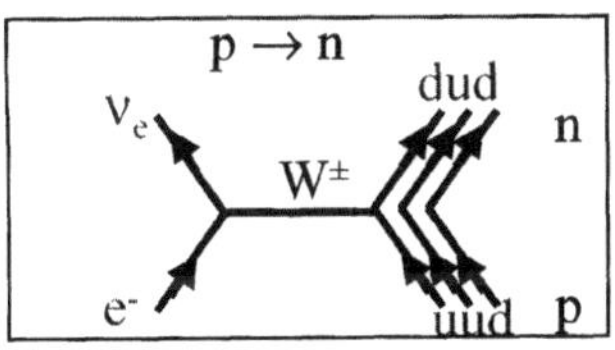

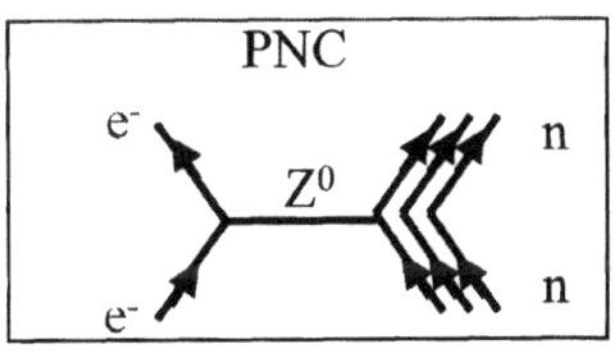

Figure 1. The charged and neutral weak interaction: the interaction occurs at the light quark level. The protons and neutrons may be embedded in a nucleus.

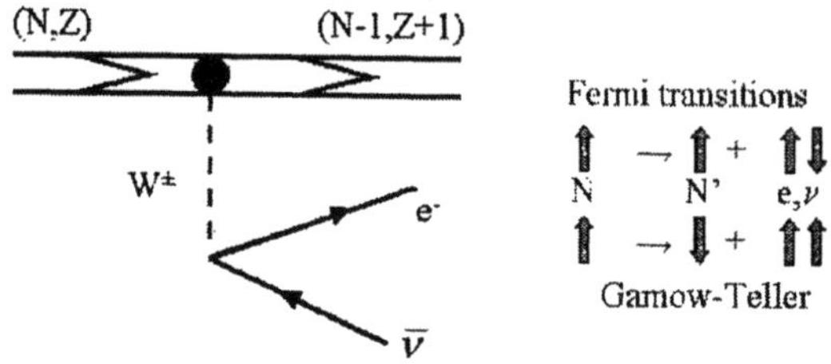

Figure 2. β-decay of a nucleus. The properties of the initial and final nuclei determine the detail of the interaction. In allowed transitions the leptons do not carry angular momentum. The coupling of the spins of the two leptons determines wether the nucleon makes a spin flip (Gamow–Teller transition) or not (Fermi transitions).

2. β-Decay

In β-decay a proton (p) or neutron (n) emits a virtual W boson which can then decay into leptons, i.e. into a positron (e^+) and neutrino (ν) or an electron (e^-) and anti neutrino ($\bar{\nu}$). The rate at which this happens is determined by phase space, nuclear structure factors and the weak interaction itself. Nuclear structure factors can be influenced by choosing a specific parent and associated daughter nucleus. In this way one can choose particular parts of the weak interaction. Well-known examples are the allowed Fermi transitions and Gamow–Teller transitions as depicted in Figure 2. The interaction in the former case contains only Dirac structures referred to as vector and scalar components, while the latter has only axial vector or tensor components. One often refers to radioactive nuclei as micro-laboratories because of this selectivity. Within the Standard Model there is no scalar or tensor component. Therefore, one way to look for new physics is to try to search for tiny components of this type in β-decay. One special type of transitions are the so-called super-allowed Fermi transitions. In these transitions the initial state and final state is a 0^+, while the valence nucleon that decays does not change it nuclear state but only its nature ($p \leftrightarrow n$). Because in these nuclei the neutron and proton core must have the same structure, the nuclear structure effect is minimized. The only remaining factors determining the transition rate (lifetime) are then the phase space, i.e. the Q-value of the decay and the strength (G_V) of the vector part of the weak interaction. In this way G_V is determined only by Q-value (entering with about the fifth power) and the decay rate (life time) of the nucleus. What one finds is that G_V has a residual dependence on nuclear structure (in particular the effect of isospin mixing) and also that indirectly, up to 2 standard deviations, this leads to an inconsistency in the Standard Model see the references in [1]. To extend the number of cases of such Fermi-transitions studies, in particular with regard to the understanding of the nuclear structure component, one has to include increasingly exotic nuclides. This is due to the requirement that the neutron and proton "valence band" must have the same quantum numbers ($Z \approx N$) while with increasing mass the neutron shells fill stronger than the proton shells. The heaviest nuclide considered for this purpose so far is ^{74}Rb [1]. It is 11 neutrons removed

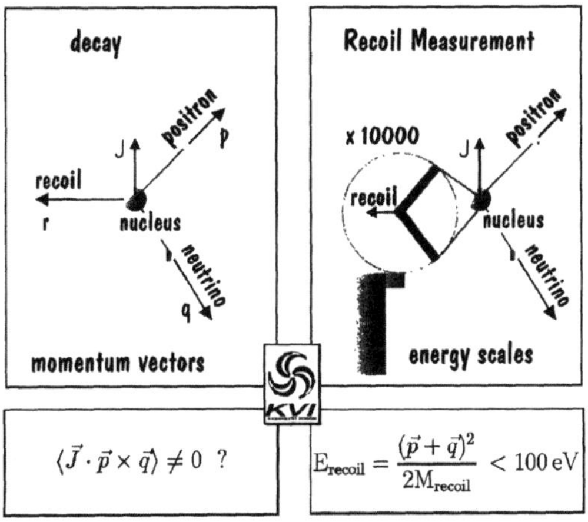

Figure 3. Correlation measurements between electron and neutrino can be replaced by electron-recoil correlations. It requires to measure the – for nuclear physics – tiny recoil energy and direction.

from the first stable Rb isotope. The use of Penning traps, e.g., to purify the isotopic samples and/or to measure Q-values has become increasingly important [2].

Much more information about the structure of weak decays is contained in the correlation between the electron and the neutrino and their spins, including the spin of the parent and daughter nucleus. Measuring the neutrino is not practical, instead one could try to measure the recoiling nucleus as depicted in Figure 3. Various correlations can be considered [3]. An overview including numerical values for the correlation coefficients can be found in [4]. Integrating over all spin degrees of freedom only the β-ν correlation is left, $W_{e\nu} = 1 + a\frac{\mathbf{p}_e \cdot \mathbf{p}_\nu}{E_e E_\nu} + b\frac{m_e}{E_e} \approx 1 + \alpha\beta_e \cos\theta_{e\nu} + b\frac{1}{\gamma_e}$. In the Standard Model $a_{F(GT)} = 1(-1/3)$ for Fermi (Gamow–Teller) transitions and the Fierz interference term $b = 0$. Integrating over all directions one finds that the recoil distribution alone allows one to test the values for a and b. An indirect method provided the best value for so far $a_F = 0.999 \pm 0.007$. It was based on the Doppler broadening of α-particle emission following β-decay of ^{32}Ar [5]. By measuring the small recoil energies directly one hopes to improves on this limit considerably. An attempt do this with an ion trapping system is underway by the Leuven group at ISOLDE [6]. Using the properties of ^{6}He a direct measurement could be made already in 1963 [7] ($a_{GT} = 0.3343 \pm 0.0030$). To improve on this value an experiment is proposed by LPC at GANIL to measure β-recoil correlations, cooling the ^{6}He sample in a Paul trap [8].

A complete different approach has been taken by groups in the USA and Canada. They purify and collect their samples in atomic traps, minimizing the effect of the trapping fields on the recoiling nucleus. In atomic physics this method (COLTRIMS in MOT's) was recently proven to be very successful [9]. The status for correlation studies in β-decay of K-isotopes was recently reported [10] and may soon improve on the current value of a_F [11]. Atomic traps may also provide a means to polarize the sample (cf. Figure 2). In this way also emission asymmetry with respect to the spin direction of the parent nucleus can be exploited (i.e. the "Wu experiment"). A first result has been reported for ^{82}Rb [12]. Behr *et al.* are making use of a "far off-resonance trap" (FORT) that was recently proposed for β-correlation studies [13], which will allow triple correlation studies (β-recoil correlations from spin polarized samples). To access these more complex correlations is one of the aims of the TRIμP programme at KVI [14].

3. The neutral interaction

In contrast to the charged weak interactions which is unambiguously signalled by β-decay, the neutral weak interaction is much more difficult to observe. The only reason one can observe it at all, is the fact that it does not conserve parity. By performing two measurements of a normal electromagnetic observable $|A_{em}|^2$ under space reflection (L/R) different results $P_{L/R} = |A_{em} \pm A_{PV}|^2$ will be found. The difference depends on the admixture of the parity violating part of the weak interaction $|A_{PV}|^2$. The difference is measured as an asymmetry $A_{L/R} = (P_L - P_R)/(P_L + P_R) = 2\Re(A_{PV}/A_{em})$. It can be enhanced by choosing a suitable weak electromagnetic transition on one hand and optimizing the strength of the weak interaction on the other. Just as the Coulomb interaction the weak interaction is proportional to the coherent sum of the individual charges of the (u, d) quarks, i.e. $Q(Z, N) = (2Z + N)Q(u) + (2N + Z)Q(d)$ for a nucleus with Z protons and N neutrons. For the electromagnetic charge with $Q_{em}(u, d) = (\frac{2}{3} - \frac{1}{3})$ one finds the trivial result $Q_{em}(N, Z) = Z$. For the weak charge one has $Q_W = \pm(1 - \frac{6\mp2}{3}\sin^2\theta_W)$ for u and d, respectively, thus $Q_W(N, Z) = Z(1 - 4\sin^2\theta_W) - N$ (to first order). With the Weinberg angle $\theta_W \approx 28°$, one finds that the weak charge depends mainly on the number of neutrons in the nucleus, i.e. $Q_W(N, Z) \approx -N$. Therefore, a heavy nucleus is preferred for these measurements. Additional factors give Z^2, so that the total dependence on Z is somewhat faster than Z^3 [15]. The most successful approach has been to study $nS_{1/2} \to n'S_{1/2}$ transitions in alkalies. Applying an electric field allows one to control the electromagnetic amplitude via Stark mixing. The best results have been obtained so far with ^{133}Cs ($Z = 55, N = 78$) atoms ($6S \to S$) [6, 16]. It was extremely important to introduce redundancy in these experiments, i.e. several different ways to bring the L/R dependence to bear and observe the L/R asymmetry ($\approx 10^{-6}$) with precision. A value of $Q_W = -72.1 \pm 0.3(\text{exp}) \pm 0.9(\text{theory})$ was deduced which would be 2.5σ below the theoretical value (including higher order – radiative – corrections) based on the Standard Model for Q_W. The importance of this observations have led various authors to recalculate the atomic and nuclear aspects of the theory, with differing results. The next heavier alkali is Francium, which would allow one to use the $7S \to 8S$ transitions. The PV effect in Fr would be more than an order of magnitude stronger than in Cs. However, there is no long-lived Fr isotope, therefore it needs to be produced continuously at an accelerator. Fr was produced and trapped in a MOT and the atomic properties of various isotopes were measured by a Stony Brook group [18]. Recently an effort to study Fr has started at Legnaro [19].

Apart from the exchange of a Z^0 between an electron and a nucleus with weak charge Q_W, parity non-conserving transitions also can occur for different reasons. For example, the weak interaction between nucleons gives rise to a peculiar static moment, the anapole moment. The electron interacts with the anapole object via exchange of photons but the interaction is parity violating. The effect of the anapole moment is often referred to as the spin dependent part of the neutral current interaction in atoms. Its effect on PNC transitions is smaller than the spin-independent part associated with Q_W. The current status has been reviewed recently showing inconsistencies between various observations, which require experimental clarification [20].

The expectation value of the triple correlation $\vec{J} \cdot \vec{p} \times \vec{q}$ depicted in Figure 3 vanishes if time reversal holds. If it is violated at any measurable level, one has direct indication for physics beyond the Standard Model. In fact any observable vanishingly small within in the Standard Model can be considered. A prime example is the possible existence of an electric dipole moment of an elementary particle or of a composite system such as an atom. This would violate both P and T. The strategy to measure a tiny dipole moment will be similar to that of measuring PNC in atoms. One needs to maximize the dipole signal. The preferred object is neutral, so that the precession in an electric field E can be measured. When one adds a magnetic field B the total precession provides a $L = (B + E)$ vs. $R = (B - E)$ measurement with $A_{L/R} = d \cdot E/(\mu \cdot B)$. Note that the electric dipole $\vec{d}$ and magnetic $\vec{\mu}$ moments are both in the direction of the spin. That the atom or nucleus could have enhanced dipole moments is a non-trivial point. Electrostatic forces between elementary particles with a certain charge and dipole moment cannot produce a finite overall dipole moment (Schiff's theorem). Only due to, e.g., relativistic effects the dipole moment of an atom could nonetheless well exceed the sum of it constituent electron-dipole moments for particular atoms and electronic configurations. In addition the nucleon–nucleon interaction could give rise to an enhanced moment, in this case depending on the nuclear structure. In fact the best upperlimit is due to a case where the nucleus dipole moment would act: for ^{199}Hg an upper limit was found with $d < 2.1 \times 10^{-28} e$ cm [21]. A review of techniques and their relation to a particular type of dipole (e.g., elementary or composite) can be found in [22]. Also for dipole searches high Z values are advantageous. In addition one searches for enhancements in the nuclear sector. It is therefore likely that also production of short-lived nuclides may be the preferred route to search for dipole moments.

One may well expect that ion and atom trapping will become increasingly important in searches for "new physics". These tools not only allow one to collect and purify samples of radioactive nuclides, they also enable the manipulation of spin degrees of freedom and the detection of slow nuclides containing the kinematic information on the neutrinos.

References

1. Ball, G. C. *et al.*, *Phys. Rev. Lett.* **86** (2001), 1454, and references therein.
2. Dilling, J. *et al.*, *Hyp. Interact.* **127** (2000), 491, and Bollen, G., *Nucl. Phys. A* **693** (2001), 3.
3. Jackson, J. D., Treiman, S. B. and Wyld, H. W., *Phys. Rev.* **106** (1957), 517.
4. Deutsch, J. and Quin, P., In: Langacker, P. (ed.), *Precision Tests of the Standard Electroweak Model*, World Scientific, 1995.
5. Adelberger, E. G. *et al.*, *Phys. Rev. Lett.* **83** (1999), 1299.
6. Delauré, B. *et al.* (2003), this volume.
7. Johnson, C. H., Pleasonton, F. and Carlson, T. A., *Phys. Rev.* **132** (1963), 1149.
8. Ban, G. *et al.*, Proceedings of APAC2000, to be published in *Hyp. Interact.*
9. van der Poel, M. *et al.*, *Phys. Rev. Lett.* **87** (2001), 1123201; Turkstra, J. W. *et al.*, *ibid*, 123202; Flechard, X. *et al.*, *ibid*, 123203.
10. Gorelov, A. *et al.*, *Hyp. Interact.* **127** (2000), 373.
11. Behr, J. A., private communication.
12. Vieira, D. J. *et al.*, *Hyp. Interact.* **127** (2000), 387.
13. Corwin, K. L. *et al.*, *Phys. Rev. Lett.* **83** (1999), 1311.
14. Wilschut, H. W. *et al.*, TRIμP proposal, 1999 (cf. http://www.kvi.nl).
15. Bouchiat, M. A. and Bouchiat, C., In: Frois, B. and Bouchiat, M. A. (eds), *Parity Violation in Atoms and Polarized Electron Scattering*, World Scientific, 1999.
16. Wood, C. S. *et al.*, *Science* **275** (1997), 1759.
17. Bennett, S. C. and Wieman, C. E., *Phys. Rev. Lett.* **82** (1999), 2484.
18. Sprouse, G. D. *et al.*, *Hyp. Interact.* **127** (2000), 381.
19. Moi, L. *et al.* (2003), this volume.
20. Haxton, W. C. and Wieman, C. E., nucl-th/0104026.
21. Romalis, M. V. *et al.*, *Phys. Rev. Lett.* **86** (2001), 2505.
22. Commins, E. D., In: P. Herczeg, C. M. Hoffman and H. V. Klapdor-Kleingrothaus (eds), *Physics Beyond the Standard Model*, World Scientific, 1999.

Hyperfine Interactions **146/147:** 83–89, 2003.

The Legnaro Francium Magneto-Optical Trap

S. N. ATUTOV[1,2], V. BIANCALANA[3], A. BURCHIANTI[3], R. CALABRESE[1], L. CORRADI[4], A. DAINELLI[4], V. GUIDI[1], B. MAI[1], C. MARINELLI[3], E. MARIOTTI[3], L. MOI[3], A. ROSSI[3], E. SCANSANI[1], G. STANCARI[1], L. TOMASSETTI[1] and S. VERONESI[3]
[1]*Dipartimento di Fisica dell'Università, INFN, Sezione di Ferrara, I-44100 Ferrara, Italy*
[2]*Permanent address: Institute of Automation and Electrometry, Novosibirsk 90, Russia*
[3]*Dipartimento di Fisica dell'Università, INFM, Unità di Siena, Italy*
[4]*INFN-Laboratori Nazionali di Legnaro, Legnaro, Italy*

Abstract. Laser cooling and trapping of radioactive atoms represent the new frontier in atomic physics and a new powerful tool in nuclear physics. We are setting up at the INFN-Legnaro National Laboratories a laser cooling facility that has as a first goal the realization of a ^{210}Fr magneto-optical trap. The general outline of the experiment and the improvements of the final trap efficiency are discussed. Some preliminary results are presented.

Key words: laser cooling and trapping, radioactive atoms, atomic spectroscopy, adsorption and desorption processes.

1. Introduction

Laser cooling and trapping of atoms and ions is today a powerful technique used in an increasing number of atomic physics experiments [1]. Traps for neutral atoms have had fantastic developments during the last decades which opened new research fields like Bose–Einstein condensation, atom optics, atom lasers, quantum computing, etc. [2]. Very cold and dense atomic samples are produced and trapped starting either from atomic beams or from gases. A cold sample of trapped atoms is the ideal tool to make high-resolution spectroscopy and to study with very high accuracy fundamental problems, such as for example atomic parity non conservation (APNC) and nuclear decay processes. Laser cooling and trapping of radioactive atoms represent in fact the new frontier in atomic and nuclear physics. Radioactive samples of ^{37,38}K [3], ^{21}Na [4], ^{208,209,210,211}Fr [5] and ^{221}Fr [6] have been cooled and trapped. Interesting possibilities can also arise when a sample of polarized radioactive atoms decays. The spatial distribution of decay alpha particles can carry information about the nuclear structure. If the decay is through β particles then electro-weak processes can be studied.

Trapping of radioactive neutral atoms imposes the solution of specific problems connected, for example, with their production rate, in general slow, and with the nuclear reactions which impose huge energy scaling to the involved particles. For example in the francium experiment a stepping down of about 17 orders of magnitude of the particle energy is necessary going through the nuclear reaction, the ion neutralization and the neutrals cooling.

Even if many trapping schemes have been proposed, the magneto-optical traps (MOT) have to be preferred has a first step. In fact they are very robust and versatile devices used in many laboratories. They make use of six orthogonal red detuned laser beams crossing in the center of a cell immersed in a quadrupole magnetic field. The atoms having a velocity below a critical value v_c, are decelerated and trapped where the laser beams cross and the magnetic field intensity is minimum. A very positive point is that the geometry gives a large freedom to the analyzing apparatus. These traps demand in the case of radioactive atoms improvements with respect to the standard approach in order to circumvent two major problems: the low atom production rates and the short atomic lifetimes.

We are setting up at the INFN-Legnaro National Laboratories a laser cooling facility devoted to high-resolution spectroscopy of poorly known radioactive isotopes and to make research and developments of experiments aimed at experimental analysis of fundamental problems. The trap will be made operative first with ^{210}Fr isotope. Francium is the heaviest of the alkali atoms and it has many of the properties of the other alkalis. In fact, despite its radioactive complications, francium is still the heaviest simple atom. This is an important issue as a good crossed check among experiments and theory is possible. Francium does not have any stable isotopes and its abundance is so small that there is at most one ounce of francium in the whole Earth at any given time as a result of the decay of other radioactive elements. It is the most unstable of the first 103 elements in the periodic table. In fact, its longest lived isotope has a half life of 22 minutes. The half-life of the ^{210}Fr isotope is about 3.2 min. ^{210}Fr has an estimated α decay branching fraction of $60 \pm 30\%$ with the remaining decays being β^+ or electron capture. After the francium discovery made by Perey during 1939 [7], relatively few experiments have been made with francium and its energy level structure is still today partially known. Liberman *et al.* [8] reported the first observation of a francium electronic resonance at the ISOLDE on-line mass separator in 1978. The ISOLDE collaboration measured the low-lying francium P level energies and many other energy levels in the S and D series [9, 10]. They also measured the isotope shift and hyperfine structure of the D_2 line for several francium isotopes, revealing the nuclear structure of $^{207-213;220-228}$Fr [11]. In 1987 Andreev *et al.* [12] reported a measurement of the francium ionization energy and extended the S and D series by detecting several Rydberg levels of ^{221}Fr using laser-resonance ionization. Quite recently, as already mentioned, ^{221}Fr [6] and ^{210}Fr [5] isotopes have been cooled and trapped. A whole series of measurements is possible with francium isotopes: search for energy levels and lifetime measurements, collision processes, molecular spectroscopy, BEC, sensitive tests of relativistic effects on the atomic structure, isotope shifts, decay processes, APNC. This last is a very hot issue as calculations show an effect 18 times larger than for cesium.

2. Experimental apparatus

The experimental apparatus has an unusual size for a laser cooling experiment. In fact as francium has to be produced by a nuclear reaction, we have to make use of a high energy ion beam sent on a suitable target. For this reason the MOT is located in the INFN-Legnaro National Laboratories where a 15 MV Tandem is available. The experimental apparatus occupies two rooms in the accelerator main hall. There is a hot room where the target is located and a cool room, where the MOT and the lasers are arranged. This disposition allows us to work next to the MOT also when the oxygen beam is on. Let us give some details. Francium atoms are produced by bombardment of a gold target with an oxygen ion beam ($100 \div 120$ MeV; charge state $+7$) according to the reaction

$$^{18}\mathrm{O} + {}^{197}\mathrm{Au} \rightarrow {}^{210}\mathrm{Fr} + 5\mathrm{n},$$

where n represents neutrons. Francium ions are produced inside the gold from where they have to be extracted and sent to the MOT with an electrostatic channel. The target is heated to a temperature near the gold melting point to maximize both francium diffusion toward the gold surface and the escape rate of francium ions. After a pathway of few meters, francium ions hit a neutralizer, kept at relatively high temperature, from which are extracted as neutral atoms and collected in a vapor cell where they are cooled and trapped by the lasers. Neutralizer materials such as yttrium and zirconium are presently under test. They have a relatively low efficiency and a careful analysis of their working conditions is important. It is also important to vary the the neutralizer location with respect the trapping region in order to optimize the neutrals transfer into the MOT.

The general layout is shown in Figure 1.

The maximum oxygen beam current is about 1 μA. The beam is focused at the target level to a few mm diameter. The target Figure 2(left) consists of a tungsten cylinder, on one end of which a gold thick (about 1 mm) sample is fixed. The target is kept at a potential of up to $+10$ kV. A conical electrode is used to focus the francium ion beam coming out from the target. Its position and the voltage have been set according to accurate simulation as shown in Figure 2(right).

A preliminary test has been made in which the francium production rate as a function of the target temperature and of the electrode voltage was studied. In Figure 3 the francium production rate is shown for a beam current equal to 50 nA. The measurement is made by collecting the francium ions onto an aluminum catcher for about 5 min. Then the catcher is rotated by 180° in front of a Si detector that counts the α particles generated by the francium decay. A ^{210}Fr production rate of the order of 5×10^4 ions per second is obtained. At the highest ion beam current we expect a rate exceeding 10^6 ions/s.

The detector was calibrated with a standard composite 3-peak alpha source (^{239}Pu, ^{241}Am, ^{244}Cm); the detected peaks correspond to francium decay energies. The decays of daughter nuclei were also observed.

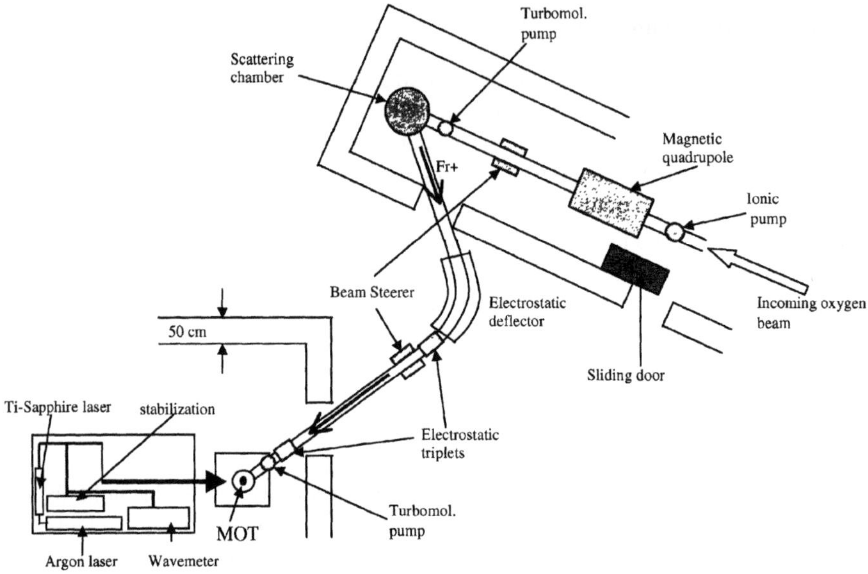

Figure 1. General sketch of the francium MOT setup.

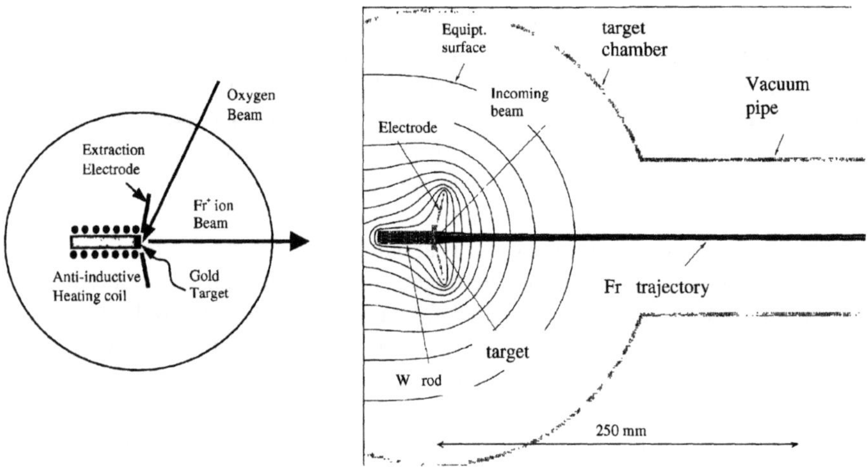

Figure 2. (*Left*) Schematic diagram of the target chamber; (*right*) Monte Carlo simulation of the ion beam for a given (+10 kV) voltage of the conical electrode.

Laser cooling exploits the resonance radiation force exerted by a red detuned laser on atoms. In the photon absorption emission processes a net momentum is transferred from photons to atoms which are pushed by light. If light is coming from all directions, atoms diffuse in a sort of gas made by photons (optical molasses) but no trap is obtained as the force does not depend on position. In a MOT this problem is solved by adding a suitable magnetic field. The MOT is formed

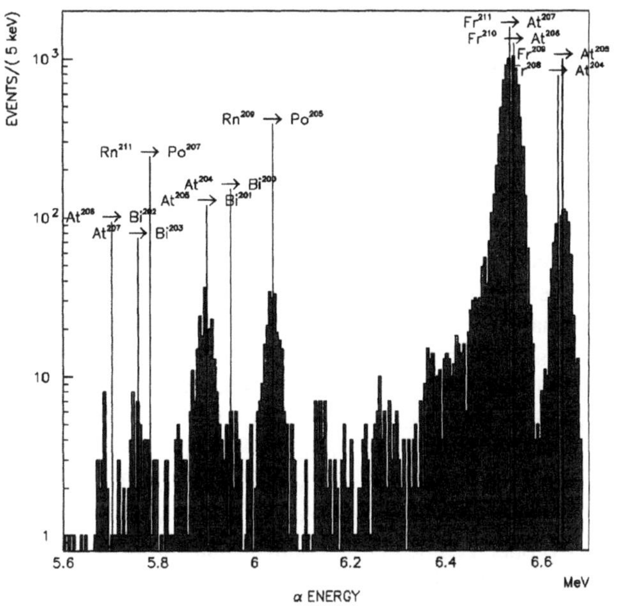

Figure 3. Measured alpha spectrum.

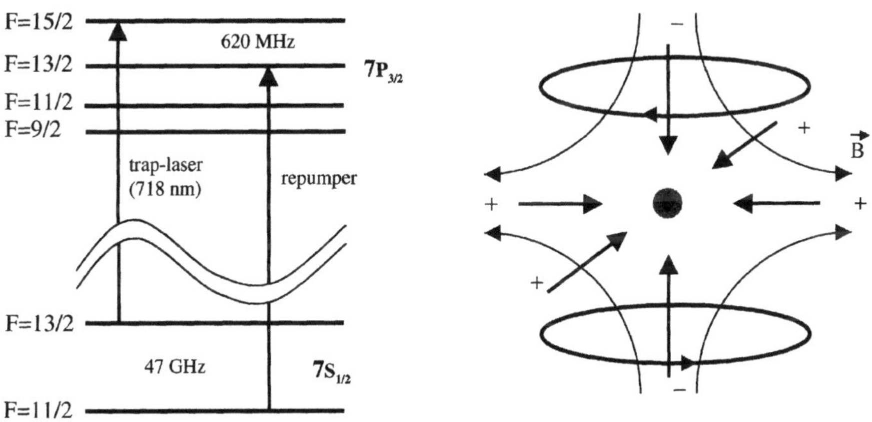

Figure 4. Francium hyperfine structure and MOT scheme.

from three orthogonal pairs of laser beams circularly polarized. They intersect at the center of a pair of coils with opposite currents. The Zeeman effect causes a shift in the atomic levels which is position dependent. The change in detuning with position leads to an imbalance in the scattering force and to a resultant force confining the atoms. In Figure 4 the francium hyperfine level structure and the MOT scheme are sketched. As francium is not a two level atom, a second laser (re-pumping laser) is necessary to avoid optical pumping to the non resonant hyperfine level, by stopping the cooling cycle and hence the trapping.

The trapping laser is a Ti:sapphire ring laser. Its wavelength is about 718 nm. The repumping laser wavelength is 718 or 817 nm (see Figure 4) and it is delivered by a diode laser. The loading of the trap is made by waiting until atoms with velocity smaller than v_c cross the trapping region and fall inside it. The loading rate R is therefore faster if v_c is maximized. v_c depends on the laser intensity, on the laser beam diameter and also on the laser bandwidth. Significant improvements have been obtained by using broad-band lasers with special spectral characteristics allowing the so called "white-light" cooling [13]. Kasevich [14] has applied this approach to the case of a lithium MOT and improvements of R larger than one order of magnitude have been obtained. A large R makes faster the trap loading and reduces the atom losses due to atom adsorpion on the cell walls or to atomic escape from the cell itself. This last problem is not a main drawback for non radioactive alkali atoms, but it is very important in case of francium that is produced at a very low rate. A drastic reduction of atom losses has to be therefore made. This is currently obtained, as proposed first by Wieman *et al.* [6], by preparing special coated cells, which prevent atoms from sticking to the wall and allow them (also because also of a proper geometry) to bounce many times on the cell walls to increase the trapping probability. The coating commonly used is known as dry film.

We plan to use a special organic coating, the PDMS (polydimethylsiloxane polimer), which has the double property to allow atoms to bounce many times at the cell walls and at the same time to release the adsorbed atoms when illuminated with light. This effect, known as LIAD (light induced atom desorption), allows one to have a light controlled atomic source [15]. This last feature gives the possibility to have very fast trap loading by keeping very slow trap decay time. Moreover there is the possibility to store the atoms in the coating for a time comparable with their half-life and then to desorb them at a given instant by using a light flash pulse. This coating has already been tested in a rubidium MOT and very promising results have been obtained [16]. Loading rates much faster than presented in literature have been measured and quite high collection efficiency has been achieved.

References

1. Metcalf, H. J. and van der Straten, P., *Laser Cooling and Trapping*, Springer-Verlag, New York, 1999.
2. Dalfovo, F. *et al.*, *Rev. Modern Phys.* **71** (1999), 463.
3. Behr, J. A. *et al.*, *Phys. Rev. Lett.* **79** (1997), 375.
4. Lu, Z.-T. *et al.*, *Phys. Rev. Lett.* **72** (1994), 3791.
5. Simsarian, J. E. *et al.*, *Phys. Rev. Lett.* **76** (1996), 3522.
6. Lu, Z.-T. *et al.*, *Phys. Rev. Lett.* **79** (1997), 994.
7. Perey, M., *C. R. Acad. Sci.* **208** (1939), 97.
8. Liberman, S. *et al.*, *C. R. Acad. Sci. B* **286** (1978), 253.
9. Duong, H. T. *et al.*, *Europhys. Lett.* **3** (1987), 175.
10. Arnold, E. *et al.*, *J. Phys. B* **23** (1990), 3511.
11. Coc, A. *et al.*, *Phys. Lett. B* **163** (1985), 66.

12. Andreev, S. V. *et al.*, *Phys. Rev. Lett.* **59** (1987), 1274.
13. Atutov, S. N. *et al.*, *Phys. Rev. Lett.* **80** (1998), 2129.
14. Anderson, B. P. *et al.*, *Phys. Rev. A* **63** (2001), 2947.
15. Atutov, S. N. *et al.*, *Phys. Rev. A* **60** (1999), 4693.
16. Atutov, S. N. *et al.*, submitted to *Phys. Rev. Lett.*

Hyperfine Interactions **146/147:** 91–95, 2003.

WITCH: A Penning Trap Retardation Spectrometer Combination for Precision Studies of the Weak Interaction

B. DELAURÉ[1], M. BECK[1], V. V. GOLOVKO[1], V. KOZLOV[1], T. PHALET[1], P. SCHUURMANS[1], N. SEVERIJNS[1], B. VEREECKE[1], S. VERSYCK[1], D. BECK[2], W. QUINT[2], F. AMES[3], K. REISINGER[3], O. FORSTNER[4], J. DEUTSCH[5], G. BOLLEN[6] and S. SCHWARZ[6]
[1]*Instituut voor Kern-en Stralingsfysica, K.U. Leuven, Belgium*
[2]*GSI-Darmstadt, Germany*
[3]*LMU Munich, Germany*
[4]*CERN, Switzerland*
[5]*UCL, Belgium*
[6]*Michigan State University, USA*

Abstract. The Weak Interaction Trap for CHarged particles (WITCH) setup is being installed at the ISOLDE-facility (CERN) to test the Standard Model of the electroweak interaction. This will be done by searching for scalar and tensor admixtures in nuclear beta decay. The beta-neutrino angular correlation is an excellent probe to determine the possible strength of those non-Standard Model contributions. The WITCH setup combines the unique storage features of a Penning trap to produce a scattering free radioactive source with a retardation spectrometer to measure the recoil energy spectrum of the daughter nuclei after beta decay with high precision. Physics beyond the Standard Model would lead to deviations from the expected spectral shape.

Key words: Penning trap, retardation spectrometer, Standard Model, weak interaction.

1. Weak interaction study

Although the V-A description of the electroweak interaction is in very good agreement with experimental data, it is believed that this Standard Model is not the ultimate theory of the weak interaction. It has too many parameters of which the value cannot been predicted by the theory itself and some fundamental questions concerning the model have not been clarified up to now. What the interaction type is concerned, the non-Standard Model admixture of scalar (S) and tensor (T)-type which are present in the most general weak interaction Hamiltonian, are only ruled out at the level of about 10% of the vector (V) interaction strength [2, 3].

In nuclear beta decay pure transitions are the best suited probes to search for non-Standard Model contributions, since a pure Fermi decay can only be of vector or scalar-type while a pure Gamow–Teller transition consists of an axial-vector or tensor contribution. The scalar and vector interaction are physically distinguishable since the 2 leptons originating from a scalar nuclear beta decay have the same helicity so that they are emitted preferentially back to back. For a vector-type interaction the helicity of the two leptons is opposite and therefore a vector decay tends to a parallel emission of neutrino and beta particle. As a consequence the recoil energy spectrum of the daughter nuclei will have its maximum at a much lower energy for a scalar decay then for a vector decay. A similar argument is valid to distinguish the tensor interaction from the axial-vector interaction.

The angular correlation between the emission directions of the beta particle and the neutrino is described by

$$W(\theta) \simeq 1 + a\frac{v}{c}\cos(\theta)\left[1 - \frac{\Gamma m}{E}b\right]. \tag{1}$$

In this equation E, v and m represent the energy, velocity and rest mass of the beta particle, $\Gamma = \sqrt{1-(\alpha Z)^2}$, b the Fierz interference coefficient and a the beta-neutrino angular correlation coefficient. Since the a and b coefficients are functions of the coupling constants of the different interaction forms [1], a measurement of these parameters can test the Standard Model. The best limits on a possible non-Standard Model admixture in nuclear beta decay have been achieved by a measurement of the beta-neutrino angular correlation coefficient of ^{32}Ar [2] (scalar interaction) and ^{6}He [3] (tensor interaction). In the WITCH experiment, which will be described below, the beta–neutrino correlation coefficient will be extracted from the shape of the measured recoil energy spectrum of the daughter nuclei after beta decay. Figure 1 shows the change in shape of two normalised spectra characterised by different a-parameters.

2. Experimental principle

In order to measure the recoil energy spectrum of the daughter nuclei after nuclear beta decay the WITCH setup uses a Penning trap as storage device for radioactive ions. Due to the recoil energy which the daughter nuclei have gained after the nuclear beta decay, the ions can escape from the trapping potential and will be guided by the magnetic field of the Penning trap to the retardation spectrometer, where the energy analysis is performed. Figure 2 shows a schematic overview of the principle of the WITCH setup.

With this approach scattering effects in the radioactive source, which is one of the limiting factors in 'classical' beta decay experiments, can be significantly reduced. Furthermore, the absence of an implantation matrix makes it possible to extract the low energetic (typically a few hundred eV) recoil ions from the well localised source which allows energy analysis and detection of these daughter nuclei. Since the injection and capture of ions in a Penning trap are not element dependent the restrictions of atom trap setups are avoided in the WITCH experiment. In this way for a certain experiment the most suited isotope candidate can be selected.

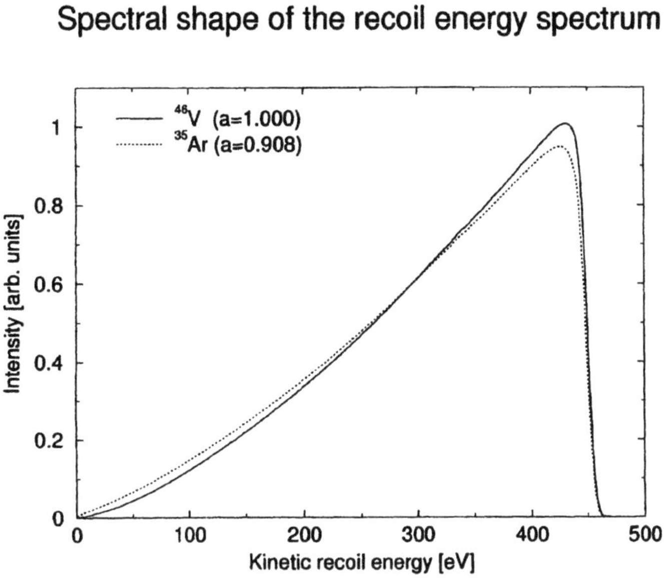

Figure 1. Differential recoil energy spectra for the decay of ^{46}V and ^{35}Ar which show the difference in shape for different a-parameter. The energy axis for endpoint energy of ^{35}Ar.

The principle of the retardation spectrometer is similar to the Mainz [4] and the Troitsk [5] spectrometers which are used to measure the electron antineutrino mass out of the beta decay of tritium. The energy analysis of the recoil ions is performed by a retardation potential which selects the ions with sufficient kinetic energy (originating from the recoil) to get over this electrostatic barrier. Those ions will be counted by a detector. By varying the retardation potential the integral recoil spectrum can be measured.

Since only the velocity component parallel to the retardation electric field is probed the analysing plane (in which the retardation potential reaches its maximum value) is placed in a low (with respect to the trap region) magnetic field. The slow variation of the magnetic field between Penning trap and analysing plane allows an adiabatic conversion of the kinetic energy of the radial ion motion into longitudinal kinetic energy such that it is really the recoil energy which is probed by the retardation potential.

3. WITCH setup

The WITCH setup will be installed at the ISOLDE-facility (CERN) where it will be connected to REXTRAP, the cylindrical Penning trap of the REX-ISOLDE-project [6]. REXTRAP will be used as a cooler and buncher for the continuous 60 keV ISOLDE-beam. After ejection from REXTRAP the ion bunches will be

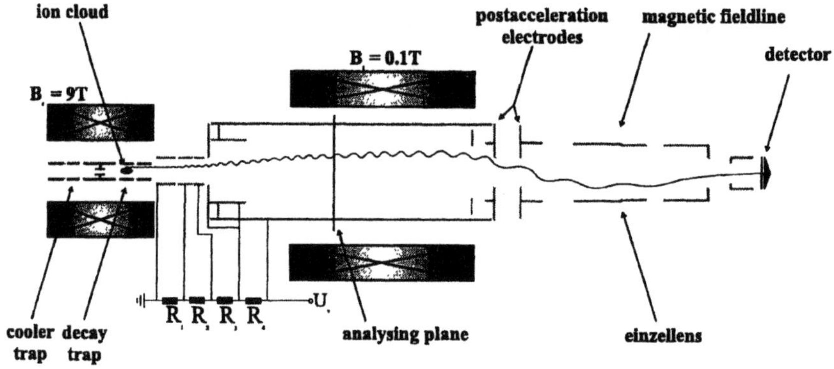

Figure 2. Schematic overview of the principle of the WITCH-setup.

guided to the vertically positioned WITCH setup, retarded and injected into the WITCH Penning trap structure.

The WITCH trap structure will consist out of two cylindrical Penning traps in a 9 T magnetic field. The first one will be used to capture and cool the ions coming from REXTRAP. After the buffergas cooling [7] in this first trap the ions will be transferred to the second trap. This Penning trap will be used to store the radioactive ions until the nuclear beta decay takes place.

In the WITCH spectrometer the retardation potential is produced by 4 cylindrical electrodes on which different voltages are applied. The magnetic field varies from 9 T in the Penning trap to 0.1 T at the position of the analysing plane. This configuration corresponds to an adiabatic conversion of 98.8% of the radial energy into axial energy. After passing the analysing plane the ions are re-accelerated to an energy of 10 keV in order to allow the detection of the recoil ions with a MCP detector. The focusing of the recoil ions onto the MCP detector is done electrostatically.

4. Time schedule and outlook

At the moment the WITCH setup is being installed at the ISOLDE facility. The construction of the experiment will be finished in June 2002. After commissioning of the full setup first experiments are planned in early 2003.

The first experiments will be focused on the search for a scalar-type weak interaction. In order to check and improve the current limits on a scalar and tensor-type interaction WITCH is aiming on a precision on the beta-neutrino angular correlation coefficient of 0.5% or better. The isotope for the first experiment is the mixed allowed beta transition of ^{35}Ar to ^{35}Cl. The next candidate is the $0^+ \rightarrow 0^+$ beta transition of ^{26m}Al. In a second stage the search for physics beyond the Standard Model will continue by focusing on a possible tensor interaction in nuclear beta decay.

Apart from Standard Model tests the WITCH setup will give the opportunity to directly detect electron capture (EC). This process will be characterised by a mono-energetic peak in the recoil energy spectrum. These peaks can be used for energy calibration of the spectrum but they will also allow to determine the Q-value of nuclear beta decays. The appearance of the different ionisation states in the recoil energy spectrum can be used to measure the charge distribution of the daughter ions both after nuclear beta decay and EC.

This research was partly funded with a specialisation fellowship of the Flemish Institute for the Stimulation of scientific-technological research in the industry (IWT).

References

1. Jackson, J. D., Treiman, S. B. and Wyld, Jr., H. W., *Nucl. Phys.* **4** (1957), 206.
2. Adelberger, E. G., Ortiz, C., Garcia, A., Swanson, H. E. *et al.*, *Phys. Rev. Lett.* **83** (1999), 1299.
3. Johnson, C. H., Pleasonton, E. and Carlson, T. A., *Phys. Rev.* **132** (1963), 1149.
4. Picard, A., Backe, H., Barth, H., Bonn, J. *et al.*, *Nucl. Instr. Meth. B* **63** (1992), 345.
5. Lobashev, V. M. and Spivak, P. E., *Nucl. Instr. Meth. A* **240** (1985), 305.
6. Habs, D., Kester, O., Rudolph, K., Thirolf, P. *et al.*, *Nucl. Instr. Meth. B* **126** (1997), 218.
7. Savard, G., Becker, S., Bollen, G., Kluge, H.-J. *et al.*, *Phys. Lett. A* **158** (1991), 247.

Hyperfine Interactions **146/147:** 97–102, 2003.

Structure and Dynamics of High-Z Ions Studied at the ESR Storage Ring

T. STÖHLKER[1,2], A. GUMBERIDZE[1,2,3], X. MA[4], H. F. BEYER[1], G. BEDNARZ[5], F. BOSCH[1], X. CAI[4], S. FRITZSCHE[6], S. HAGMANN[2,7], C. KOZHUHAROV[1], O. KLEPPER[1], D. LIESEN[1], P. H. MOKLER[1], D. SIERPOWSKI[5], Z. STACHURA[8], M. STECK[1], A. SURZHYKOV[1], S. TOLEIKIS[1], A. WARCZAK[5] and Y. ZOU[9]

[1] *GSI-Darmstadt, Germany*
[2] *IKF, University of Frankfurt, Germany*
[3] *Tbilisi State University, Georgia*
[4] *IMP, Lanzhou, China*
[5] *Institute of Physics, Cracow University, Poland*
[6] *University of Kassel, Germany*
[7] *Kansas State University, Kansas, USA*
[8] *INP, Cracow, Poland*
[9] *Fudan University, Shanghai, China*

Abstract. The ESR storage ring offers unique conditions for precision investigations in the realm of atomic structure and collisions research at highest nuclear charges. For atomic structure studies, a very recent experiment will be discussed aiming on a precise determination of the electron–electron QED contribution to the groundstate ionization potential in He-like uranium. For the regime of collision dynamics, the close interplay between dynamics and atomic structure is outlined. Here, for the case of the $2p_{3/2} \rightarrow 1s_{1/2}$ transition in H-like uranium, a strong interference between the E1 and M2 transition amplitudes has been identified which gives rise to a remarkable modification of the angular distribution of emitted photons.

Key words: QED, Lamb shift, recombination, high-Z.

1. Introduction

The study of strong field effects occurring in relativistic collisions involving high-Z projectiles is a corner stone of the current atomic physics program at the ESR. This overview first reports on atomic physics experiments addressing QED effects for the ground-state in heliumlike uranium. Here the goal of the experiment is to reach a sensitivity which provides a valuable test of second-order radiative corrections, a goal which has not yet been reached up to now for H-like high-Z ions. Secondly we discuss an interference between the El and M2 transition amplitudes taking place for the decay of the $2p_{3/2}$ state in H-like uranium. To the best of our knowledge, such an effect has not been observed before up to now for any other atomic system.

We emphasize, that this observation was only possible by means of the radiative electron capture process in relativistic collisions, a process which leads to strongly aligned excited projectile states.

2. Two-electron QED studies for helium-like uranium

Helium-like ions are the simplest atomic multi-body systems. Very recently the theoretical as well as the experimental investigations of these fundamental systems have obtained a considerable improved precision [1, 3]. The experimental progress manifests itself in a novel approach and exploits Radiative Recombination (RR) transitions into the vacant $1s$ shell of bare and H-like ions [3]. In this process an electron may undergo a direct transition into a bound state of the stationary ion via the emission of a photon carrying away the energy difference between the initial and final electron state, i.e. $\hbar\omega = E_{\rm kin} + E_B$ (compare Figure 1). The difference in the centroid energies for such radiative recombination transitions into the vacant K-shell of bare and H-like high-Z ions is equal to the difference in the ionization potential between the hydrogen-like and the helium-like ions formed by the recombination process. It gives exactly the two-electron contribution to the ground-state energy of the heliumlike ions. Most important, all one-electron contributions to the binding energy such as the finite-nuclear size corrections and the one-electron self-energy cancel out completely in this type of experiment [1]. These first experiments on the ground-state of He-like ions conducted at Super-EBIT were very successful and provided to the first time results which are at the threshold to become sensitive to higher order QED effects [3, 4]. However, at high nuclear charges such as $Z = 83$ or even $Z = 92$ it turned out that the production efficiency for bare ions is not sufficient and the results suffered from poor by counting statistics. Very recently we applied this technique at the electron cooler device at the ESR storage ring for the case of He-like uranium. The aim of this experiment is to measure precisely the two-electron contribution to the ionization potential in He-like uranium of 2.2 keV with an accuracy better than 5 eV [1]. Since

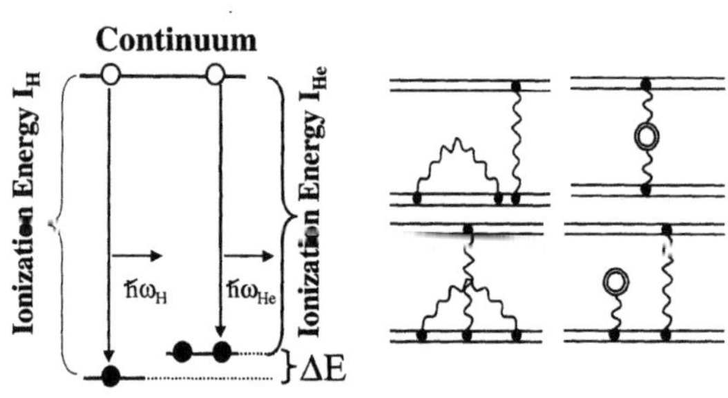

Figure 1. Left side: schematic presentation of the RR process of free electrons into the initially bare and H-like ions. The energy difference $\Delta E = \hbar\omega_{\rm H} - \hbar\omega_{\rm He}$ gives exactly the two-electron contribution to the ionization potential in He-like ions. Right side: Feynman diagrams for the two-electron self-energy and vacuum polarization.

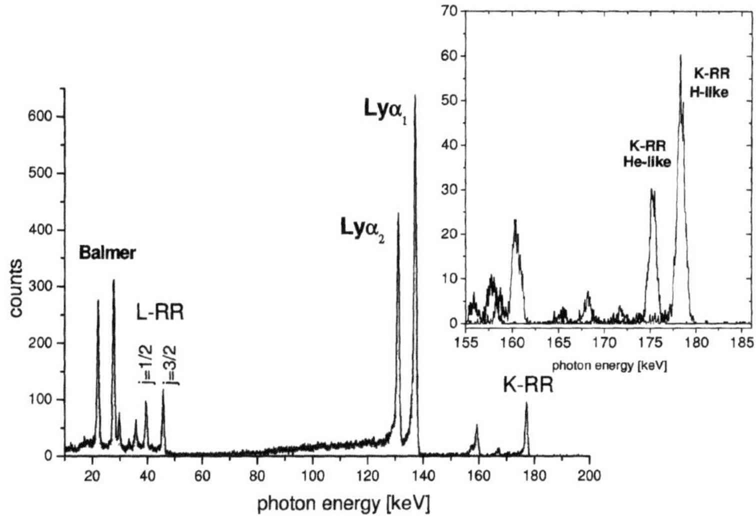

Figure 2. X-ray spectrum for H-like uranium as observed for decelerated ions at the electron cooler of the ESR storage ring. For details see text.

two-electron QED effects are calculated to contribute 7 eV, such an experimental study would therefore provide the very first test of higher-order QED corrections (higher order in α) in the domain of high-Z ions (some of the relevant Feynman diagrams are displayed in Figure 1).

For the experiment, the setup available at the electron cooler section was used. There, the X-rays emitted via RR of the cooler electrons into the ground state of the bare- or H-like ions can be detected by a solid state detector which views the interaction region at an observation angle close to 0° [2]. In particular the experiment benefits from the recently established deceleration technique [5]. At low energies, all uncertainties associated with Doppler corrections are strongly reduced compared to high-energy beams. Also, for decelerated ions the bremsstrahlung intensity caused by the cooler electrons is strongly reduced (due to the comparably small cooler voltage of 27 kV and current of 100 mA). Consequently, very clean conditions for X-ray spectroscopy are present at the cooler section. This is depicted in Figure 2 where a preliminarily calibrated, coincident X-ray spectrum is plotted as observed for initially bare uranium ions at an energy of 43 MeV/u. Note that the strongly reduced bremsstrahlung intensity allowed us for the first time to observe even RR transition into the L-shell, located at the low-energy part of the spectrum. In the spectrum, the X-ray transitions for RR into the ground state, which are of particular interest for the current study, show up in the high energy part of the spectrum. Most important, the tails of the Lyman-α transition lines, caused by cascade feeding of the L-shell levels [2], are consequently not present in the case of the RR photon emission. In the inset of Figure 2 the energy regime relevance for K-RR transitions is displayed. There the line intensities as measured for initially bare-

and H-like uranium are compared. From a first preliminarily analysis, a statistical accuracy for the splitting of close to 9 eV can already be stated. Keeping in mind that a total beam time of 4 days has been used for collection of the data, a second more extended run should provide us with the desired experimental accuracy of 4 eV or even below.

3. The decay of the $2p_{3/2}$ state in H-like uranium

For the particular case of the $2p_{3/2}$ state in H-like uranium an interference between the E1 and M2 transition amplitudes has been observed recently [8] which alters significantly the photon angular distribution of the Lyman-α_1 ($2p_{3/2} \rightarrow 1s_{1/2}$) transition. This finding is in particular surprising because the M2 decay branch contributes less than 1% to the total decay rate (see Figure 3).

In the past, the angular distribution of this decay was the subject of intense experimental and theoretical studies in order to elucidate the magnetic substate population of Radiative Electron Capture (REC) in relativistic collisions of bare uranium with low-Z target atoms [6, 7]. If the magnetic sublevels with different absolute magnetic quantum numbers μ are populated non-statistically, alignment occurs leading to an anisotrope emission pattern. For the particular case of the $2p_{3/2}$ state this angular distribution is given by [7]

$$W(\theta) \propto 1 + \beta_{20}\left(1 - \frac{3}{2}\sin^2\theta\right), \tag{1}$$

where θ is the angle between the direction of the excitation photon and the beam direction (emitter frame) while β_{20} denotes the anisotropy coefficient. In the case of the $2p_{3/2}$ level, the anisotropy coefficient can be expressed as:

$$\beta_{20} = \frac{1}{2}\frac{\sigma(3/2, \pm 3/2) - \sigma(3/2, \pm 1/2)}{\sigma(3/2, \pm 3/2) + \sigma(3/2, \pm 1/2)}. \tag{2}$$

Here $\sigma(3/2, \mu_n)$ describes the population of substate μ_n of the $2p_{3/2}$ level. The well known expression (2) takes into account only the dominant electric dipole (E1) term whereas the weaker magnetic quadrupole component (M2) is neglected. Based on expression (2), the theoretical and observed angular distributions were compared in detail and a remarkable variance was found [6, 7]. This deviation was surprising because REC is otherwise one of the best studied processes for high-Z, hydrogen-like ions for which an excellent agreement between theory and experiment is typically found. By using the density matrix theory one can consider the magnetic (M2) decay branch in a coherent way. As a result, the alignment coefficient parameter β_{20} must now be replaced by the product $\beta_{20} \cdot f(\mathrm{E1}, \mathrm{M2})$ where $f(\mathrm{E1}, \mathrm{M2})$ is a *structure* function which takes into account the interference effect. For H-like uranium the structure parameter $f(\mathrm{E1}, \mathrm{M2})$ reaches a value of as large as 1.28 [8]. In Figure 3 we compare the experimental results for H-like uranium (solid points) with the corresponding theoretical findings (full line) and the results

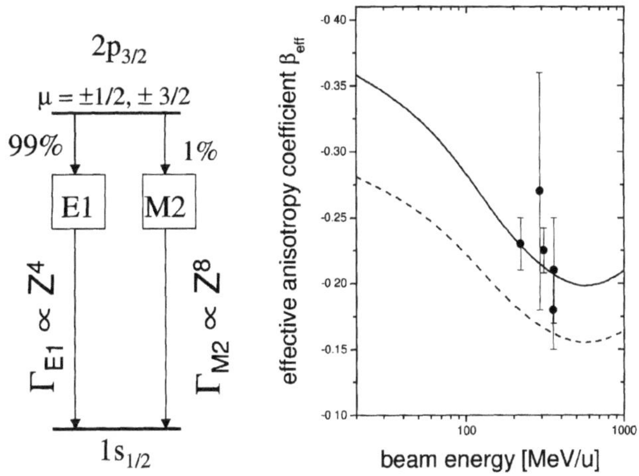

Figure 3. Experimental and theoretical alignment coefficient for the $2p_{3/2}$ state populated by REC into bare uranium as function of beam energy. Solid line: predictions taken the correct structure parameter $f(\text{E1, M2}) = 1.28$. Dashed line: $f(\text{E1, M2}) = 1$ (for details see text).

obtained assuming $f(\text{E1, M2}) = 1$ (dashed line), i.e. neglecting the interference term. From the figure it is evident that the former departure of the theoretical results from the experimental values is removed when taking the interference term is taken into account. This proves the importance of the interference between the E1 and M2 decay branches for the decay of $2p_{3/2}$ state.

4. Summary

We report on a first experiment at the storage ring ESR which aims at a precise determination of the two-electron contribution to the ionization potential in He-like uranium. Although the data analysis is still in progress, the preliminary data show that for the case of high-Z ions, the count rate statistics will no longer be the accuracy limiting factor. In contrast, at Super-EBIT, limitations on counting statistics prevent experiments beyond $Z = 83$. Moreover, the interference between the E1 and M2 transition amplitudes observed for the decay of $2p_{3/2}$ state in H-like uranium demonstrates the strong interplay between atomic collision and structure research. This observation was only possible by means of REC, a process leading to strongly aligned excited states. This also proves that although we are dealing with the simplest atomic systems, i.e. a H-like ion, unexpected phenomena may occur which refine our knowledge about the physics at strong Coulomb fields.

References

1. Persson, H., Salomonson, S., Sunnergren, P. and Lindgren, I., *Phys. Rev. Lett.* **76** (1996), 204.
2. Beyer, H. F. *et al.*, *Z. Phys. D* **35** (1995), 169.

3. Marrs, R. E., Elliott, S. R. and Stöhlker, Th., *Phys. Rev. A* **52**, (1995), 3577.
4. Stöhlker, Th., Elliott, S. R. and Marrs, R. E., *Hyp. Interact.* **99** (1996), 217.
5. Stöhlker, Th. *et al.*, *Phys. Rev. A* **58** (1998), 2043.
6. Stöhlker, Th. *et al.*, *Phys. Rev. Lett.* **79** (1997), 3270.
7. Eichler, J., *Nucl. Phys. A* **572** (1994), 147.
8. Surzhykov, A., Fritzsche, S., Gumberidze, A. and Stöhlker, Th., *Phys. Rev. Lett.* **88** (2002), 153001.

Hyperfine Interactions **146/147:** 103–108, 2003.

Fast-Beam Laser Spectroscopy of Helium-like Silicon

E. G. MYERS, M. REDSHAW and B. ROEDER
Department of Physics, Florida State University, Tallahassee, Florida 32306-4350, USA

Abstract. Doppler-tuned laser spectroscopy has been applied to a foil-stripped 1 MeV/u silicon ion beam to measure the $1s2s\ {}^1S_0$–$1s2p\ {}^3P_1$ intercombination interval in ${}^{28}\mathrm{Si}^{12+}$. To obtain sufficient transition probability the Si^{12+} beam was merged co-linearly with over 2 kW of continuous-wave laser light at 1319 nm inside an ultra-high finesse build-up cavity. The measurement differentiates between current theories for the $n = 2$ levels of helium-like ions.

Key words: helium-like ions, laser spectroscopy.

1. Introduction

The $n = 2$ levels of helium-like ions, see Figure 1, provide interesting tests of relativistic atomic many-body theory [1–3]. Although the underlying interaction (QED) is well-understood, the best procedure to obtain accurate results remains the subject of much theoretical effort. At $Z = 14$ relativistic calculations [4–6] obtain the structure energies up to order $(Z\alpha)^4$ a.u. and there is now interest in investigating QED corrections of this order. As for the Lamb shift in hydrogen-like ions, explicit QED effects are largest for s-states.

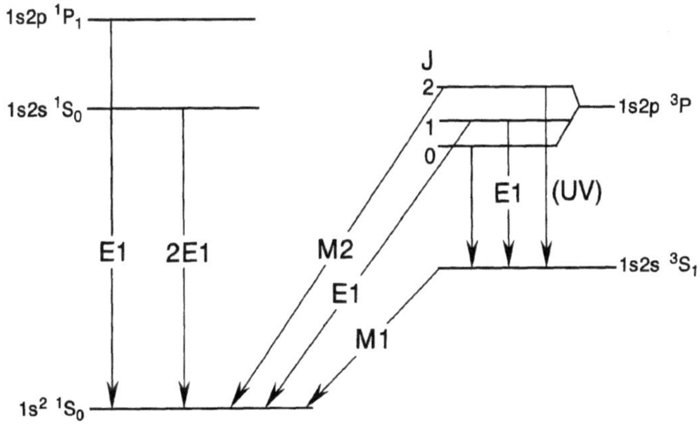

Figure 1. Schematic of the $n = 2$ energy levels of helium-like ions showing the principal radiative decay modes.

However, at least for $Z > 10$, existing calculations predict the results of X-ray spectroscopy satisfactorily, e.g., see [7, 8], and of ultra-violet spectroscopy of the $1s2p\,^3P$–$1s2s\,^3S$ transitions [9, 10], at current experimental precision. Laser spectroscopy can achieve much higher precision [11] and has been applied to the $1s2s\,^3S$–$1s2p\,^3P$ transitions for Z up to 5. Extension to higher Z is difficult because the transitions move further into the ultra-violet. Instead, as has been shown for helium-like nitrogen [12], a path for extending laser techniques to higher Z helium-like ions is to use the $1s2s\,^1S_0$–$1s2p\,^3P_1$ intercombination transition. This is a relativistically allowed E1 transition and lies in the infra-red for $Z < 40$.

For the $1s2s\,^1S_0$–$1s2p\,^3P_1$, the laser induced transition probability for a given laser intensity varies as $\sim Z^{-6}$. The rapidly increasing radiative decays of the various levels shown in Figure 1 cause signal-to-background ratios to fall even faster [11]. To achieve a good signal to background ratio, high power pulsed lasers could be used. However, because cw techniques allow higher ultimate precision, and for economy, we chose to work cw and increase the laser power at the interaction region using a high-finesse build-up cavity (BUC). In Si^{12+} the wavelength of the $1s2s\,^1S_0$–$1s2p\,^3P_1$ interval is approx. 1.383 μm. This makes a convenient match with the 1.319 μm (second strongest) laser line of Nd^{3+}:YAG, allowing for the $\sim$5% Doppler shift with a foil-stripped ion beam at 1 MeV/u.

2. Experiment

The set-up is shown in Figure 2. Beams of $^{28}Si^{5+}$ and $^{28}Si^{6+}$ ions were obtained from the Florida State University FN tandem accelerator at energies of approx. 29.1 MeV. The ions were analyzed in a 90° bending magnet and then passed through a 10 or a 4 μg cm^{-2} carbon foil. This yields a Si^{12+} fraction of up to 7.5% [13] of which of the order of 1% are formed in the metastable $1s2s\,^1S_0$ level, mean lifetime 11.5 ns [14]. The Si^{12+} charge state was then deflected 5° horizontally by a dipole magnet, bringing it co-linear with the standing wave inside the BUC.

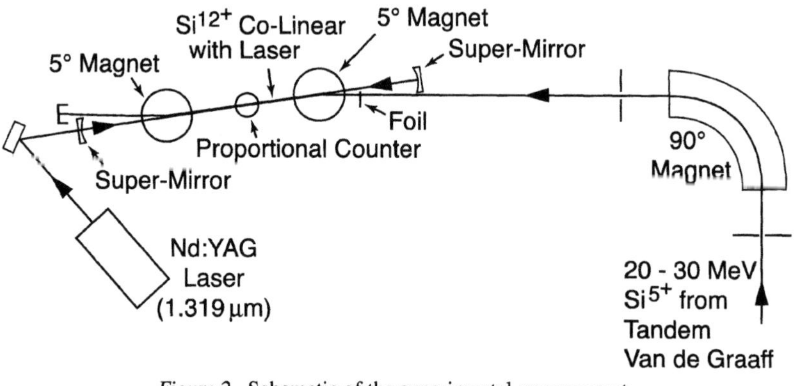

Figure 2. Schematic of the experimental arrangement.

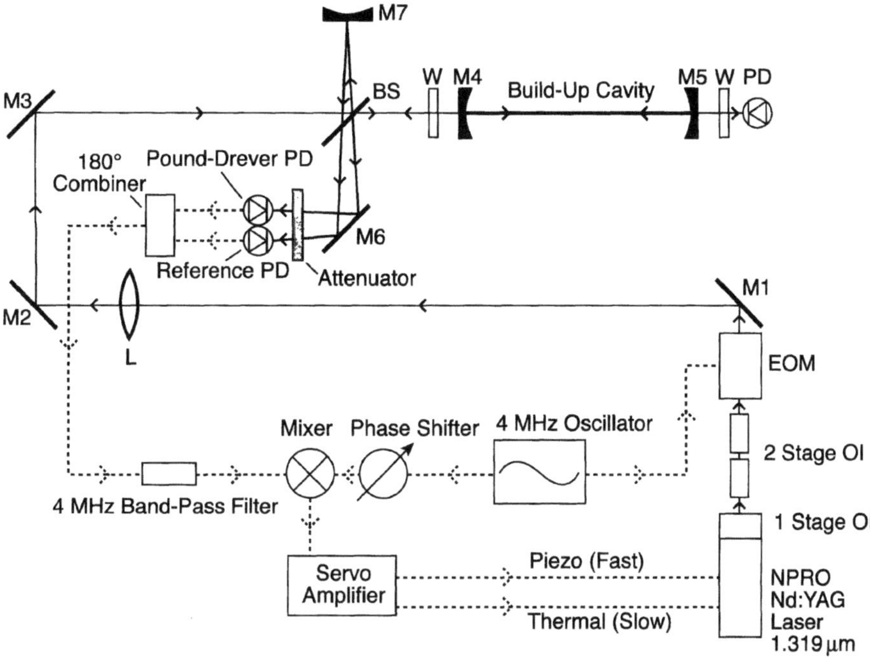

Figure 3. Schematic of the laser and build-up cavity optics: NPRO stands for non-planar ring oscillator, OI – opto-isolator, EOM – electro-optic modulator, M1 to M7 – mirrors, L – lens, PD – photo-diode. Dotted lines indicate electrical signals.

When the co-propagating laser beam was Doppler-shifted into resonance with the $1s2s\,^1S_0$–$1s2p\,^3P_1$ transition, population was transferred to the $1s2p\,^3P_1$ level, mean lifetime 6.36 ps [15]. The signal was the increase in X-ray yield, detected by a proportional counter, due to the subsequent decay of $1s2p\,^3P_1$ to the groundstate. The resonance was scanned by varying the beam velocity by scanning the field in the 90° magnet [16]. The laser induced signal was detected above the strong background from other K X-rays by modulating the power in the BUC at 50 Hz.

The BUC [17] used 1 m radius-of-curvature superpolished mirrors with transmissions of 25 ± 5 ppm, spaced 90 cm apart using invar rods inside the vacuum chamber. The cavity was excited by a 210 mW diode-pumped mono-lithic Nd:YAG laser. The laser was locked to one of the 3 kHz FWHM cavity resonances using the Pound–Drever–Hall scheme [18], see Figure 3. Because of the inherent frequency stability of this laser a simplified servo-amplifier based on the work of De Riva *et al.* [19] sufficed. The circulating intracavity power in the BUC was estimated from the transmitted power to be 2.4 kW.

A search in beam energy over the region predicted by [4] revealed a resonance of the expected height and width, see Figure 4. The observed resonance width, equivalent to 1.03 ± 0.3 cm^{-1} FWHM, is consistent with the (Lorentzian) natural width of 0.834 cm^{-1}, the estimated transition probability, which leads to power

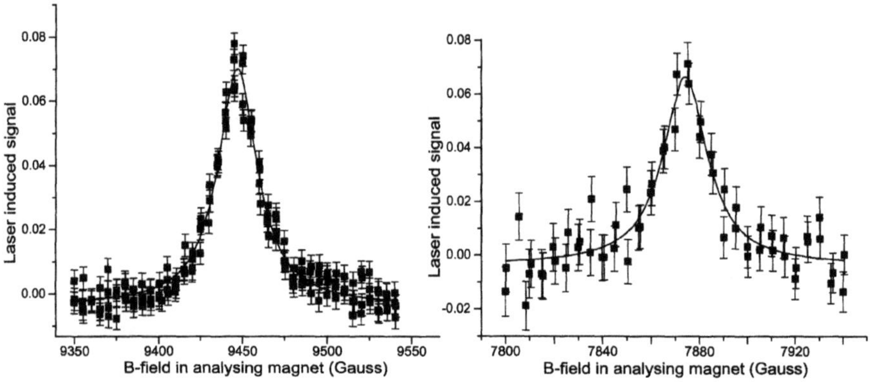

Figure 4. Doppler-tuned resonances of the $^{28}Si^{12+}$ $1s2s\,^1S_0$–$1s2p\,^3P_1$ intercombination transition, with Si^{5+} (*left*), and Si^{6+} (*right*) ions incident on a nominally 4 μg cm^{-2} carbon foil. The horizontal axis is the magnetic field in the analysing magnet used to determine the beam velocity. The vertical axis is the fractional increase in count rate due to the laser. The solid lines are least-squares fits with Lorentzians.

broadening (due to depopulation of $1s2s\,^1S_0$) of approx. 10%, and a Gaussian contribution of 0.3 cm^{-1}, due to an estimated beam energy spread after the foil of 50 keV FWHM [20].

The statistical precision of the data in Figure 4 is consistent with determining the wavenumber of the intercombination transition to 1 ppm. In the present experiment the precision is limited by the uncertainty in the beam velocity (and hence Doppler shift), due to uncertainty in the magnet calibration and energy lost in the foil. To determine both of these, we used the well known $^{19}F(p, \alpha\gamma)\,^{16}O$ nuclear resonance occurring at a proton energy of 872.11(20) keV [21], with both gas and solid hydrocarbon targets. Using the 90° magnet fields corresponding to the laser and gas target nuclear resonances, plus the measurement of ^{19}F energy loss in the foil, we determined the beam velocity corresponding to the centroid of the Si^{12+} laser resonances, see Equation (1) of [22]. Hence, using the relativistic Doppler formula and the measured laser wavenumber, we derived the $1s2s\,^1S_0$–$1s2p\,^3P_1$ wavenumber.

3. Results

Our result is compared with theory in Table I. The best agreement is with [4] which used an all-orders RMBPT and QED corrections from [23]. The remaining discrepancy is equivalent to 0.025(8) $(Z\alpha)^4$ a.u. The calculations of [6], who used RCIT, are in strong disagreement, apparently due mainly to their QED corrections. Agreement is improved if these are replaced with those of Drake, as shown in the last line of Table I. (We allow for the difference in the definition of the QED contribution as discussed in [6].) Further work in both the structure and QED parts of

Table I. Experimental result for the $^{28}Si^{12+}$ $1s2s\,^1S_0$–$1s2p\,^3P_1$ intercombination interval compared with recent theory. Units are cm^{-1}

This experiment	7230.5(2)
Plante, Johnson and Sapirstein [4]	7231.1
Cheng, Chen, Johnson and Sapirstein [6]	7264.7
Drake [23]	7251.8
Cheng *et al.* [6], Drake [23] (QED)	7228.9

the theory is indicated. The precision of the experiment can be improved, perhaps an order of magnitude, by using a co- and counter-propagating beam technique [16, 20], or by precision time-of-flight measurement of the beam velocity.

Acknowledgements

We thank the Oxford EBIT group for the loan of the wavemeter and for discussions on build-up cavities. Important contributions of J. K. Thompson, R. Harry, J.-S. Cho, J. Parnell, R. K. Nadaskay and of other students and staff of the Florida State University Superconducting Accelerator Laboratory are gratefully acknowledged. This work was partially supported by NSF grants PHY-9970991 and PHY-0098142, NATO CRG-960003, and the State of Florida.

References

1. Sapirstein, J., *Rev. Mod. Phys.* **70** (1998), 55.
2. Lindgren, I., *Int. J. Quant. Chem.* **57** (1996), 683.
3. Persson, H., Salomonson, S., Sunnergren, P. and Lindgren, I., *Phys. Rev. Lett.* **76** (1996), 204.
4. Plante, D. R., Johnson, W. R. and Sapirstein, J., *Phys. Rev. A* **49** (1994), 3519.
5. Chen, M. H., Cheng, K. T. and Johnson, W. R., *Phys. Rev. A* **47** (1993), 3692.
6. Cheng, K. T., Chen, H. M., Johnson, W. R. and Sapirstein, J., *Phys. Rev. A* **50** (1994), 247.
7. Marrs, R. E., Elliott, S. R. and Stöhlker, Th., *Phys. Rev. A* **52** (1995), 3577.
8. Chantler, C. T., Paterson, D., Hudson, L. T., Serpa, F. G., Gillaspy, J. D. and Takács, E., *Phys. Rev. A* **62** (2000), 042501.
9. Kukla K. W. *et al.*, *Phys. Rev. A* **51** (1995), 1905.
10. Howie, D. J. H., Silver, J. D. and Myers, E. G., *J. Phys. B* **29** (1996), 927.
11. Myers, E. G., In: S. G. Karshenboim *et al.* (eds), *The Hydrogen Atom: Precision Physics of Simple Atomic Systems*, Springer, 2001, p. 179.
12. Myers, E. G., Thompson, J. K., Gavathas, E. P., Claussen, N. R., Silver, J. D. and Howie, D. J. H. *Phys. Rev. Lett.* **75** (1995), 3637.
13. Shima, K., Kuno, N., Yamanouchi, M. and Tawara, H., *At. Data Nucl. Data Tables* **51** (1992), 173.
14. Derevianko, A. and Johnson, W. R., *Phys. Rev. A* **56** (1997), 1288.
15. Johnson, W. R., Plante, D. R. and Sapirstein, J., *Adv. At. Mol. Opt. Phys.* **35** (1995), 255.

16. Thompson, J. K., Howie, D. J. H. and Myers, E. G., *Phys. Rev. A* **57** (1998), 180.
17. Fee, M. S., Chu, S., Mills, Jr., A. P., Chichester, R. J., Zuckerman, D. M., Shaw, E. D. and Danzmann, K., *Phys. Rev. A* **48** (1993), 192.
18. Drever, R. W. P., Hall, J. L., Kowalski, F. V., Hough, J., Ford, G. M., Munley, A. J. and Ward, H., *Appl. Phys. B* **31** (1983), 97.
19. De Riva, A. M. *et al.*, *Rev. Sci. Instrum.* **67** (1996), 2680.
20. Myers, E. G., Margolis, H. S., Thompson, J. K., Farmer, M. A., Silver, J. D. and Tarbutt, M. R., *Phys. Rev. Lett.* **82** (1999), 4200.
21. Ajzenberg-Selove, F., *Nuc. Phys. A* **300** (1978), 1.
22. Myers, E. G. and Tarbutt, M. R., *Phys. Rev. A* **61** (1999), 10501(R).
23. Drake, G. W. F., *Can. J. Phys.* **66** (1988), 586.

Hyperfine Interactions **146/147:** 109–113, 2003.

Physics with Highly-Charged Ions in an EBIT

J. R. CRESPO LÓPEZ-URRUTIA[1], B. BAPAT[1], I. DRAGANIĆ[1], B. FEUERSTEIN[1], D. FISCHER[1], H. LÖRCH[1], R. MOSHAMMER[1], J. ULLRICH[1], R. D. DUBOIS[2] and Y. ZOU[3]
[1] *Max-Planck-Insitut für Kernphysik, Saupfercheckweg 1, D-69117 Heidelberg, Germany*
[2] *University of Missoury, Rolla, USA*
[3] *Fudan University, Shanghai, PR of China*

Abstract. After the commissioning of the Freiburg electron beam ion trap, experiments on dielectronic recombination of the low-lying resonances in He-like Ar^{16+} have been carried out at high resolution. Forbidden transitions ("coronal lines") of highly charged argon ions in the optical range have been measured with an accuracy around 1 ppm. Ions extracted from FreEBIT have been used to perform collision experiments using the Cold Target Recoil-Ion Momentum Spectroscopy (COLTRIMS) technique.

Key words: ion sources, highly charged ions, spectroscopy, ion–atom collisions, ion–electron collisions, recoil ion momentum spectroscopy.

1. Optical spectroscopy

The new electron beam ion trap facility FreEBIT was developed to study the physics of highly charged ions with special interest in visible spectroscopy, hyperfine structure studies and collision processes. It was designed as a combination of ion trap and ion source [1] which can deliver beams of highly charged ions to perform kinematically complete collision experiments [2]. Simultaneously the EBIT can be used for spectroscopy studies of HCIs in the X-ray, soft X-ray, VUV, UV, and visible region. The main interest is to obtain data relevant for plasma diagnostics and astrophysics, as well as fundamental information on the hyperfine structure of hydrogenic ions. The design of the apparatus allows a convenient optical access to the central trap region due to its wide radial magnet bore openings. Two quartz lenses are mounted inside of one of those; these *in vacuo* lenses relay the image of the trap to a plane just outside the quartz vacuum window. From there, a set of mirrors and lenses rotate the orientation of the image from horizontal to vertical and project it onto the entrance slit of a grating monochromator. This instrument can reach a dispersion of 0.34 nm/mm at the focal plane. Cryogenically cooled CCD detectors with 2000×800 pixels on a 30 mm $\times$ 8 mm surface are used to detect the incident radiation with very low noise levels. With the two available monochromators, four gratings and two CCD detectors, the energy region from 190 nm to 850 nm can be covered with very good resolution and quantum efficiency.

In the first mode of observation, the trap is imaged onto the narrow entrance slit, to approach the resolution limit of the instrument. Narrow spectral lines from the trapped ions are thus obtained. By positioning spectral lamps at the position of the first real image of the trap that appears just outside the quartz window, calibration spectra are recorded before and after each exposure. Afterwards, the grating is slightly rotated. A new exposure with its corresponding calibrations is then recorded. The entire process is repeated as many as thirty times. In this way, the statistical limitations posed by the small number of pixels illuminated across the width of the spectral line are overcome, and the sampling of the line profile is improved by recording several hundred data points for each line profile. Each spectrum is evaluated by fitting individual Gauss functions to the calibration lines (roughly 10 lines on each spectrum) and then using a least-squares-algorithm to obtain a second degree polynomial for the dispersion function. Since the deviation from the linear dispersion for this type of grating monochromator is small, a quadratic function provides already a very good approximation to the real dispersion. Typical parameters are: At 3.5 pixels FWHM, and a linear dispersion of 0.0086 nm/pixel, a calibration line has a $3 \cdot 10^{-2}$ nm FWHM. The integral number of counts on the line is about 20 000. The centroid of this line can be so determined to better than $1/100$ of the line profile when enough data points are recorded and the line profile is well understood. That provided, a $2 \cdot 10^{-4}$ nm centroid determination is achieved. The wavelength calibration data are taken from sources where they are given with an accuracy of 10^{-4} to 10^{-5} nm. With ten calibration lines spawning the central region of the detector (about 15 nm), the standard deviation of the calibration lines from the fitted dispersion curve is $1.6 \cdot 10^{-4}$ nm, agreeing well with the statistical uncertainty of the individual centroid determination. The fitted curve is more accurate than the individual calibration line by a factor roughly proportional to the square root of the number of calibration lines (10) minus the number of degrees of freedom for the fitting function (3). In this case we obtain a numerical result of $6 \cdot 10^{-5}$ nm, and are therefore confident about the calibration at the level of 10^{-4} nm. Given the fact that the calibration procedure is repeated at different grating positions, non-linearities in the detector response of the individual pixels (or pixel columns, in the usual binning mode applied) and the pixel geometry (for a CCD camera a very small figure) become negligible. The effect from deviations of the line profile from an ideal Gauss has been studied by varying the intervals around the line centre for the fitting procedure in order to take into account or neglect varying amounts of background, scattered light and so on which can affect the line wings. Usually such tests led to centroid shifts of 0.004 pixels on average, or $3.5 \cdot 10^{-5}$ nm, although in some cases the shift can be as big as $1.5 \cdot 10^{-4}$ nm under very unrealistic choices of the fitting interval. This systematic effect could cause an estimated error of 10^{-4} nm under conservative assumptions. Another possible source of systematic deviations is the location of the spectral lamp in the optical setup. We have performed stepwise changes in this location, resulting in shifts of the order of 10^{-4} nm, consistent with the statistical quality of

Table I. Experimental wavelengths of some forbidden transitions (preliminary)

Ion	Transition	Wavelength (nm)	Exp. error (nm)
Ar^{9+} (ArX)	$2s^2 2p^5\ ^2P_{3/2}$–$^2P_{1/2}$	553.3260	0.0006
Ar^{13+} (ArXIV)	$2s^2 2p\ ^2P_{3/2}$–$^2P_{1/2}$	441.2567	0.0004
Ar^{14+} (ArXV)	$2s^2\ ^3P_3$–$^3P_2$	594.3884	0.0006

the data. Temperature drifts as possible sources of systematic errors are ruled out by our calibration procedure. The time analysis of the data shows only statistical departures from the average wavelength value, and no systematic drifts whatsoever. Possible influences from the Zeeman or Stark effect on the wavelength can be ruled out by simple estimates.

Measurements for different highly charged argon ions were carried out in order to obtain a systematic series for systems with 4, 5, 8 and 9 electrons. Also some interesting krypton forbidden transitions were observed. All these lines are of potential application for diagnostic purposes (e.g., Doppler shift of solar flares). These (preliminary) results are shown in Table I. The experimental error is mainly determined by the statistics of the measured lines in comparison with the calibration lines, and finally by the quality of the calibration lines. With a total error of roughly 1 ppm, these results belong to the few at this level of accuracy for highly charged ions. We are not aware of any other data with an error bar of less than 7 ppm for visible radiation from highly charged ions. In the X-ray region, very few measurements have an accuracy of better than 20 ppm (again for highly charged ions). Therefore, the present data are well suited to test relativistic atomic physics structure codes for few-electron systems with an unprecedented precision.

Another different type of measurement was carried by using a wide open slit. In this way, a multi-wavelength imaging of the spatial distribution of the ions in the trap was obtained. Together with the Doppler-width data obtained in the first set of measurements, detailed information about the thermal velocity and distribution of the ions are obtained, which are necessary to derive cross sections for excitation and recombination processes in the trap.

2. Dielectronic recombination

Dielectronic recombination studies have been carried out with highly charged argon ions by using an electron beam of variable energy and recording the photons emitted during the radiative stabilization of the doubly excited states populated by the DR process. The photon energy is measured with a high-purity germanium detector with a resolution of roughly 150 eV. The beam energy was scanned from 1.5 to 4.5 keV. By using very low beam currents, the energy spread of the electron beam, which is mainly caused by its own space charge, was reduced to levels as

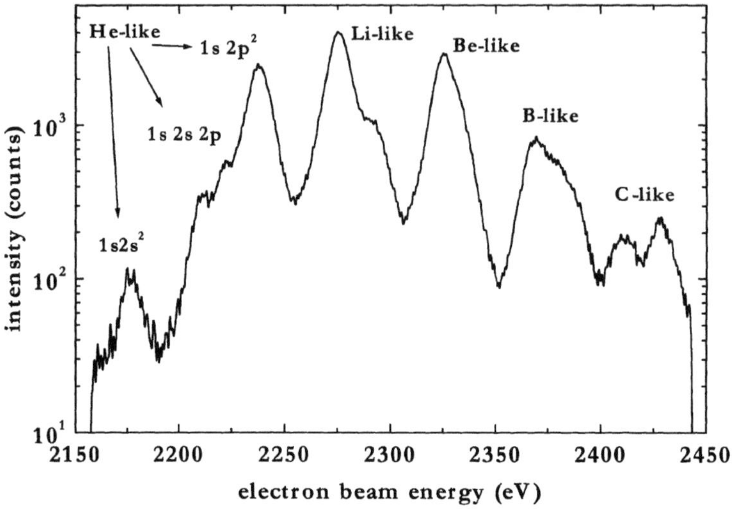

Figure 1. Photon yield of the 2–1 transitions after excitation of the KLL dielectronic resonances of highly-charged Ar ions as a function of electron beam energy.

low as 8 eV FWHM. This allowed us to resolve the structure of the KLL resonance and to separate the contributions from He, Li, Be and B-like satellites (see Figure 1). The absolute resonance energy was determined by varying the electron beam current from 3 to 50 mA and extrapolating the space charge shift to zero current, with an estimated absolute error of about 0.5 eV. In the preliminary analysis of the data two important features appear clearly resolved: resonantly excited two-electron-one-photon transitions and the two-photon deexcitation continuum.

3. Ion–atom collisions

Extracted ions (Ne^{8+}, Ne^{7+}) were transported to a COLTRIMS spectrometer to study state-selective charge exchange processes in ion–atom collisions with an excellent energy resolution of 0.7 eV [3]. In future kinematically complete measurements of multi-electron transfer reactions accompanied by subsequent autoionization of effectively populated highly excited projectile states can be anticipated yielding detailed information about the many electron dynamics in slow highly charged ion–atom collisions. At the new location of FreEBIT at the Max-Planck-Institut für Kernphysik in Heidelberg, where the facility has been transferred in July 2001, several beamlines are under construction to use the extracted highly charged ions in addition for surface studies and for high precision experiments after retrapping specific ions in an external ion trap. A laser spectroscopy laboratory is under construction which will allow high resolution hyperfine structure measurements of hydrogen-like ions.

Acknowledgements

We thank the Deutsche Forschunggemeinschaft for financial support under contract No. UL 166/2-1 and from the Leibniz Award 1999.

References

1. Crespo López-Urrutia, J. R. *et al.*, *Phys. Scr. T* **80** (1999), 502; Crespo López-Urrutia, J. R. *et al.*, *Hyp. Interact.* **127** (2000), 497; Crespo López-Urrutia, J. R. *et al.*, *Phys. Scr. T* **92** (2001), 110.
2. Ullrich, J. *et al.*, *J. Phys. B* **30** (1997), 2917 (topical review).
3. Fischer, D. *et al.*, this volume; Fischer, D. *et al.*, *J. Phys. B* **35** (2002), 1369.

Hyperfine Interactions **146/147:** 115–119, 2003.

Systematic Calculations of Total Atomic Binding Energies

P. INDELICATO[1], G. RODRIGUES[1], J. P. SANTOS[2], P. PATTÉ[3], J. P. MARQUES[3] and F. PARENTE[3]
[1]*Laboratoire Kastler-Brossel, Unité Mixte de Recherche du CNRS n° C8552, École Normale Supérieure et Université Pierre et Marie Curie, Case 74, 4 place Jussieu, F-75252 Paris CEDEX 05, France; e-mail: paul.indelicato@spectro.jussieu.fr*
[2]*Departamento de Física, Faculdade de Ciências e Tecnologia, Universidade Nova de Lisboa, Monte de Caparica, 2825-5164 Caparica, Portugal, and Centro de Física Atómica da Universidade de Lisboa, Av. Prof. Gama Pinto 2, 1649-003 Lisboa, Portugal*
[3]*Departamento Física da Universidade de Lisboa, and Centro de Física Atómica da Universidade de Lisboa, Av. Prof. Gama Pinto 2, 1649-003 Lisboa, Portugal*

Abstract. We have calculated total atomic binding energies of 3- to 91-electron ions of all atoms with $Z = 3$ to 118, in the Dirac–Fock model, for applications to atomic mass determination from highly-charged ions. In this process we have determined the ground-state configuration of many ions for which it was not known. We also provide total electronic correlation including Breit correlation for iso-electronic series of beryllium, neon, magnesium and argon, using the multiconfiguration Dirac–Fock approach.

Key words: atomic binding energies, fundamental state configurations.

1. Introduction

Recent work on highly ionized atoms using Penning traps [1, 3] requires precise values for electronic binding energies in order to determine atomic masses from the ionic ones obtained experimentally with traps. The relative uncertainties of such experimental results can vary from 10^{-7} to 10^{-10}, depending on the ionic species handled, on the lifetime of the nucleus and of the experimental apparatus. Although for the highest uncertainties the theoretical Dirac–Fock value is more than sufficient, the lowest ones are approaching a region that requires the inclusion of electronic correlation. In this paper we describe a new calculation of the total binding energy of all ions with nuclei having a charge $3 \leqslant Z \leqslant 118$ from neutral to 91-electron configurations. We also describe calculation of correlation energy for the Be-like, Ne-like, Mg-like and Ar-like iso-electronic sequence for $4 \leqslant Z \leqslant 118$.

2. Calculation

Since we have to deal with very heavy ions, we use an *ab initio* relativistic method. The multi-configuration Dirac–Fock method (MCDF) [2] is particularly well suited for the task as it enables to deal easily with arbitrarily complicated outer shell structure. It has been used recently to provide accurate binding energy for Cs ions [6]. In the present work we use the MCDF method to calculate the ground configuration total energy of all ions with 3 to 91 electrons, for $3 \leqslant Z \leqslant 118$. In many cases, for high-Z ions, the ground configuration is not known, and we thus had to evaluate energies for several configurations to find the lower one. This was done without correlation and thus in some cases could provide a ground configuration different from the real one. However, this would provide a negligible change in the total binding energy, with respect to experimental precision.

The binding energy is evaluated with a full self-consistent treatment of the Breit interaction. One-electron self-energy and self-energy screening using the Welton approximation [4, 5] are included. Vacuum polarization is also taken into account including Källén et Sabry, Wichmann and Kroll and iterated Uehling potential. The main missing correction is correlation, which represents roughly 1 eV/electron (see [6]). We have thus done a MCDF calculation of the total correlation energy for a number of closed shell ions (Be-like, Ne-like, Mg-like and Ar-like) over the whole range of Z studied in this work.

3. Results and discussion

The differences between Dirac–Fock binding energies for neutral atom and for ion of structure [A], for a few elements that have been recently measured at SMILETRAP are presented in Table I. One can note that the difference between our values and the Hartree–Fock Slater values of Scofield is 1.4 times the experimental error bar which is around 50 eV. The properties of the ground configuration for a subset of our results are presented in Table II. Even for high-Z, all ions with charge larger than 3+ have a relativistic shell structure.

With the MCDF method, correlation is treated by minimization of an energy functional, which is the mean value of the Hamiltonian within the virtual orbital space spanned by a selected number of configurations, including the effect of the electron–electron interaction to all-orders, but in an incomplete way. For the present work we follow the procedure recently published for Cs ions [6]. From a comparison between MCDF and RMBPT calculation, it was demonstrated that, for the iso-electronic sequence of closed shell ions with no more than 18 electrons (Be, Ne, Mg, Ar), the MCDF method can provide correlation energy with accuracy ranging from 0.1 eV (including up to the $n = 7$ shell for [Be] and [Ne]) to 1 eV (including up to the $n = 6$ shell [Mg] and [Ar]) when all possible single and double excitations from all occupied shell are accounted for. This leads to $\approx$600 (for [Be]) to $\approx$5000 fully relaxed configurations (for [Ar]).

Table I. Difference in binding energy between ion of structure [A] and neutral atom (eV). Scofield: Hartree–Fock Slater calculation (unpublished)

Atom	Ion	Binding energy	Scofield
Ge	[Ne]	−7982.51	
Ge	[F]	−10159.27	
Se	[Ne]	−10118.76	
Se	[F]	−12655.84	
Hg	[Ni]	−56377.66	−56800.2

Table II. Structure of some ground configurations for $66 \leqslant Z \leqslant 118$. The ground configuration for atomic numbers in boldface have been newly determined in the present work

Element	Number of electrons	Configuration	$^{2S+1}L_j$
Tb	66	$[Xe]4f^{10}6s$	$^6I_{17/2}$
	67 ... 118	$[Xe]4f^{11}$	$^4I_{15/2}$
Dy	66	$[Xe]4f^{10}6s^2$	$^5I_8$
	67	$[Xe]4f^{11}6s$	$^5I_8$
	68 ... 118	$[Xe]4f^{12}$	$^3H_6$
Er	68	$[Xe]4f^{12}6s^2$	$^3H_6$
	69	$[Xe]4f^{13}6s$	$^3F_4$
	70 ... 118	$[Xe]4f^{14}$	$^1S_0$
Tm	69	$[Xe]4f^{13}6s^2$	$^2F_{7/2}$
	70 ... 71	$[Xe]4/f^{14}6s$	$^2S_{1/2}$
	72 ... 118	$\mathbf{[Xe]4f^{14}5d}$	$\mathbf{^2S_{3/2}}$
Yb	70 ... 71	$[Xe]4f^{14}6s^2$	$^1S_0$
	72 ... 118	$\mathbf{[Xe]4f^{14}5d^2}$	$\mathbf{^3F_2}$
Lu	71	$[Xe]4f^{14}5d6s^2$	$^2D_{3/2}$
	72	$\mathbf{[Xe]4f^{14}5d^26s}$	$\mathbf{^4F_{3/2}}$
	73 ... 118	$\mathbf{[Xe]4f^{14}5d^3}$	$\mathbf{^4F_{3/2}}$
Hf	72	$[Xe]4f^{14}5d^26s^2$	$^3F_2$
	73	$[Xe]4f^{14}5d^36s$	$^5F_1$
	74 ... 118	$\mathbf{[Xe]4f^{14}5d^4}$	$\mathbf{^5D_0}$
W	74	$[Xe]4f^{14}5d^46s^2$	$^5D_0$
	75 ... 76	$\mathbf{[Xe]4f^{14}5d^56s}$	$\mathbf{^7S_3}$
	77 ... 118	$\mathbf{[Xe]4f^{14}5d^6}$	$\mathbf{^5D_4}$
Th	90	$[Rn]6d^27s^2$	$^3F_2$
	91	$\mathbf{[Rn]5f^26d7s}$	$\mathbf{^5K_5}$
	92	$\mathbf{[Rn]5f^36d}$	$\mathbf{^5L_6}$
	93 ... 118	$\mathbf{[Rn]5f^4}$	$\mathbf{^5I_4}$

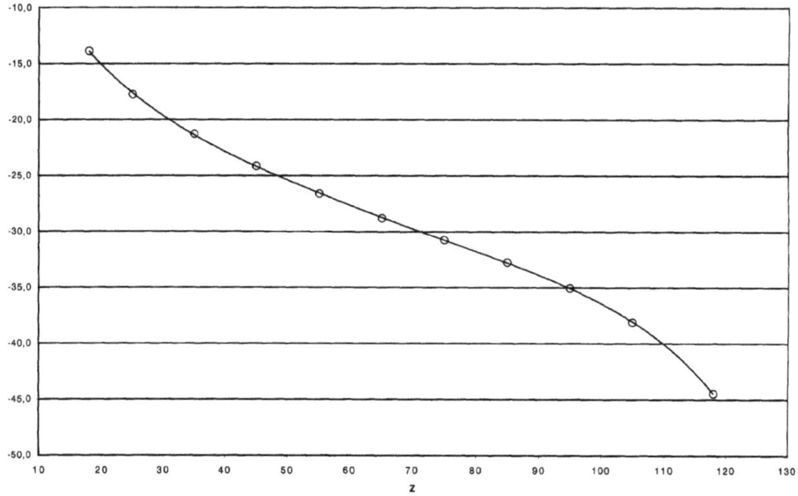

Figure 1. Total correlation energy (in Dirac–Fock level) for the Ar iso-electronic series for elements with atomic number $18 \leqslant Z \leqslant 118$ (eV). The open circles represent the calculated value and the line the fit with a five coefficient polynomial.

For the above iso-electronic series, the total electronic correlation is calculated for some elements with atomic number comprised between neutral and $Z = 118$. The electronic correlation for the remaining elements of this interval was interpolated using a 5th degree polynomial. The estimated uncertainty for this calculation is of the order of the eV. The results obtained for the argon iso-electronic sequence are shown in Figure 1.

4. Conclusion

We have shown that the use of *ab initio* relativistic technique is necessary to provide high-accuracy binding energy data that are required for the obtention of accurate atomic masses. Because it can accommodate arbitrary outer-shell structure, the MCDF method is for the moment better suited for that task than relativistic many-body perturbation theory, even though it cannot compete for the moment, as was shown in [6], when very accurate correlation energy for many-electron ions are needed.

Acknowledgements

This research was supported in part by JNICT (Portugal) under project Praxis/C/FIS/10030/98, and by the TMR Network EUROTRAPS Contract Number ERBFM-RXCT970144. We wish to thank T. Fritioff for providing us their Hg results before publication.

References

1. Carlberg, C., Fritioff, T. and Bergström, I., *Phys. Rev. Lett.* **83**(22) (1999), 4506.
2. Desclaux, J. P., In: E. Clementi (ed.), *Methods and Techniques in Computational Chemistry*, Vol. A, STEF, Cagliary, 1993.
3. Douysset, G., Fritioff, T., Carlberg, C., Bergström, I. and Björkhage, M., *Phys. Rev. Lett.* **86**(19) (2001), 4259.
4. Indelicato, P. and Desclaux, J. P., *Phys. Rev. A* **42**(9) (1990), 5139.
5. Indelicato, P., Gorceix, O. and Desclaux, J. P., *J. Phys. B* **20** (1987), 651.
6. Rodrigues, G. C., Ourdane, M. A., Bieron, J., Indelicato, P. and Lindroth, E., *Phys. Rev. A* **63**(1) (2001), 012510.

Hyperfine Interactions **146/147:** 121–125, 2003.

Relativistic Transition Energies and Radiative Transition Rates for Forbidden Transitions in the $1s^2 2s^2 2p^4$ Atomic Configuration for $20 \leqslant Z \leqslant 30$

J. P. MARQUES[1], F. PARENTE[1] and P. INDELICATO[2]

[1] *Centro de Física Atómica e Departamento de Física da Universidade de Lisboa, Campo Grande, Ed. C8, Piso 6, 1749-016 Lisboa, Portugal; e-mail: jmpm@fc.ul.pt*

[2] *Laboratoire Kastler-Brossel, École Normale Supérieure et Université P. et M. Curie, Unité Mixte de Recherche du CNRS n° C8552 Case 74; 4, pl. Jussieu, 75252 Paris CEDEX 05, France*

Abstract. The Multiconfiguration Dirac–Fock (MCDF) method was used to compute energy levels and forbidden transition probabilities for the $2p^4$ ground configuration of the oxygen isoelectronic sequence for $20 \leqslant Z \leqslant 30$. Hyperfine quenching is shown to have no influence on the lifetime of the metastable $^3P_0$ level.

1. Introduction

An accurate knowledge of fundamental atomic quantities is important not only for atomic physics experiments, but also for fields like astrophysics and plasma and fusion physics. Measurement of radiative lifetimes of metastable states provide valuable information for understanding emission lines in astrophysical objects, plasma modelling, solar physics and in atomic physics theory.

The study of forbidden transitions can bring valuable insight in the understanding of astronomical spectra, mainly because for highly charged ions, transitions from metastable levels occur at much longer wavelengths than other transitions in the same ion [1]. Of particular interest are the lines belonging to the $2p^4$ configuration because they are present in solar flares [2] and have been also observed in tokamaks [3, 4]. Transition energies and probabilities within the $2p^4$ configuration of the oxygen isoelectronic sequence have been object of particular attention in the last decades. The SUPERSTRUCTURE code of Eissner *et al.* [5] was used by Kastner *et al.* [6], who included only the $2p^4$, $2p^6$ and $2p^3\,3p$ configurations, and Bathia *et al.* [7], who accounted for correlation within the $n = 2$ complex only. Most recently [8], the same code was used to calculate radiative rates for forbidden transitions in the oxygen isoelectronic sequence for $Z \leqslant 28$. Of particular interest were the Multi-Configuration Dirac–Fock calculations of Cheng *et al.* [9], who included all configurations within the $n = 2$ complex to account for electron correlation and intermediate coupling and also selected QED contributions. Edlén [10] expressed the energy difference between the observed values in the O-like spectra and the theoretical values of Cheng *et al.* [9] in terms of a polynomial in Z, thus providing a set of "recommended" energy values. Fischer and Saha [11] calculated energy levels, as well as electric quadrupole and magnetic dipole transition probabilities, for the ground-state configuration of the oxygen isoelectronic sequence, using a multiconfiguration Hartree–Fock method with Breit–Pauli corrections, and claiming a better agreement with observation than other theoretical results, except for the $^1D_2$ level. Baluja and Zeippen [1] performed a similar calculation using configuration interaction wave functions in the computer program CIV3, including relativistic corrections in the Breit-Pauli approximation.

In the Multiconfiguration Dirac Fock (MCDF) method implemented by Desclaux and Indelicate [12] the total wave function for an atomic state is calculated with the help of the variational principle. The wave function is written as a linear combination of configuration state functions (CSF), which are eigenfunctions of the total angular momentum $\mathbf{J}^2$ and its projection J_z. The CSF are antisymmetric products of one electron wave functions expressed as linear combinations of Slater determinants of Dirac 4-spinors. The function thus obtained is an eigenfunction of the no-pair relativistic Hamiltonian, the total energy of the atomic system being the corresponding eigenvalue. Exact one-electron radiative corrections (self-energy and vacuum polarization) are added afterwards. More details can be found in [12–14].

2. Results and discussion

2.1. ENERGIES

We used the most recent version of the MCDF code of Desclaux and Indelicato [12] to compute energy levels for the $2p^4$ ground configuration of the oxygen isoelectronic sequence for $20 \leqslant Z \leqslant 30$. Inclusion of electronic correlation is very important [11]. In our calculation electronic configurations up to $n = 3$ were included, leading to more than 700 jj configurations. We found that the correlation energy accounts for 10% of $^3P_0$–$^3P_2$ transition energy for $Z = 30$ and 3% for $Z = 20$. We focused our attention on the $20 \leqslant Z \leqslant 30$ atomic number region because $^3P_0$ and $^3P_1$ level excitation energies cross at $Z = 24$ (see Figure 1). As a consequence, for $Z > 24$, the $^3P_0$ level can decay only via a Electric quadrupole (E2) transition to the ground level, thus suggesting that hyperfine quenching could play a decisive role on the lifetime of this level.

Excitation energies for $^3P_2$, $^3P_1$, $^3P_0$, $^1D_2$ and $^1S_0$ levels were computed. These results, for selected Z values, are listed in Table I, together with experimental and theoretical results obtained by other authors. In the present calculation the wave function was optimized for each level inside the calculation. The Gaunt part of the Breit interaction was made self-consistent, while retardation and QED effects, namely vacuum polarization and self-energy were treated as perturbations. The

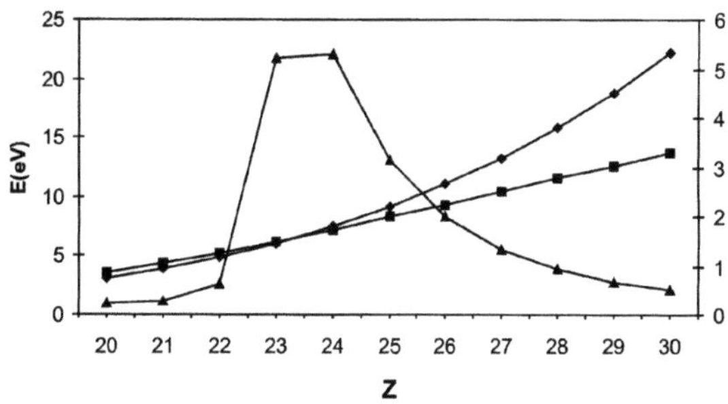

Figure 1. Excitation energies for $^3P_0$ (■) and $^3P_1$ (♦) levels (left axis) and $^3P_0$ (▲) lifetime (right axis) plotted against Z.

Table I. Excitation energies (in cm $^{-1}$) for the $2p^4$ ground configuration

Z	Level	This work	Exp. [10]	Calc. [8]	Calc. [1]	Calc. [11]
20	$^3P_1$	24404	24465	24467	24411	24541
	$^3P_0$	28822	28880	28651	28569	28892
	$^1D_2$	89523	88202	90300	89371	90607
	$^1S_0$	179146	178568	178786	178220	179705
22	$^3P_1$	39190	39277	39425	39294	39503
	$^3P_0$	42357	42309	41956	41840	42404
	$^1D_2$	110009	108717	111038	110082	111435
	$^1S_0$	216041	215509	215544	215121	216724
24	$^3P_1$	60253	60375	60834	60563	60896
	$^3P_0$	58214	58050	57416	57287	58239
	$^1D_2$	136233	134991	137675	136649	138186
	$^1S_0$	263308	262926	262908	262523	264336
26	$^3P_1$	89267	89439	90495	89889	90486
	$^3P_0$	75434	75188	73950	73888	75428
	$^1D_2$	170018	168848	172149	170885	172776
	$^1S_0$	325227	325149	325326	324893	327017
28	$^3P_1$	128075	128308	130396	129337	130237
	$^3P_0$	93072	92836	90484	90688	92989
	$^1D_2$	213337	212262	216573	214959	217304
	$^1S_0$	406483	406819	407805	407163	409616
30	$^3P_1$	178668	178973			182299
	$^3P_0$	110504	110437			110273
	$^1D_2$	268284	267325			274032
	$^1S_0$	511792	512557			517000

Table II. Transition probabilities (in s^{-1})

	$^3P_0$–$^3P_2$ E2	$^3P_0$–$^3P_1$ M1	$^3P_1$–$^3P_2$ M1	$^1S_0$–$^3P_2$ E2	$^1S_0$–$^3P_1$ M1	$^1D_2$–$^3P_2$ M1	$^1D_2$–$^3P_1$ M1
	$Z = 20$						
Present	1.226E−2	4.461E+0	3.170E+2	1.808E+0	7.954E+3	7.613E+2	1.046E+2
Ref. [8]	1.256E−2	4.453E+0	3.195E+2	1.768E+0	8.052E+3	7.858E+2	1.018E+2
Ref. [1]	1.244E−2	4.475E+0	3.213E+2	1.687E+0	8.095E+3	7.640E+2	9.890E+1
	$Z = 22$						
Present	5.505E−2	1.580E+0	1.287E+3	3.503E+0	2.247E+4	2.312E+3	2.222E+2
Ref. [8]	5.545E−2	1.387E+0	1.296E+3	3.440E+0	2.298E+4	2.380E+3	2.171E+2
Ref. [1]	5.477E−2	1.392E+0	1.304E+3	3.231E+0	2.304E+4	2.318E+3	2.111E+2
	$Z = 24$						
Present	1.883E−1		4.564E+3	6.222E+0		6.450E+3	
Ref. [8]	1.878E−1		4.591E+3	5.993E+0		6.643E+3	
Ref. [1]	1.850E−1		4.624E+3	5.594E+0		6.473E+3	
	$Z = 26$						
Present	5.029E−1		1.442E+4	1.037E+1		1.683E+4	
Ref. [8]	4.999E−1		1.449E+4	9.712E+0		1.738E+4	
Ref. [1]	4.907E−1		1.461E+4	9.052E+0		1.692E+4	
	$Z = 28$						
Present	1.085E+0		4.130E+4	1.620E+1		4.151E+4	
Ref. [8]	1.082E+0		4.142E+4	1.499E+1		4.302E+4	
Ref. [1]	1.095E+0		4.185E+4	1.412E+1		4.182E+4	
	$Z = 30$						
Present	1.974E+0		1.087E+5	2.422E+1		9.739E+4	

values in the column labelled "Exp." in Table I are Edlén's [10] "recommended" values.

The results of this work are in general in a better agreement with the experiment than previous published calculations.

2.2 TRANSITION PROBABILITIES

We computed the probability of all magnetic dipole (M1) and electric quadrupole (E2) transitions, within the $2p^4$ complex, leading to the $^3P_2$ and $^3P_1$ levels. The aim of this calculation was to check the effect of the hyperfine quenching on the lifetime of the $^3P_0$ level, mainly for $Z > 24$, where this level can decay only via a E2 transition to the ground level.

To compute the transition probabilities fully relaxed orbitals were used, i.e., non-orthogonality between initial and final state orbitals were taken into account.

We found that, for $Z = 24$, where the hyperfine interaction is expected to be greater, this effect changes the lifetime of the $^3P_0$ level from 5.310 s to 5.308 s, while for $Z = 28$ there is no change. This results from the fact that the hyperfine matrix elements are too small and because the intensity of the $^3P_0$–$^3P_2$ E2 line increases strongly with Z, as can been seen in Table II.

Acknowledgements

This work was supported in part by the European Community under TMR Contract No. FMRX-CT97-0144 (EUROTRAPS).

References

1. Baluja, K. L. and Zeippen, C. J., *J. Phys. B* **21** (1988), 1455.
2. Edlén, B., *Phys. Scr.* **26** (1982), 71.
3. Hinnov, E., Suckewer, S., Cohen, S. and Sato, K., *Phys. Rev. A* **25** (1982), 2293.
4. Peacock, N. J., Stamp, M. F. and Silver, J. D., *Phys. Scr. T* **8** (1984), 10.
5. Eissner, W., Jones, M. and Sana, H. P., *Comput. Phys. Commun.* **8** (1983), 270.
6. Kastner, S. O., Bathia, A. K. and Cohen, L., *Phys. Scr.* **15** (1977), 259.
7. Bathia, A. K., Feldman, U. and Doscheck, G. A., *Astron. Astrophys.* **80** (1979), 22.
8. Galavís, M. E., Mendonza, C. and Zeippen, C. J., *Astron. Astrophys. Suppl. Ser.* **123** (1997), 159.
9. Cheng, K. T., Kim, Y. K. and Desclaux, J. P., *At. Data Nucl. Data Tables* **24** (1979), 111.
10. Edlén, B., *Phys. Scr.* **28** (1983), 51.
11. Froese Fischer, C. and Saha, H. P., *Phys. Rev. A* **28** (1983), 3169.
12. Desclaux, J. P., *Relativistic Multiconfiguration Dirac–Fock Package,* Vol. A, Cagliary, STEF, 1993.
13. Indelicato, P., *Phys. Rev. A* **51** (1995), 1132.
14. Marques, J. P., Parente, F. and Indelicato, P., *J. Phys. B* **34** (2001), 3487.

Hyperfine Interactions **146/147:** 127–131, 2003.

Correlation and Relativistic Effects on Landé g_J Factors of Atomic Ions

P. INDELICATO[1], A.-M. MÅRTENSSON-PENDRILL[2], W. QUINT[3] and J.-P. DESCLAUX[4]

[1] *Laboratoire Kastler-Brossel, Unité Mixte de Recherche du CNRS n° C8552, École Normale Supérieure et Université Pierre et Marie Curie, Case 74, 4 place Jussieu, F-75252 Paris CEDEX 05, France; e-mail: paul.indelicato@spectro.jussieu.fr*

[2] *Physics, Göteborg University and Chalmers University of Technology SE-41296 Göteborg, Sweden*

[3] *GSI, Planckstraße 1, Darmstadt, D-64291, Germany*

[4] *15 Chemin du Billery, F-38360, Sassenage, France*

Abstract. We investigate relativistic effects on the Landé g_J factors of atomic ions using Multiconfiguration Dirac–Fock technique and Relativistic Many-Body perturbation theory. The role of Breit interaction, negative energy continuum and correlation effects is studied in Li-like, B-like, N-like ions and Ca^+ ground-states. We also investigate Ti-like ions which have a long-lived excited state with $J = 4$. Those ions are all good candidates for employing the continuous Stern–Gerlach effect to measure their g_J-factor and provide accurate tests of relativistic many-body calculations.

Key words: Landé factors, highly-charged ions.

1. Introduction

The use of Penning traps has provided new experimental possibilities to do high-accuracy measurements of Landé g_J-factors. The ground-state Landé factor of hydrogen-like carbon has been measured to an accuracy of 1 ppb by microwave spectroscopy via the continuous Stern–Gerlach effect [7]. Other ions have been studied like Ca^+, Ba^+ (including metastable states) with laser-microwave double resonance spectroscopy. In this paper we investigate the size of different relativistic and correlation contributions to the g_J-factor of several ions in the ground state, and to the g_J-factor of Ti-like ions in a very metastable state with a high-angular momentum [13]. We discuss the possibility of employing the continuous Stern–Gerlach effect for a measurements of the g_J-factor of those highly charged few-electron systems.

2. Theoretical considerations

Performing relativistic many-body calculations is much more complicated than the equivalent non-relativistic case. The only fully clean way of doing those calculation is to use quantum electrodynamics (QED). Yet because of the difficulties of the evaluation of QED diagrams, and the large number of them that contributes to correlation at low-Z (perturbation parameter for correlation is $1/Z$), such a calculation cannot be realized in practice for light elements. One must then use an approximate Hamiltonian, derived from QED, and use either Relativistic Many-Body Perturbation theory (RMBPT) or a variational method like Multiconfiguration Dirac–Fock (MCDF). For evaluating quantities other than energy levels, like Landé g_J factors, an additional difficulty is that the approximate Hamiltonian derived from QED is a no-pair approximation [5, 8], that is required to calculate energies. Yet this approximation is not good enough for some one-body operators, e.g., like the operator used for evaluating M1 transition probabilities [9]. Also the magnetic and retardation part of the electron–electron interaction could play an important role, which complicates even more the picture.

Here we investigate correlation and relativistic effects in several atoms and ions to investigate the importance of different effects and motivate new experiments. Up to now the evaluation of QED radiative corrections to Landé factors has been performed only for hydrogen-like ions [1, 14] where they lead to extraordinarily good tests of QED when compared to experiment [6, 7]. In this work we have used the MCDF code developed by two of us (J.P.D. and P.I.) to evaluate Landé factors following the formalism described in [2]. We also performed RMBPT calculations following the method described in [11] to evaluate the accuracy of the approximations.

3. Results and discussion

3.1. LITHIUM-LIKE IONS

We have performed Multiconfiguration Dirac–Fock calculations of the Landé g_J-factors of three-electron ions. Relativistic effects are included ab-initio. Full self-consistent treatment of the Breit interaction allows us to include the Breit interaction effects to all orders. In this calculation we have been able to obtain a fully-optimized wave function including all single, double and triple excitations to all levels with $n \leqslant 6$ and $\ell \leqslant 5$, which represents 4152 jj configurations. Note that limiting the calculation to double excitations leads to only 468 jj configurations. The results for $Z = 4$ are represented in Figure 1. We compare results obtained with an offspring of GRASP [3] by Bieron (Coulomb only, configuration active space approximation [10]), recent high-accuracy variational calculations [20], experiment [19] and our MCDF calculations (Coulomb, full Breit, pair or no-pair approximation) and RMBPT results [11]. The results for the MCDF calculation with full account of the Breit interaction, the high-precision variational calculation and the RMBPT values are all in agreement with experiment. Calculations for Z up to 20 are being performed. They will provide high-quality tests of QED and many-body effects for three-electron ions when experiments using the continuous Stern–Gerlach effect [6, 7] will be performed.

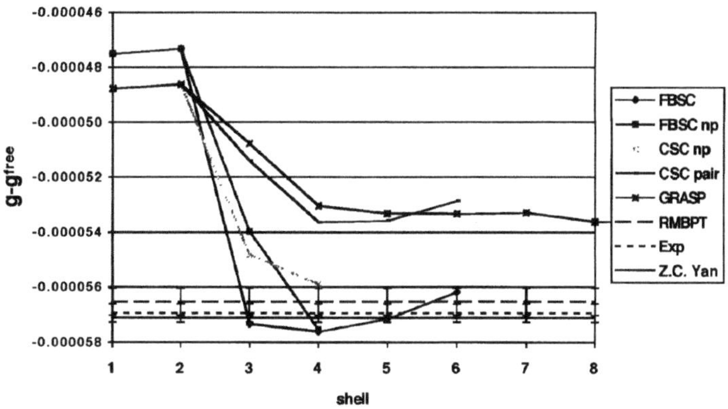

Figure 1. Landé g_J factors for Be^+. FBSC: full Breit interaction used in the SCF process (this work). CSC: only the Coulomb interaction is used in the self-consistent field (this work). When np is added, no pair approximation has been used to obtain the wave function. GRASP: MCDF calculation using Oxford General Relativistic Atomic Structure Package [3, 10]. Exp: Experimental value from [19]. Z. C. Yan: Hylleraas-type non-relativistic calculations from [20].

3.2. EVALUATION OF THE Ca^+ LANDÉ FACTOR

The Landé factor of Ca^+ has been measured with high-accuracy at Mainz in a Penning trap [16]. We have performed a RMBPT calculation including Relativistic Random Phase Approximation (RPA) contributions. We also did a Relativistic Configuration Interaction (RCI) calculation starting from a Dirac–Fock wave function and adding single excitations of s and p orbitals (d orbital effects were found negligible). Using the Dirac–Fock potential we generated a basis set using B-splines, and included (discretized) continuum wave functions in the set of RCI orbitals. Both methods provide similar results.

Higher-order correlation (double excitation for RCI, beyond RPA for RMBPT) and Breit correlation remain to be investigated. The results are presented in Table I.

3.3. OTHER IONS

Highly-charged ions can exhibit interesting properties in their Landé factors. One can study cases in which the active electron is not an s electron, and investigate relativistic effects. We study B-like ions ($p_{1/2}$ ground state), N-like ions ($p_{3/2}$ ground state) and a long-lived metastable state [13] of Ti-like ($3d^4$, $J = 4$) ions, a very interesting sequence [4, 12, 15, 17, 18]. The results are presented in Table II. As it is only an exploratory work, we performed only Dirac–Fock calculations. Yet in the case of the N-like sequence we took into account the intra-shell correlation (i.e., using $1s^22s^22p + 1s^22p^3$ configurations), which is expected to give a large effect.

Table I. Different approximations to the Ca^+ Landé factor

Method	RCI basis set size	$g_J - g_{\text{free}}$	g_J
		Coulomb only	
Dirac–Fock		-4.851×10^{-5}	2.0022708
RCI	20	-5.342×10^{-5}	2.0022659
RCI	25	-5.305×10^{-5}	2.0022662
RCI	30	-5.399×10^{-5}	2.0022653
RMBPT		-5.280×10^{-5}	2.0022665
		with Breit contribution	
Dirac–Fock		-5.159×10^{-5}	2.0022677
Experiment [16]		$-6.263(9) \times 10^{-5}$	2.00225664(9)

Table II. Landé factors in B-like ([B]), N-like ([N]) and Ti-like ([Ti]) ions. IS: intra-shell correlation. The change of sign in the QED corrections comes from the fact that $p_{1/2}$ and $p_{3/2}$ orbitals have opposite relativistic angular momentum number κ – see, e.g., [2]

	B $J = 1/2$	[B] Xe $J = 1/2$	N $J = 3/2$	[N] Xe $J = 3/2$	[Ti] Ta $J = 4$
Coulomb	0.666623	0.641814	1.999861	1.357132	1.060885
Breit contr.	0.000037	0.000750	−0.000022	0.002431	0.002074
QED (Coul.)	−0.000773	−0.000773	0.002319	0.000871	0.000167
QED (Breit)	0.000000	0.000000	0.000000	0.000005	0.000004
IS Coul.	−0.000016	−0.001404			
IS Breit	−0.000025	−0.000467			
Total	0.665846	0.639920	2.002159	1.360439	1.063131

4. Conclusion

We have demonstrated a number of relativistic correlation effects on Landé factors and provided some accurate predictions for some of them. Many interesting studies on few-electron ions or ions with complex structure can be undertaken, that will be a challenge for theory, but should lead to new insight on the relativistic many-body problem. The continuous Stern–Gerlach effect is well suited to investigate experimentally all those ions, including some of them in a long lived metastable state.

Acknowledgements

This research was supported in part by the TMR Network Eurotraps Contract Number ERBFMRXCT970144.

References

1. Beier, T., Lindgren, I., Persson, H., Salomonson, S., Sunnergren, P., Häffner, H. and Hermanspahn, N., *Phys. Rev. A* **62**(3) (2000), 032510.
2. Cheng, K. and Childs, W., *Phys. Rev. A* **31**(5) (1985), 2775.
3. Dyall, K. G., Grant, I. P., Johnson, C. T., Parpia, F. A. and Plummer, E. P., *Comp. Phys. Commun.* **55** (1989), 425.
4. Feldman, U., Sugar, J. and Indelicato, P., *J. Optic. Soc. Am. B* **8**(1) (1990), 3.
5. Grant, I. P., *Relativistic Atomic Structure Calculations*, Vol. 2, Plenum Press, New York, 1988, p. 132.
6. Häffner, H., Beier, T., Hermanspahn, N., Kluge, H.-J., Quint, W., Stahl, S., Verdú, J. and Werth, G., *Phys. Rev. Lett.* **85**(25) (2000), 5308.
7. Hermanspahn, N., Häffner, H., Kluge, H.-J., Quint, W., Stahl, S., Verdú, J. and Werth, G., *Phys. Rev. Lett.* **84**(3) (2000), 427.
8. Indelicato, P., *Phys. Rev. A* **51**(2) (1995), 1132.
9. Indelicato, P., *Phys. Rev. Lett.* **77**(16) (1996), 3323.
10. Indelicato, P., Lindroth, E., Beier, T., Bieron, J., Costa, A., Lindgren, I., Marques, J., Martins, M., Ourdane, M., Parente, F., Patt, P., Rodrigues, G., Salomonson, S. and Santos, J., *Hyp. Interact.* **132**(1–4) (2001), 347.
11. Lindroth, E. and Ynnerman, A., *Phys. Rev. A* **47**(2) (1993), 961.
12. Morgan, C., Serpa, F., Takacs, E., Meyer, E., Gillaspy, J., Sugar, J., Roberts, J., Brown, C. and Feldman, U., *Phys. Rev. Lett.* **74**(10) (1995), 1716.
13. Parente, F., Marques, J. and Indelicato, P., *Europhys. Lett.* **26**(6) (1994), 437.
14. Persson, H., Salomonson, S., Sunnergren, P. and Lindgren, I., *Phys. Rev. A* **56**(4) (1997), R2499.
15. Porto, J. V., Kink, I. and Gillaspy, J. D., *Phys. Rev. A* **61**(5) (2001), 054501.
16. Tommaseo, G., Pfeil, T., Revalde, G. and Werth, G., In: *The Seventh European Conference on Atomic and Molecular Physics*, Berlin, 2001.
17. Utter, S. B., Beiersdorfer, P. and Brown, G. V., *Phys. Rev. A* **61**(3) (2000), 030503(R).
18. Watanabe, H., Crosby, D., Currell, F. J., Fukami, T., Kato, D., Ohtani, S., Silver, J. D. and Yamada, C., *Phys. Rev. A* **63**(4) (2001), 042513.
19. Wineland, D. J., Bellinger, J. J. and Itano, W. M. *Phys. Rev. Lett.* **50**(9) (1983), 628.
20. Yan, Z.-C., *Phys. Rev. Lett.* **86**(25) (2001), 5683.

Hyperfine Interactions **146/147:** 133–137, 2003.

Magnetic Sublevel Population Studied for H- and He-like Uranium in Relativistic Ion–Atom Collisions

A. GUMBERIDZE[1,2,9], T. STÖHLKER[1,2], G. BEDNARZ[3], F. BOSCH[1], S. FRITZSCHE[5], S. HAGMANN[6], D. C. IONESCU[1,7], O. KLEPPER[1], C. KOZHUHAROV[1], A. KRÄMER[1,7], D. LIESEN[1], X. MA[1,8], R. MANN[1], P. H. MOKLER[1], D. SIERPOWSKI[3], Z. STACHURA[4], M. STECK[1], S. TOLEIKIS[1] and A. WARCZAK[3]

[1] *GSI-Darmstadt, Germany*
[2] *IKF, University of Frankfurt, Germany*
[3] *Institute of Physics, Cracow University, Poland*
[4] *INP, Cracow, Poland*
[5] *University of Kassel, Germany*
[6] *Kansas State University, Kansas, USA*
[7] *TU, Dresden, Germany*
[8] *IMP, Lanzhou, China*
[9] *Tbilisi State University, Georgia*

Abstract. An experimental study for K-shell excitation of helium-like uranium in relativistic collisions with low-Z gaseous target is presented. Within this experiment information about the population of the magnetic sublevels has been obtained via a photon angular differential study of the decay photons associated with the excitation process. The preliminary results presented show, for the particular case of the $^3P_1$ level, a surprisingly strong population of the magnetic sublevels with $\mu = \pm 1$.

1. Introduction

The experimental information about Coulomb excitation of one- and few-electron projectiles occurring in relativistic atomic collisions is very scarce. The lack of data must be attributed to the experimental difficulties which arise from the fact that excitation is not accompanied by projectile charge exchange. As a consequence, this process can only be studied in single pass experiments by measuring the photon production in coincidence with primary beams of low intensity [1, 2]. Very recently, an alternative experimental approach has been introduced at the storage ring ESR where projectile excitation has been studied by detecting the projectile X-ray emission in anti-coincidence with charge exchange [3]. By using this technique we now started to extend our earlier investigations to a more detailed angular differential study for the collision systems U^{91+}, $U^{90+} \rightarrow N_2$ at 217 MeV/u.

2. Experimental arrangement

The experiment was performed at the experimental storage ring (ESR) at GSI-Darmstadt by using H-, He-like uranium ions at an energy of 217 MeV/u. For X-ray detection, observation angles in the range between $\approx$10° and 150° were used at the atomic physics photon detection chamber of the internal target of the ESR (for details see [4]). Here, projectile X-rays produced in collisions of the stored ion beams with a N_2 jet-target were detected by an array of solid state detectors (for details see [4]), covering observation angles in the range between 13 and 150° with respect to the beam axis.

In the experiment, the projectile X-ray emission was measured in coincidence with down-charged ions as well as in a single mode, i.e. without any coincidence requirement. As a representative example we depict in Figure 1 X-ray spectra recorded for $U^{91+} \rightarrow N_2$ collisions at the forward angle of close to 10°. In the spectra, the transitions arising from electron capture ($K\alpha$ transitions in He-like uranium) and these from excitation ($Ly\alpha$ transitions in H-like uranium) can clearly be distinguished by both the transition energies as well as by the coincidence requirement. Indeed, no $K\alpha$ transitions are observed in the anti-coincidence spectrum. This also proves that the MWPC detector used for particle detection operates with a detection efficiency very close to 100%.

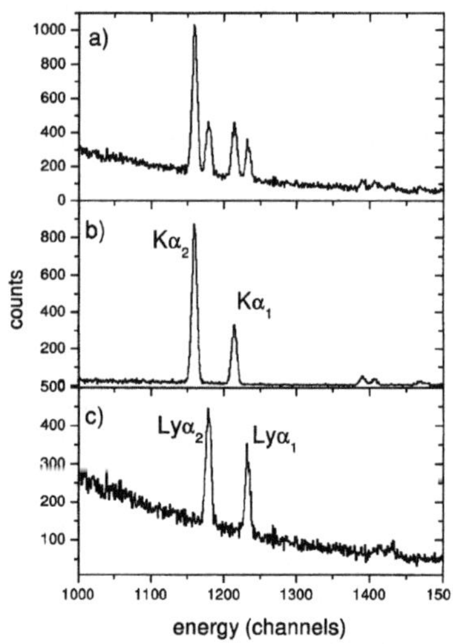

Figure 1. X-ray spectra recorded for 217 MeV/u $U^{91+} \rightarrow N_2$ collisions at the forward angle of close to 10° ((a): total emission spectrum without coincidence requirement; (b): photons in coincidence with electron capture; (c): photons in anti-coincidence with electron capture).

3. K-shell excitation of heliumlike uranium

In the following we concentrate on the formation of magnetic-sublevels by Coulomb excitation as well as by electron capture. Information about this topic can be obtained from the study of the angular distribution of the photons associated with these processes. For the particular case of E1 transitions, the photon angular correlation has the form:

$$W(\theta) = A_0 + A_2 P_2(\cos\theta) \propto 1 + \beta_{20}\left(1 - \frac{3}{2}\sin^2\theta\right). \tag{1}$$

Here θ is the angle between the de-excitation photon and the axis defined by the projectile motion (projectile frame) while P_2 is the second-order Legendre polynomial. The angular correlation is completely determined by the anisotropy coefficient β_{20}. In general, $W(\theta)$ is symmetric about 90° in the projectile frame and isotropic if the intermediate state has $j_n = 1/2$ as it is the case for the Lyα_2 ($2p_{1/2} \rightarrow 1s_{1/2}$) transition. For the particular case of the $2p_{3/2}$ transition, however, one may also determine β_{20} from the alignment $\mathcal{A}_2$ of the state which is defined as

$$\mathcal{A}_2 = \frac{\sigma(3/2, \pm 3/2) - \sigma(3/2, \pm 1/2)}{\sigma(3/2, \pm 3/2) + \sigma(3/2, \pm 1/2)} = \frac{1}{\alpha}\beta_{20}, \tag{2}$$

where $\sigma(j = 3/2, \mu)$ is the population of the magnetic substate with $\mu = \pm 1/2$, $\pm 3/2$. For the $2p_{3/2} \rightarrow 1s_{1/2}$ transition, $\alpha = 1/2$. Quite similar expressions can be found for the case of the E1 decay of the $[1s_{1/2}, 2p_{1/2}]^3P_1$ and the $[1s_{1/2}, 2p_{3/2}]^1P_1$ states in the He-like systems. Here, however, magnetic sublevels with the quantum numbers of $\mu_{J=1} = 0, \pm 1$ must be considered.

In our current experiment we strongly benefit from the fact that the Lyα_2 transition arising from the decay of the $2s_{1/2}, 2p_{1/2}$ levels is known to be precisely isotropic. Consequently, it provides an ideal tool to measure a possible anisotropy of the close-spaced Lyα_1 or Kα transitions. In Figure 2 the preliminary results for the emission pattern of the Lyα_1 and the Kα transitions are shown, normalized to the Lyα_2 intensity. In all cases no alignment is observed and the magnetic sublevels are therefore populated statistically. In the case of the Lyα_1 transition ($2p_{3/2} \rightarrow 1s_{1/2}$) induced by excitation this finding seems to be in agreement with theoretical predictions [5]. Moreover, also for the case of Kα_1 emission $[1s_{1/2}, 2p_{3/2}]^1P_1$, ${}^3P_2 \rightarrow 1^1S_0$, caused by electron capture an isotropic emission pattern is found.

For K-shell excitation, however, we observed a markedly difference between the H- and the He-like species. In Figure 3, the Kα_1/Kα_2 intensity ratio, as measured for K-shell excitation of He-like uranium in 217 MeV/u $U^{91+} \rightarrow N_2$ collisions, is plotted as a function of the observation angle. For comparison the corresponding intensity ratio (solid squares) as measured for capture into H-like uranium is shown in addition. In contrast to electron capture, the data for excitation exhibit a pronounced deviation from a constant intensity ratio. Note, that for the case of K-shell excitation of high-Z He-like ions, only the $[1s_{1/2}, 2p_{3/2}]^1P_1$ contributes to the Kα_1 transition whereas the Kα_2 intensity arises from the decay of the

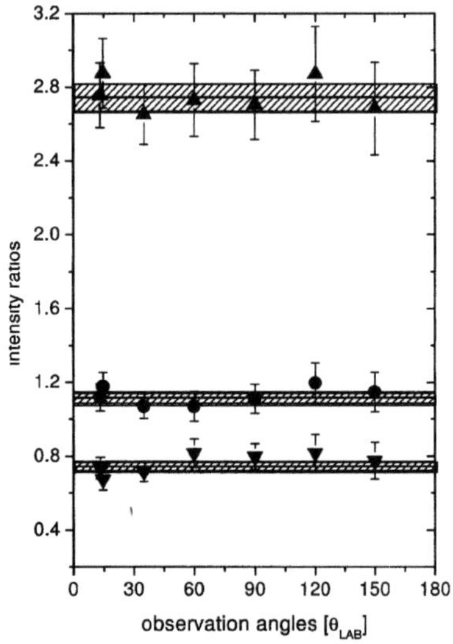

Figure 2. The intensities of $K\alpha_2$- (up-triangles), $K\alpha_1$- (solid circles), $Ly\alpha_1$-transitions (down-triangles) normalized to the $Ly\alpha_2$ line as function of observation angle (preliminary result). The full lines refer to the corresponding mean values and the shaded areas give the associated uncertainties.

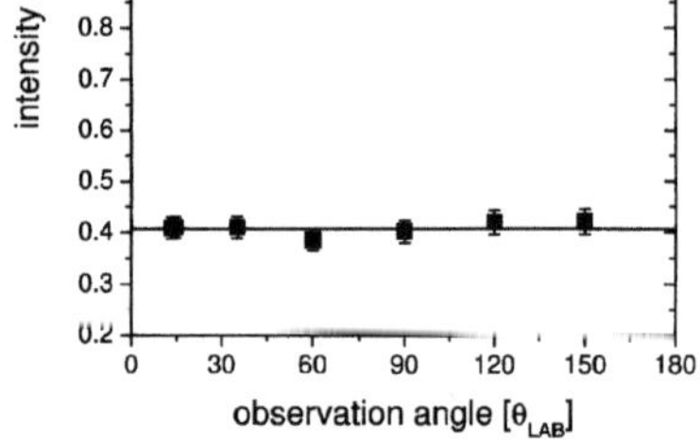

Figure 3. Preliminary result for the $K\alpha_1/K\alpha_2$ intensity ratio (solid circles) measured for K-shell excitation of He-like uranium in collisions with N_2 at 217 MeV/u. The solid line refers to a least square fit of Equation (1) to the experimental data. The solid squares show corresponding intensity ratio as measured for capture into H-like uranium.

$[1s_{1/2}, 2p_{1/2}]^3P_1$ level only. In both cases an alignment of the different sublevels is possible. At present, it has not been clarified which of the two states causes the observed anisotropic intensity ratio. However, at the current state of data analysis there are strong indications of a positive alignment of the $[1s_{1/2}, 2p_{1/2}]^3P_1$ level. This means that the magnetic sublevels with $\mu = \pm 1$ are preferably populated in the collision.

4. Summary

An experimental study for K-shell excitation of helium-like uranium is presented for relativistic collisions with low-Z gaseous target. Here, the information about the population of the magnetic sublevels has been obtained via a photon angular differential study of the decay photons associated with the excitation process. The preliminary results presented show for the particular case of the $^3P_1$ level a preferred population of the magnetic sublevels with $\mu = \pm 1$. This finding cannot be explained on the basis of theoretical predictions for excitation of the H-like species. This is remarkable because we are dealing with a high-Z ion where the electron-electron interaction is commonly assumed to be of minor importance. Consequently, additional theoretical studies are required which explicitly take into account both the relativistic effects associated with the strong fields of high-Z ions as well as the multi-electron aspect of the He-like system studied.

References

1. Stöhlker, Th. *et al.*, *Phys. Rev. A* **57** (1998), 845; *Phys. Lett. A* **238** (1998), 43.
2. Ludziejewski, T. *et al.*, *Phys. Rev. A* **61** (2000), 052706-1.
3. Krämer, A. *et al.*, *Physica Scripta T* **80** (1999), 424.
4. Stöhlker, Th. *et al.*, *AIP Conference Proc.* **506** (2000), 389.
5. Ionescu, D. C. and Stöhlker, Th., *Phys. Rev. A* (2003), 022705.
6. Stöhlker, Th. *et al.*, *Phys. Rev. Lett.* **79** (1997), 3270.

Hyperfine Interactions **146/147:** 139–143, 2003.

Lifetime Measurement of the Metastable 2^3P_0 State in Helium-like ^{197}Au

S. TOLEIKIS[1], E. BERDERMANN[1], H. F. BEYER[1], F. BOSCH[1], M. CZANTA[1], A. GUMBERIDZE[1], C. KOZHUHAROV[1], D. LIESEN[1], X. MA[1], T. STÖHLKER[1], B. MANIL[2], P. INDELICATO[2], A. SIMIONOVICI[3], A. WARCZAK[4], Z. STACHURA[5], R. MARRUS[6], D. SCHNEIDER[7], R. W. DUNFORD[8] and Y. ZOU[9]

[1] *GSI, Darmstadt, Germany*
[2] *Université P. et M. Curie, Paris, France*
[3] *ESRF, Grenoble, France*
[4] *IFUJ, Cracow, Poland*
[5] *INP, Cracow, Poland*
[6] *University of California, Berkeley, USA*
[7] *LLNL, Livermore, USA*
[8] *ANL, Argonne, USA*
[9] *Jhiaotong University, Shanghai, China*

Abstract. The determination of atomic lifetimes offers the possibility to investigate various aspects of the atomic structure. In the case of helium-like ^{197}Au with nuclear spin 3/2, due to hyperfine-quenching, one can determine the 2^3P_0–2^3P_1 fine-structure splitting from the lifetime measurement of the metastable 2^3P_0 state. This lifetime has been measured at the GSI accelerator facility with the beam-foil time-of-flight technique and has taken advantage of particle-X-ray coincidences using a charge state spectrometer in conjuction with a newly developed CVD-diamond particle detector. A preliminary analysis has given a value of 22.45 ± 0.66 ps for the lifetime of the 2^3P_0 state in helium-like gold. This is in good agreement with theoretical predictions.

Key words: fine-structure splitting, highly charged ions, beam-foil spectroscopy, hyperfine quenching, lifetimes of atomic states.

1. Introduction

The radiative decay of the 2^3P_0 state of helium-like ions with non-zero nuclear spin is mainly to the 1^1S_0 ground state and occurs by a rare mechanism in atomic physics often referred to as hyperfine quenching [1]. Normally, this 0–0 transition is rigorously forbidden by the angular momentum selection rules, and in ions with zero nuclear spin this state will decay by a fully-allowed electric dipole transition to the 2^3S_1 state. However, in ions with non-zero nuclear spin the hyperfine interaction will mix a small amount of the 2^3P_1 state into the 2^3P_0 wave function. Because the decay rate of the 2^3P_1 state to the 1^1S_0 ground state is high, the radiative decay of the mixed state is dominated by the hyperfine quenching mode.

The calculation of the decay rate of the mixed state to the ground state involves several atomic parameters, including the decay rate of the 2^3P_1 state and the 2^3P_0–2^3P_1 fine-structure splitting. The fine-structure splitting is of particular interest, because it is determined by the electron–electron interaction and the two-electron ion is the simplest atomic system in which it can be studied. Because of the rapid scaling of the relativistic part of this interaction with atomic number Z, measurements at high Z provide a particularly sensitive test of the relativistic theory.

The theoretical approach for two-electron ions at high Z is basically different compared to a low-Z system. At low Z the calculations of the fine-structure generally include the effects of relativity by a perturbation expansion in the parameter $(Z\alpha)^2$ and the calculations for helium are accurate at the ppm level, whereas calculations in the high-Z region claim accuracy of only a few percent [2, 3], because here perturbation theory does not work, as the expansion parameter is not a small quantity any more (in the case of gold: $Z\alpha = 0.577$). Accurate experimental measurements performed at high Z can therefore provide benchmark values for theoretical work.

The experimental approach is also different. While in low-Z systems the fine-structure splitting is directly measurable using radio-frequency spectroscopy or laser spectroscopy, the extension of direct methods to higher Z is experimentally not feasible and indirect methods must be used. In the experiment we report here, the fine-structure splitting is inferred from the measured decay rate of the hyperfine-quenched decay of the 2^3P_0 state. The use of this method to make precise measurements of the fine-structure splitting was first demonstrated in Ag^{45+} [1] and subsequently applied to Nb^{39+} [4] and Gd^{62+} [5]. The current measurement on Au^{77+} described in this paper extends the method to ions with the highest Z so far investigated.

2. Experimental method

The experiment was performed at the Gesellschaft für Schwerionenforschung (GSI) heavy-ion synchrotron facility located in Darmstadt, Germany. We have used the same basic beam-foil time-of-flight technique employed in previous experiments [1, 4, 5]. A schematic of the setup is shown in Figure 1. A hydrogen-like gold beam (Au^{77+}) was extracted from the heavy-ion synchrotron (SIS) accelerator with an energy of 194.8 MeV/u ($\beta = 0.5621$). This beam passes through a Ni target foil of thickness 1.5 mg/cm^2 hereby producing mainly excited helium-like ions by single electron capture. It is estimated that about 1% of the helium-like ions are in the 2^3P_0 state of interest. A pair of Ge(i) X-ray detectors are located outside the vacuum chamber on opposite sides of the ion beam behind the target foil. The X-rays emitted by the decay of the excited helium-like ions are viewed

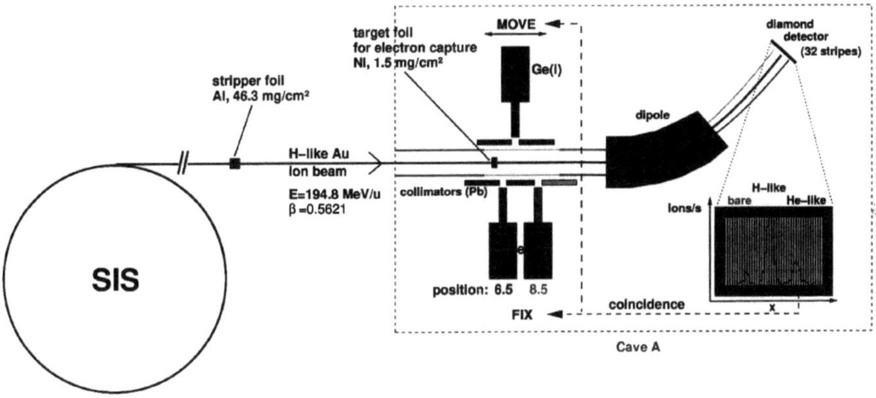

Figure 1. Experimental setup.

by the detectors through a pair of lead collimators. These collimators prevent the detectors from viewing prompt X-rays emitted by the beam at the target.

The position of one detector is fixed while the other detector is moveable. A precision translator is used which establishes the change in the position of the detector with an accuracy $\leqslant 1\ \mu$m. The raw data is obtained from a variation of the distance between the target foil and the moveable detector and the measurement of the ratio of counts from the 2^3P_0–1^1S_0 transition in the moveable detector relative to the counts in the fixed detector. The measurement of the ratio allows normalization directly to the ion population in the excited state of interest and eliminates many systematic errors associated with normalization to the integrated beam current. Thereby, a decay curve can be traced out and a measurement of the decay length can be obtained.

In this experiment we have employed for the first time a particle detector mounted downstream of a charge-state spectrometer consisting of a quadrupole doublet and a bending magnet which allows the separation of the different charge states of the gold ion beam [6]. The particle detector is a newly-developed 32-fold strip CVD-diamond detector with a detection area of 60×40 mm^2. The advantages of this new type of particle detector are its time resolution below 50 ps and its single-particle count rate capability of up to 10^8 ions/s [7]. Using this detector, we have been able to measure the X-rays in coincidence with helium-like ions counted in the diamond detector. This enables us to eliminate spurious X-rays observed from a variety of other sources. This advantage is illustrated in Figure 2 where two sample energy spectra of the moveable detector are shown. The left spectrum shows a raw energy spectrum without coincidence and the right spectrum shows an energy spectrum in coincidence with helium-like ions observed in the diamond detector. The improvement in signal/noise ratio is obvious.

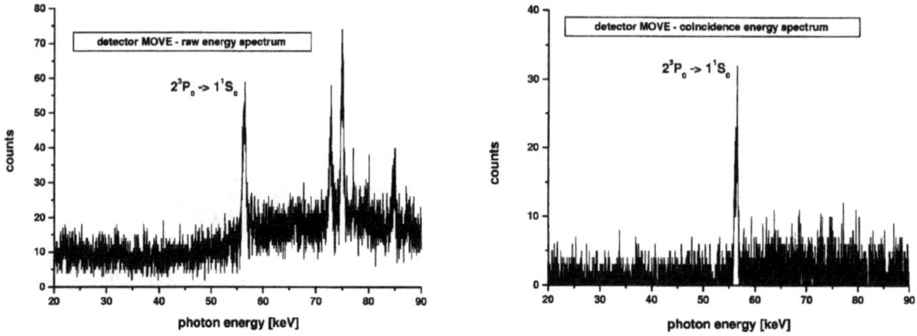

Figure 2. X-ray spectra obtained with the moveable Ge(i) detector without (*left*) and with coincidence condition.

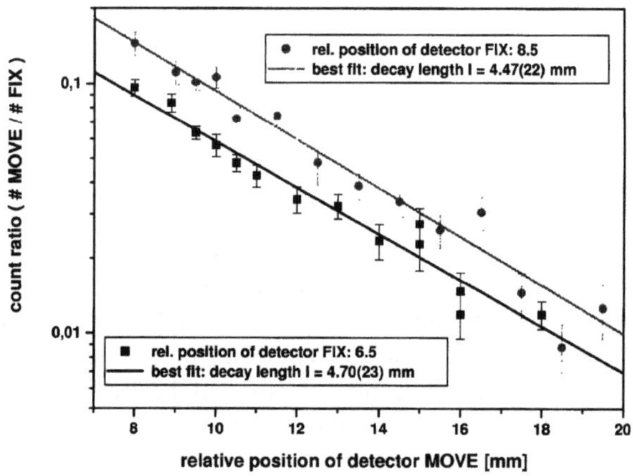

Figure 3. Measured decay curves for two normalization positions.

3. Preliminary results

For each detector the count rate of the 2^3P_0–1^1S_0 transition is determined from a fit to the area under the peak in the final coincidence energy spectrum with a Gaussian plus background. The result of a preliminary analysis of the obtained data is shown in Figure 3.

Here the count ratio of the moveable detector relative to the fixed detector is plotted as a function of the relative position of the moveable detector. The error bars of the ratios correspond to the statistical uncertainty. Decay curves from two different normalization positions have been traced out, covering almost three decay lengths. The decay length has been determined for each decay curve from a fit with a single exponential without background. The results for the decay lengths are also plotted in Figure 3. With this decay length l the lifetime τ is determined by

$$\tau = \frac{l}{c\beta\gamma},$$

Table I. Comparison of the lifetime τ of the 2^3P_0 state between experiment and theory. Our experimental value is the weighted mean of the two decay curves

	τ [ps]	
Experiment (preliminary)	(22.45 ± 0.66)	
Johnson *et al.* (1997) Bohr–Weisskopf	23.04	[8]
Indelicato (2001) Bohr–Weisskopf	23.35	[9]
Indelicato (2001) no Bohr–Weisskopf	22.61	[9]

where c denotes the speed of light, β the velocity of the ions and γ the Lorentz factor.

The preliminary result for the lifetime of the 2^3P_0 state and a comparison with recent calculations are summarized in Table I. The given error for the lifetime is associated with the fitting procedure and is the only major contribution to the total error. Other possible errors like the uncertainty of the velocity of the ions, delayed cascade feeding or the misalignment of the setup are assumed to contribute to $\leqslant 0.1$ ps to the total error. The preliminary result for the lifetime of the 2^3P_0 state is in good agreement with theoretical predictions. No value for the fine-structure splitting can be given as the evaluation is still in progress.

References

1. Marrus, R., Simionovici, A., Indelicato, P., Dietrich, D. D., Charles, P., Briand, J. P., Finlayson, K., Bosch, F., Liesen, D. and Parente, F., *Phys. Rev. Lett.* **63** (1989), 502.
2. Drake, G. W. F., *Can. J. Phys.* **66** (1988), 586.
3. Gorceix, O., Indelicato, P. and Desclaux, J. P., *J. Phys. B* **20** (1987).
4. Simionovici, A., Birkett, B. B., Briand, J. P., Charles, P., Dietrich, D. D., Finlayson, K., Indelicato, P., Liesen, D. and Marrus, R., *Phys. Rev. A* **48** (1993), 1695.
5. Indelicato, P., Birkett, B. B., Briand, J. P., Charles, P., Dietrich, D. D., Marrus, R. and Simionovici, A., *Phys. Rev. Lett.* **68** (1992), 1307.
6. Bräuning, H. *et al.*, GSI-2000-1, 2000, p. 91.
7. Berdermann, E., Fischer, B. E., Schlögl, M., Stelzer, H. and Voss, B., GSI-Preprint-2000-09.
8. Johnson, W. R., Cheng, K. T. and Plante, D. R., *Phys. Rev. A* **55** (1997), 2728.
9. Indelicato, P., private communication, 2001.

Hyperfine Interactions **146/147:** 145–150, 2003.

Nuclear and Electron Polarization Contributions to the HFS of Hydrogen- and Lithium-like Ions

M. TOMASELLI[1,2], T. KÜHL[2], W. NÖRTERSHÄUSER[2], G. EWALD[2], R. SANCHEZ[2], A. GLUZICKA[2], S. FRITZSCHE[3] and L. C. LIU[4]

[1] *Technische Universität Darmstadt, D-64289 Darmstadt, Germany; e-mail: m.tomaselli@gsi.de*
[2] *Gesellschaft für Schwerionenforschung mbH (GSI), D-64291 Darmstadt, Germany*
[3] *University of Kassel, D-34132 Kassel, Germany*
[4] *T.-Division, Los Alamos National Laboratory, Los Alamos, NM 87545, USA*

Abstract. The Dynamic Correlation Model (DCM) has been used to calculate nuclear ground-state wave functions of nuclei with one particle/hole in the closed shells. The strong mixing amplitudes between the valence particle/hole and the intrinsic vacuum states (valence hole coupled to core excitations) characterize the dynamic calculations of the hyperfine-structure splitting energy of the hydrogenlike ions which are in good agreement with measured values if the QED corrections are neglected. New experiments on the hyperfine-structure splitting energies of lithium-like ions could help in clarifying this still open point.

Key words: hydrogen-like atoms, lithium-like atoms, Bohr–Weisskopf effect, QED, nuclear magnetization.

1. Introduction and theory

Hydrogenlike high-Z atoms offer fascinating possibilities for testing the interplay of nuclear and atomic structure. They have recently become available for experiments at GSI in Darmstadt [1] and at the LLNL in Livermore [2–4] and have been used to test the theory of quantum electrodynamics (QED) in extremely strong electric and magnetic fields. Measuring the ground-state hyperfine structure (HFS) of these hydrogenlike systems is a sensitive method to explore QED and nuclear contributions to the electron energy. Experiments at GSI included laser spectroscopic investigations of $^{209}Bi^{82+}$ and $^{207}Pb^{81+}$ at the heavy-ion storage ring [1]. These measurements, with a relative accuracy $< 10^{-3}$, allowed the first test of QED in the strong electromagnetic field of highly charged heavy ions. At Livermore, hydrogenlike $^{165}Ho^{66+}$ [2], $^{185}Re^{74+}$, and $^{187}Re^{74+}$ [3], $^{203}Tl^{81+}$, and $^{205}Tl^{81+}$ [4] have been produced and stored in a high-energy electron-beam ion trap (SuperEBIT) by an energy variable electron beam, axially compressed by a strong magnetic field. In this contribution we revisite the calculations performed for the hyperfine splitting by using the Dynamic Correlation Model (DCM) [5]. The DCM describes the structure of open-shell nuclei with an odd number of valence particles in terms

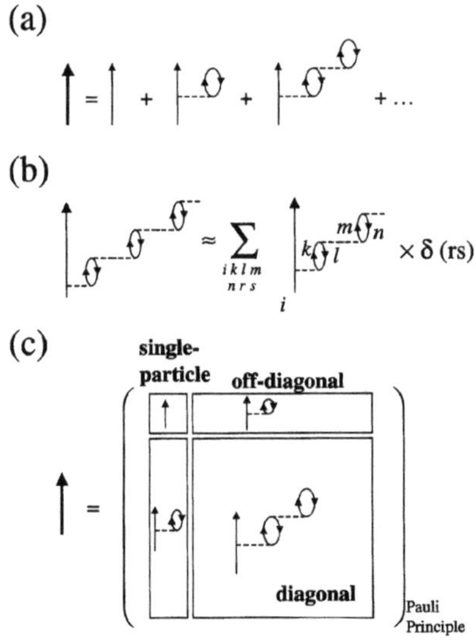

Figure 1. Dressed particles in the DCM: (a) perturbation theory, (b) dynamic linearization, (c) DCM eigenvalue equations.

of clusters: the valence and the core cluster. In contrast to Hartree–Fock approaches the DCM includes a coupling mechanism acting between the core and the valence particles, which modifies the Hartree–Fock description and polarizes the core via particle–hole excitations ($2\hbar w$) of protons and neutrons. Thus, the valence particle becomes "dressed" in the sense that it coexists with complex excitations of the core. From a Hartree–Fock point-of-view, these dressed particles correspond at most to a particular solution of an extended, nonlinear HF Hamiltonian. A symbolic representation of the equations governing the dynamics of the system is given in Figure 1(*left*). DCM corrections applied to hyperfine splitting calculations show good agreement with experimental data [1–4] if the QED corrections are neglected, however, by adding these corrections a systematic deviation between theory and experiment [6–8] has been observed. In order to clarify this open point an analysis of the non-linear terms contributing to the HFS was provided in [7] by performing a term-to-term comparison with the results of their theoretical models [9, 10]. Since the calculated Bohr–Weisskopf term (ε) is larger than that obtained in perturbation theories [9–11], the reason for this systematic disagreement could be the modification of the magnetization and the charge distribution by the additional DCM terms. When compared with hydrogen-like ions, further information about the hyperfine interaction of the nucleus with the electronic cloud of high-Z ions can be obtained by studying the ground-state splitting of few-electron ions. The simplest systems which have a closed K-shell are lithium-like ions. For these systems, the hyperfine splitting arises first of all from the coupling of the $2s$ electron with the nuclear angular momentum I since the closed K-shell does not contribute to the total (electronic) angular momentum J. Moreover, a filled $1s^2$ shell yields the advantage to reduce the influence of both the nucleus and of quantum electrodynamical effects. On the other hand, difficulties arise from the correlated motion of the electrons and from a (possibly) enhanced polarization of the interaction among the electrons and the nucleus. For the ground state hyperfine transition for lithium-like bismuth a similar situation was found as for hydrogen-like bismuth. In the zero-order, the hyperfine interaction of the nuclear moment with the motion of the electrons is represented by a dipole field with a $1/r^3$ behaviour. This behaviour leads to the typically strong dependence of the hyperfine splitting on nuclear parameters. Like for the (extended) nuclear charge distribution, a point model for the magnetic interaction represents, of course, a rather crude model and should be replaced by an extended nuclear magnetization distribution. But while in the traditional approach the different measured contributions are obtained additively, an improved treatment should include also nucleon–nucleon correlation to provide a consistent description of the interaction of the electron cloud with the nucleus. In the traditional approach to theoretical hyperfine splitting studies, the different contributions from the electron–electron interaction, QED effects, and the extended nucleus are obtained *additively*. This also applies to the coupling of the electrons with the motion of the nucleons as is summarized by the Bohr–Weisskopf effect. An alternative view point is taken by the dynamic correlation model (DCM) which includes both the coupling of the internal nuclear motion as well as the electronic interaction and the creation and annihilation of virtual electron–positron pairs within the same framework.

2. Results and discussion

In Table I the DCM results are compared to ground-state HFS splittings calculated for a point nucleus and with results which take the finite spatial distribution of the nuclear charge into account (Breit–Schawlow correction). Additionally, QED

Table I. Ground state hyperfine structure splittings for a point-nucleus (E_{PN}), including Breit–Schawlow (E_{BS}), DCM, and QED corrections. The QED contributions include vacuum polarization and self-energy. All values are in eV

	$^{165}Ho^{66+}$	$^{185}R^{74+}$	$^{187}Re^{74+}$	$^{203}Tl^{80+}$	$^{205}Tl^{80+}$	$^{207}Pb^{81+}$	$^{209}Bi^{82+}$
E_{PN}	2.3007	3.0103	3.0411	3.0184	2.9890	1.3998	5.8395
E_{BS}	2.1957	2.7976	2.8263	3.3073	3.3374	1.2528	5.1922
E_{tot}^{DCM}	2.1649	2.7192	2.7449	3.2130	3.2770	1.2166	5.0832
ΔE_{QED}	−0.0103	−0.0142	−0.0143	−0.0176	−0.0177	−0.0067	−0.0280
E_{tot}	2.1546	2.7050	2.7306	3.1954	3.2213	1.2099	5.0552
E_{exp}	2.1646 (6)	2.7187 (18)	2.7449 (18)	3.21351(25)	3.24409(29)	1.2159 (2)	5.0841 (4)
Ref.	[2]	[3]	[3]	[4]	[4]	[1]a	[1]b

Table II. Calculated Bohr–Weisskopf terms ε in different theories (SO: Spin–Orbit interaction)

Ref.	$^{165}Ho^{66+}$	$^{185}Re^{74+}$	$^{187}Re^{74+}$	$^{203}Tl^{80+}$	$^{205}Tl^{80+}$	$^{207}Pb^{81+}$	$^{209}Bi^{82+}$
[10]	0.0099	0.0118	0.0119	0.0174	0.0174	0.0429	0.0131
[9] (no SO)	0.0085	0.0120	–	0.0177	0.0177	0.0419	0.0133
[9] (with SO)	0.0089	0.0122	–	0.0179	0.0179	–	0.0118
[11]	0.0086	0.013	–	0.020	0.020	0.036	0.011
DCM	0.0140	0.0280	0.0288	0.0285	0.0295	0.0289	0.0210

radiative corrections [9] are listed and combined with the DCM results. The pure DCM splittings agree remarkably well with the experimental values, which are given in the last row of the table, while the wavelengths obtained after adding the QED contributions show a systematic shift to larger wavelengths. Again, in Table II, a systematic deviation is observed: the ε values obtained with the DCM are considerably larger then the ε calculated within perturbative calculations. However the magnetization distributions used to calculate ε in perturbation theories are obtained from a single-particle or single-hole model that does not reproduce the experimental magnetic moments. In the DCM other terms then the single particle/hole terns contribute to the magnetization; without these terms the experimental magnetic moments cannot be obtained. Similar correlation terms should be provided in perturbation calculations so that the integration of the modified magnetic distribution yields the experimental magnetic moment. In the DCM these nuclear correlation terms also modify the Coulomb central potential derived from the charge distribution, thus changing the Breit–Schawlow contributions. Consequently, although we start the calculations with Dirac spinors that are solutions for a Fermi charge distribution, the final spinors are modified by the correlations. The corrections include "de facto" virtual mesons, which can decay into e^+e^- pairs (see Figure 1(*right*), where this effect has been symbolically given for odd-hole nuclei). Similar modifications are also considered by QED calculations. Hence, the disagreement between λ_{tot} and λ_{exp} in Table I could be the result of double-counting. The Bohr–Weisskopf ε calculated in the DCM might include effects that are considered by QED as well. Another possibility is that nuclear correlations, which are of obvious importance for the size of the magnetic moments, lead to additional QED terms that are currently not included in the calculated corrections. Thus, the central question is how to compare the DCM terms and the radiative corrections. The importance of this problem is evident: In the extreme single-particle model QED and nuclear-magnetization corrections for high-Z atoms are on the same order of magnitude. Hence, the feasibility of testing the QED corrections depends strongly on the accuracy of the model used to evaluate the nuclear magnetization (see Figure 2). Table III displays the individual and total contributions to the hyperfine splitting for lithium-like bismuth from previous and our computations. In our model, taking into account the QED corrections, the hyperfine structure

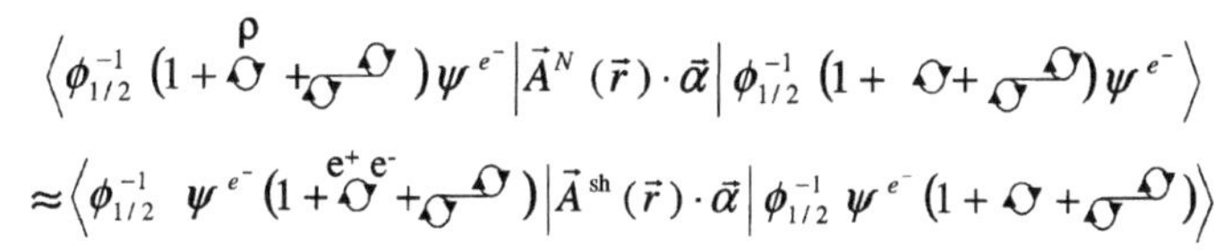

Figure 2. Symbolic equivalence between many body and single particle approximations.

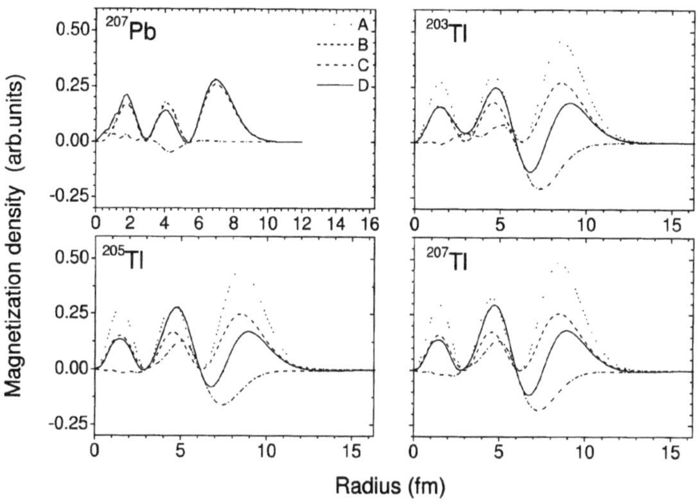

Figure 3. Calculated magnetization distribution for ^{207}Pb and ^{203,205,207}Tl. Line A shows the single-particle result, line B has been normalized with the spectroscopic factor of the single particle, line C shows the contributions of diagonal and off-diagonal elements in DCM and line D is the total magnetization.

Table III. Comparison of different hfs calculations for Li-like bismuth with experimental HFS splitting

Contribution (meV)	[11]	[12]	[13]
one-electron	958.50(5)	958.49	958.51
charge distr.	−113.8(3)	−108.42	−113.61
mag. distr.	−13.9(2)	−18.68	−14.1
total QED	−4.44	−4.06	−4.81
e–e interaction	−29.45 (4)	8.43	−34.4
boiling of the QED vacuum [15]	−.−	0.06	−7.69
total theory	796.9 (2)	792.8	783.9 (3.0)
measurement [14]		820 (26) meV	

splitting for lithium-like $^{209}Bi^{80+}$ ions is reduced by about 1.3 % if compared with previous calculations.

References

1. (a) Klaft, I. *et al.*, *Phys. Rev. Lett.* **73** (1993), 2425; (b) Seelig, P. *et al.*, *Phys. Rev. Lett.* **81** (1998), 4824.
2. Crespo López-Urrutia, J. R., Beiersdorfer, D. W., Savin, P. and Widmann, K., *Phys. Rev. Lett.* **77** (1996), 826.
3. Crespo López-Urrutia, J. R., Beiersdorfer, P., Widmann, K., Birkett, B. B., Martensson-Pendrill, A.-M. and Gustavsson, M. G. H., *Phys. Rev. A* **57** (1998), 879.
4. Beiersdorfer, P. *et al.*, *Phys. Rev. A* **64** (2001), 032506.
5. Tomaselli, M., *Phys. Rev. C* **37** (1988), 349; *Ann. Phys.* **205** (1991), 362; *Phys. Rev C* **48** (1993), 2290; Tomaselli, M., Liu, L. C. *et al.*, *J. Opt. B* **5** (2003), 395.
6. Tomaselli, M., Schneider, S. M., Kankeleit, E. and Kühl, T., *Phys. Rev. C* **51** (1995), 2989.
7. Tomaselli, M., Kühl, T., Seelig, P., Holbrow, C. and Kankeleit, E., *Phys. Rev. C* **58** (1998), 1524.
8. Tomaselli, M., Kühl, T., Nörtershäuser, W., Dax, A., Marx, D., Sanchez, R., Wang, H. and Fritzsche, S., *Phys. Rev. A* **65** (2002), 022502.
9. Shabaev, V. M., Tomaselli, M., Kühl, T., Artemyev, A. N. and Yerokhin, V. A., *Phys. Rev. A* **56** (1997), 252.
10. Gustavsson, M. G. H., Forssén, Chr. and Mårtensson-Pendrill, A.-M., *Hyp. Interact.* **127** (2000), 347.
11. Shabaev, V. M. *et al.*, *Phys. Rev. A* **57** (1998), 149.
12. Bouchard, S. and Indelicato, P., *Eur. Phys. J. D* **8** (2000), 59.
13. Tomaselli, M., Fritzsche, S., Kühl, T. and Winter, H., *Hyp. Interact.* **127** (2000), 315.
14. Beiersdorfer, P. *et al.*, *Phys. Rev. Lett.* **80** (1998), 3082.
15. Tomaselli, M. *et al.*, *Can. J. Phys.* **80** (2002), 1347.

Hyperfine Interactions **146/147:** 151–159, 2003.

Studies of Lifetimes in an Ion Storage Ring Using Laser Technique

DANIJELA ROSTOHAR[1], ANNA DERKATCH[1], HENRIK HARTMAN[2], LARS-OLOV NORLIN[3], PEDER ROYEN[1], PETER SCHEF[1] and SVEN MANNERVIK[1]

[1]*Physics Department, Stockholm University, AlbaNova S-10691 Stockholm, Sweden*
[2]*Atomic Spectroscopy; Department of Physics, Box 118, S-22100 Lund, Sweden*
[3]*Royal Institute of Technology AlbaNova, S-10691 Stockholm, Sweden*

Abstract. The laser-probing method for lifetime measurements of metastable levels, performed by applying the Fast Ion Beam Laser (FIBLAS) method to ions stored in a storage ring, has been developed by the Stockholm group. Recently, we have applied this method to lifetime measurements of close lying metastable levels. In this paper we discuss experimental studies of ions with complex structure and present the first experimentally obtained lifetimes of selected metastable levels in complex systems as Fe^+, Eu^+ and La^+.

1. Introduction

Lifetimes of metastable states, which are in the range of ms to s, find applications both in fundamental atomic structure studies and in relation to astrophysical observations. Laboratory lifetime investigations of highly forbidden transitions are usually difficult since the ions usually are exposed to rapid collisional quenching and special methods for trapping under ultra-high vacuum (UHV) conditions are needed. The long trapping time offered by a storage ring makes it a powerful tool for measurements of long-lived states. The laser-probing technique, recently developed by S. Mannervik and co-workers, which utilities the spectroscopic advantage of the FIBLAS method characterised by sub-Doppler resolution, has been applied to ions stored for a long time in a storage ring. The method has properties of high resolution and high sensitivity and can be used for lifetime measurements of metastable levels, with lifetimes from a few ms up to a few seconds. The results obtained in lifetime measurements of metastable levels in Xe^+, Ca^+ and Sr^+ and comparison with measurements performed in different traps were discussed in paper [1]. Here we discuss the application of the laser probing technique to complex systems for which high resolution and high sensitivity is needed. We present a summary of results obtained in recent investigations of complex systems as Fe^+, Eu^+ and La^+. The particular interest for selection of these ion species will be briefly discussed during the presentation of the obtained results.

2. Experiment

2.1. THE CRYRING FACILITY

The experiments were performed at the ion storage ring CRYRING at the Manne Siegbahn Laboratory in Stockholm [2]. In previous papers, the facility was described in more details [5, 7], and only a short overview is given here. A Nielsen type of hot-cathode, low-voltage electron impact ion source (MINIS) was used for ion production. The ion beam was extracted from the source by electrostatic electrodes at a maximum voltage of 40 kV. After mass separation by an analysing magnet and passage through a radio-frequency quadrupole (RFQ), ions were injected into the storage ring. The ion beam current of the stored ion beams, measured by a calibrated current transformer, was typically 0.5–1 μA. The storage time of singly charged ions at UHV condition in CRYRING (pressure below 10^{-11} mbar) is limited by neutralisation in collisions with particles of the rest gas. The lifetime of the stored ion beam was recorded by counting the number of neutral particles after one of the dipole magnets. The particle detector consisted of a BaF_2 scintillator and a PM tube [3]. The lifetime of the stored ion beam was estimated to be 30 s and 18 s for the Fe^+ and Eu^+ beam, respectively.

2.2. THE CONCEPT OF LASER-PROBING TECHNIQUE

Only a small fraction of all ions produced by the ion source and stored in the ring are in metastable levels. By laser pulses applied at different delay times after injection of ions one can investigate how many ions have remained in the metastable level. The laser frequency is chosen to correspond to a transition from the metastable level to some higher lying level while the number of emitted photons when this level decays is monitored. This prompt fluorescence is proportional to the population of the metastable state. The probing method is destructive and new ions must be injected before the next probe pulse can be applied. The laser pulse duration is chosen to let most of the metastable state population to be pumped out.

As spontaneous decay of the metastable state can occur anywhere in the ring, the detection efficiency for spontaneous decay will be small because the detector covers only a small part of the ring circumference. The high efficiency of the laser probing method is achieved by localising the interaction area of the ions in metastable state and the laser light in front of the detector. This is obtained by applying an additional acceleration of the ions by a Doppler Tuning Device (DTD) [4, 5] by which a local Doppler shift is obtained.

Laser light of the desired wavelength ranges was provided by a laser system consisting of a Coherent 699-29 Autoscan Ring Dye Laser pumped by a Coherent Innova 400-25 Argon Ion Laser. The laser light was transported into the storage ring by a set of mirrors and focussed to a spot diameter of a few millimetres at the interaction region. Since the laser light is merged with the ion beam in collinear geometry, the instrumental line widths of the recorded spectra will be improved by the kinematic compression effect [6].

The laser-induced fluorescence was observed through a mesh in one of the electrodes in the DTD and focussed by a lens system on a single photon counting photomultiplier equipped with a coloured glass filter to block scattered laser light. Typical background rates including scattered laser light are 10 s^{-1}.

2.3. EXPERIMENTAL PROCEDURE

The data points in the decay curve, recorded during the lifetime measurements correspond to different ring cycles. Thus the measurements would be sensitive to a situation where the injected intensity of ions may vary from injection to injection. Also the production of ions in the metastable level by the ion source may vary. During the lifetime measurement five different curves are stored to get full information about possible instabilities (see Figure 1). The points in the *lifetime raw data* curve correspond to the fluorescence counts read in the time window at different delay from ion injection in different ring cycles. The width of the time window, i.e. the time when the shutter is open, has been chosen to assure that more

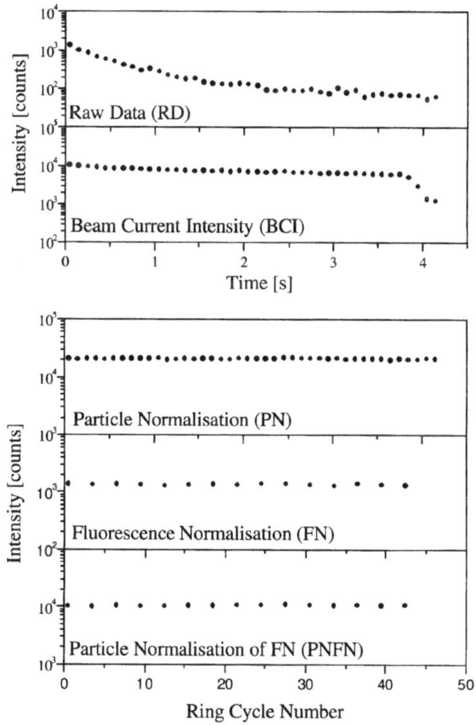

Figure 1. A typical data set for lifetime measurements by the laser probing technique.

than 90% of ions in metastable state are pumped out. The *beam current intensity* in the corresponding time window shows the decay of the stored ion beam current. The *lifetime raw data* and *beam current intensity* curves give information about the lifetime of the selected metastable level and the shape of ion beam intensity decay. Information about the stability of the production and storage of all ions, and in particular, the ions in metastable state are given by the next three curves. The *particle normalisation* curve is obtained by reading the number of counts from the BaF_2 detector for every ring cycle during a fixed externally gated time window of a few hundred milliseconds. The position of this window is chosen after inspection of the ion beam current curve for the particular stored beam. To monitor the stability in production of ions in metastable state, laser induced fluorescence at a *fixed* time delay is recorded every fourth cycle. This signal is stored in the fourth curve. The corresponding particle normalisation is also recorded. It should be noted that flat *particle normalisation* and *fluorescence normalisation* curves show that the production of ions and ions in metastable state by the ion source was stable.

The number of stored ions in metastable state can be changed by collisional interaction with the residual gas due to excitation, deexcitation and neutralisation [7]. Besides injection instabilities, these processes could also be sources of systematic errors in the determination of radiative lifetimes. In order to subtract contributions from ions in metastable state produced in collision processes, a separate repopulation curve is determined [5]. This curve is recorded in a similar way as the decay curve. In the beginning of every ring cycle, the ions in the metastable state produced by ion source are, however, quenched by a laser pulse (typically a hundred milliseconds long). A second laser pulse will be applied with different delays from the injection in order to investigate how many ions in metastable state are created by collisional interaction. The contribution from repopulation can subsequently be accounted for by subtracting the repopulation curve from the lifetime curve.

To distinguish between radiative decay and depopulation (deexcitation and neutralisation) due to the collisions with atoms and molecules of the residual gas, the lifetime measurement was also performed as a function of rest gas pressure. Plotting the decay rate of the metastable level versus relative pressure (Stern–Vollmer plot), the radiative decay will be obtained by extrapolation to zero pressure.

The ion beam decay curve can be recorded by feeding the signal from the particle detector into a multichannel scaler (MCS) triggered at ion injection. This curve usually shows the existence of initial loss of ions, which is much faster than later in the cycle where ions are lost mainly due to the neutralisation. To reduce the influence of this initial loss, the measurement should be started with a delay after injection. This delay time was typically chosen to be 50–100 ms. Longer delay costs loss in the number of ions in metastable state. Corrections for the remaining effect will be obtained in the data analysis.

2.4. LIFETIME MEASUREMENTS FOR COMPLEX IONS BY LASER-PROBING TECHNIQUE

In our lifetime measurements the ions are excited into the metastable states directly in the ion source. The fraction of ions in metastable state produced in the ion source is difficult to determine and it can only be roughly estimated to be less than 1%. In the case of Fe^+, this fraction of ions in metastable state will be distributed over 62 possible metastable levels. This fact explains that the signals in the Fe^+ measurements were weak and that, for example, the $3d^5\ 4s^2 a^6 S_{5/2}$ level in the best case gave a signal-to-noise ratio of 10 [9]. To illustrate the weakness of this particular signal, Figure 2 shows a comparison of the resonance signals obtained for Ca^+ and Fe^+, respectively. The high detection efficiency of laser-probing method, however, is more than a factor of 10^3 higher than for passive observation of spontaneous decay from the metastable level, so lifetime investigations of metastable levels in Fe^+ has still been successful.

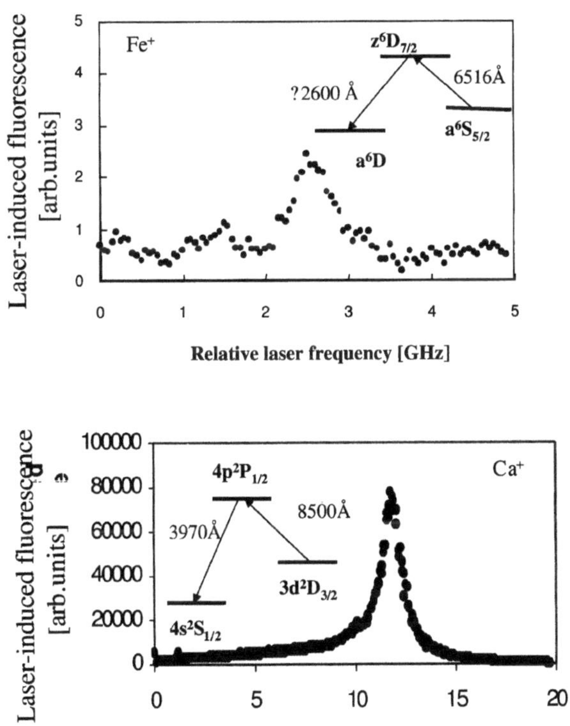

Figure 2. Laser-induced fluorescence spectra obtained using the excitation schemes shown in insets. The signal was 1000 time stronger in the Ca^+ case compared with the signal obtained in the Fe^+ case.

No repopulation was observed for the levels studied in Fe^+ [9, 10] and Eu^+ [11]. The explanation could be that, because of unfavourable signal-to-noise ratio, the repopulation cannot be resolved from the background. Also, the repopulation, as a process induced by collisional excitation with the rest gas, is not selective and in the case of Fe^+ it will be spread over all 62 metastable levels. In the lifetime investigation of the a^1G_4 and b^1D_2 in La^+ [12], repopulation was observed and metastable ions produced by collisional quenching were subtracted. Here it was found that the repopulation of the two levels was different.

Collisional destruction rates were extracted from lifetime measurements performed at the base pressure (10^{-11} mbar) and at raised average pressure. Since at raised pressure two counteracting processes are enhanced: the collisional destruction and repopulation, the estimation of the collisional deexcitation rate can be difficult. For the metastable levels in Fe^+ collisional destruction was only observed for the $3d^6(^3H)4s\ b^2H_{11/2}$ level. Using a Stern–Vollmer plot this rate was estimated to be 0.028 s^{-1} at base pressure [10]. Inclusion of this effect changed the extracted radiative lifetime of this level by 10%. Due to the fact that the investigated levels have similar energy, the observed collision effect was assumed to be approximately the same and was used to subtract the collision destruction effect for lifetimes of all other levels in Fe^+. For the investigated levels in Eu^+ no significant extra contribution from collisional destruction was observed as the average pressure in the ring was increased by 50%. The contribution from collisional destruction for the a^1G_4 level in La^+ was estimated to be 2%. Similarly as in the case of Fe^+, this observed collision rate was used to subtract the collision destruction effect for the b^1D_2 level in La^+.

As mentioned above the ion beam decay curve usually exhibits an initial loss of ions, which has a slope much steeper than the later part of the particle curve where the slope is determined by the neutralisation rate. The strength of this initial peak has been found to depend on the ion beam intensity but also on the specific settings of ring parameters. The ion beam current of stored La^+ ions was about 20 times lower than the corresponding beam current for measurements with Eu^+ and Fe^+ and the transient peak was not visible here (see Figure 3). According to this fact we conclude that the intensity and the slope of transient part can be changed by changing the ring settings and decreasing the beam intensity. Since the extra loss is of instrumental character it will be equal for both ground state ions and ions in the metastable state. A way to compensate for this extra loss is to divide the fluorescence decay curve with this extra decay component extracted from the ion beam current intensity curve (curve No. 2 at Figure 1). This procedure has been utilised in the analysis of Eu^+ and Fe^+ and the correction resulted in an increase of the lifetime up to 10–20%.

Since the probing procedure is sequential, normalisation to the number of ions in metastable state produced in the ion source should be performed. The production of ions in metastable state was very stable in experiments performed with Fe^+ and Eu^+ ions, and the effect of this normalisation was negligible. For La^+, however,

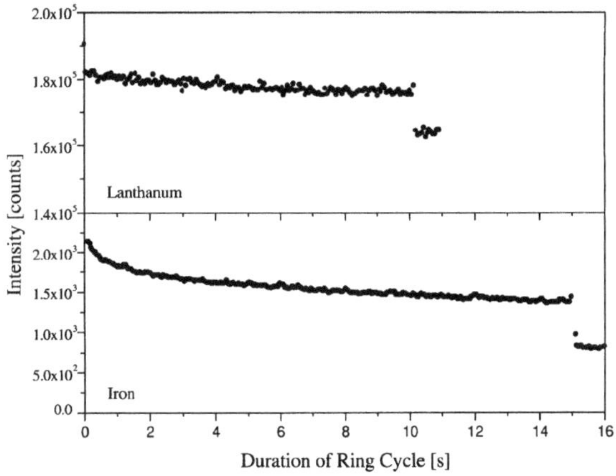

Figure 3. Comparison of the ion beam current curves for the La^+ and Fe^+ ion beams obtained with the BaF_2 detector. The data analysis in the La^+ case was started from time corresponding to second point on the curve.

the production of ions in metastable state was decreasing slightly during the time while the total production of ions was stable (see Figure 4). Here normalisation to the *fluorescence normalisation* curve gave a significant (but small) correction to the decay rate of the two investigated levels.

3. Results for complex ions

3.1. LIFETIME MEASUREMENT OF METASTABLE LEVELS IN Fe^+

High cosmic abundance makes iron very important for astronomical investigations. Therefore Johansson recently started an international project for collection of radiative decay rate data denoted as FERRUM [8]. In collaboration with his group we have made the first experimental investigation of decay rates for the $3d^5\,4s^2a^6S_{5/2}$, $3d^6(^3D)4s\;b^4D_{7/2}$, $3d^6(^3G)4s\;a^4G_{9/2}$ and $3d^6(^3H)4s\;b^2H_{11/2}$ levels in Fe^+. More details about these particular measurements are reported in [9] and [10] and here we only summarise the results.

The lifetime of the $3d^54s^2a^6S_{5/2}$, $3d^6(^3D)4s\;b^4D_{7/2}$, the $3d^6(^3G)4s\;a^4G_{9/2}$ and $3d^6(^3H)4s\;b^2H_{11/2}$ levels were measured by probing with the laser wavelengths locked to 6516, 6456, 5426 and 5536 Å, respectively. After data analysis, which was made in several steps, the lifetimes were determined to be 230 (30) ms, 530 (30) ms, 670 (20) ms and 3.8 (0.3) s, respectively. As discussed in the previous section, the high density of metastable levels and the low population of each state, was a strong challenge for the laser probing technique.

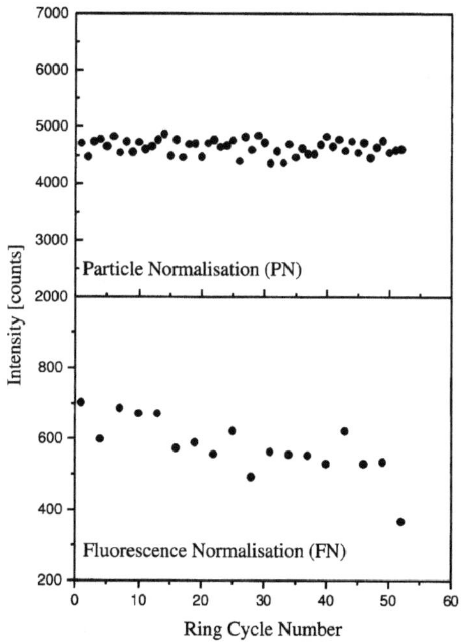

Figure 4. The *Particle Normalisation* and *Fluorescence Normalisation* curve recorded in the lifetime investigations of metastable levels in La^+. The production of La^+ metastable ions was decreasing during the time while the production of ions was stable.

3.2. LIFETIME MEASUREMENT OF METASTABLE LEVELS IN Eu^+

Singly charged europium has a very complex atomic structure due to the open f-shell. Many possible energy levels and crowed optical spectra make rare earth elements very complicated for experimental and theoretical analysis. No experimental or theoretical investigations of forbidden transitions have been reported until now.

The first experimental lifetime investigation of metastable levels in Eu^+ was done by applying the laser probing method. The lifetime of the a^9D_J ($J = 2$–5) levels were all found to be around 1 s. The procedure of data analysis and error estimation, as well as more experimental details, were discussed in [11].

3.3. LIFETIME MEASUREMENT OF METASTABLE LEVELS IN La^+

Lanthanum, with atomic number 57, is the first of the rare earth in the lanthanide series and its electronic structure is less difficult to handle in theoretical investigations as compared with europium, which is positioned in the middle of the series. This was our motivation for performing lifetime measurements of the a^1G_4 and b^1D_2 levels in La^+ [12]. As an integrated part of this investigation, calculations were performed by E. Biémont. The experimentally obtained lifetimes of the a^1G_4

and b^1D_2 metastable levels are 5.2 (0.2) s and 2.1 (0.3) s, respectively and corresponding theoretical results were 5.09 s and 2.28 s. The lifetime of the a^1G_4 level is the longest lifetime measured so far by the laser-probing technique in the ring.

4. Summary

In this paper we have discussed the laser probing technique of a stored ion beam as a method for experimental determination of lifetimes of metastable levels in ions with complex electronic structure. We have presented recent results obtained in lifetime investigations of metastable levels in Fe^+, Eu^+ and La^+. The results prove that the laser probing technique is a powerful tool for lifetime studies of ions with complex atomic structure.

References

1. Mannervik, S., *Hyp. Interact.* **127** (2000), 237.
2. Abrahamsson, K. *et al.*, *Nucl. Instrum. Methods Phys. Res. B* **79** (1993), 269.
3. Kerek, A., Klamra, W., Norlin, L.-O., Novak, D., Westman, S., Lidberg, J. and Mannervik, S., *Proceedings of the 6th European Particle Accelerator Conference (EPAC 98)*, Institute of Physics Publishing, Bristol and Philadelphia, 1998, p. 1577.
4. Short, R. T., Mannervik, S., Larsson, M., Sigray, P. and Sonnek, D., *Phys. Rev. A* **39** (1989), 3969.
5. Lidberg, J., Al-Khalili, A., Norlin, L.-O., Royen, P., Tordoir, X. and Mannervik, S., *Nucl. Instrum. Methods Phys. Res. B* **152** (1999), 157.
6. Kaufman, S. L., *Opt. Comm.* **17** (1976), 309.
7. Mannervik, S., Lidberg, J., Norlin, L.-O. and Royen, P., *Phys. Rev. A* **56** (1997), R1075.
8. Sikström, C. M., Schultz-Johanning, M., Kock, M., Li, Z.-S., Nilsson, H., Johansson, S., Lundberg, H. and Raassen, A. J. J., *J. Phys. B* **32** (1999), 5687.
9. Rostohar, D., Derkatch, A., Hartman, H., Johansson, S., Lundberg, H., Mannervik, S., Norlin, L.-O., Royen, P. and Schmitt, A., *Phys. Rev. Lett.* **86** (2001), 1466.
10. Hartman, H., Derkatch, A., Johansson, S., Lundberg, H., Mannervik, S., Norlin, L.-O., Rostohar, D. and Royen, P., *Astron. Astrophys.* **397** (2003), 1143.
11. Rostohar, D., Andersson, K., Derkatch, A., Hartman, H., Mannervik, S., Norlin, L.-O., Royen, P., Schmitt, A. and Tordoir, X., *Physica Scripta* **64** (2001), 1.
12. Derkatch, A., Manervik, S., Norlin, L.-O., Rostohar, D., Royen, P., Schef, P. and Biémont, E., *Phys. Rev. A* **65** (2002), 062508.

Hyperfine Interactions **146/147:** 161–170, 2003.

Strong Relaxation and Correlation Effects on the $2p^5 3s$–$2p^6$ Spectrum of Neutral Neon

C. Z. DONG[1,2], L. Y. XIE[1], X. X. ZHOU[1], X. W. MA[2] and S. FRITZSCHE[3]
[1]*Department of Physics, Northwest Normal University, Lanzhou 730070, China*
[2]*Center of Theoretical Nuclear Physics, National Laboratory of Heavy-Ion Accelerator of Lanzhou, Lanzhou 730000, China*
[3]*Department of Physics, University of Kassel, D-34119 Kassel, Germany*

Abstract. The effects of correlation and relaxation on the $2p^5 3s\ {}^{1,3}P_1$–$2p^6\ {}^1S_0$ resonance and intercombination transitions of neutral neon have been studied systematically using the Multi-configuration Dirac–Fock (MCDF) method. Results are presented for the energies and lifetimes of the low-lying ${}^{1,3}P_1$ levels as well as the according transition probabilities. Lifetimes for the two metastable ${}^3P_{0,2}$ levels are obtained also from the calculation of the ${}^3P_2$–${}^1S_0$ magnetic-quadrupole (M2) and from the ${}^3P_0$–${}^3P_1$ magnetic-dipole (M1) and ${}^3P_0$–${}^3P_2$ electric-quadrupole (E2) transitions, respectively. Comparison is made for our theoretical results with previous computations and measurements; they are found to be in good agreement with recently evaluated data.

Key words: MCDF method, correlation effects, atomic relaxation, transition probabilities.

1. Introduction

Although at the (heavy) end of the second-row elements in the periodic table, neon is still an abundant element in the universe. Hence, spectral information about neon in rather different environments has attracted astrophysical interest for many years to understand, for example, the details of stellar nucleosynthesis. Moreover, neon – at different stages of ionization – is also frequently used in many laboratory and industrial applications [1]. But although quite a number of (theoretical and experimental) investigations have been carried out for neon during the last four decades, up to the present, there is still a lack of accurate transition data, even for lines among the low-lying levels.

The $1s^2 2s^2 2p^6\ {}^1S_0$ ground state of neutral neon is known to represent a very stable (and chemically inert) configuration. The lowest excited levels of neon then arise if one electron from the $2p$ shell *becomes* a $3s$ electron which may *couple* with the remaining electron cloud to any of the (four) ${}^3P_{0,1,2}$ and ${}^1P_1$ levels. From these excited levels, however, only the two ${}^3P_1$ levels can decay (back) to the ${}^1S_0$ ground state via an electric-dipole allowed resonance or (at least) intercombination transition. The other two levels, ${}^3P_2$ and ${}^3P_0$ in contrast, are *meta-stable* with lifetimes of several (ten) seconds, which must decay either by quenching or via *forbidden lines.* While the $^3P_0$ level can decay to both, the $^3P_1$ level via magnetic-dipole (M1) or the $^3P_2$ via an electric-quadrupole (E2) transition, the $^3P_2$ level has to undergo a magnetic-quadrupole (M2) transition back to the ground-state if no additional fields or collisions occur.

During the last decades, a number of theoretical [2–4] and experimental studies [5, 6] have been carried out on the $2p^53s$–$2p^6$ transitions in neon, especially for the $2p^53s\ ^{1,3}P_1$–$2p^6\ ^1S_0$ resonance and intercombination lines. For these two levels, for example, Curtis *et al.* [5] measured the lifetimes by using beam-boil techniques, while Zhong *et al.* [6] explored the absolute (optical) oscillator strengths by using a high-resolution electron-energy-loss spectrometer (EELS). Till today, however, accurate measurements of the lifetimes of the noble gases (often) suffer from various systematic shifts which affect also the experimental accuracy. By comparing the data from different measurements [5, 7, 8], rather sizeable deviations from each other are found in the literature. For the magnetic-quadrupole transitions of the metastable $^3P_2$ level, moreover, theoretical studies are still scarce [3]. To improve the present understanding of these low-lying levels of atomic neon, below we present and discuss theoretical data from a systematically enlarged multiconfiguration Dirac–Fock (MCDF) calculation.

In the analysis of the $2p^53s$–$2p^6$ spectrum of atomic neon, major difficulties arise from the rather strong effects of correlation and the rearrangement of the electron density, when these densities are compared for the ground and the excited levels. These two effects often influence the transition probabilities, in particular, for most forbidden and weak lines. To improve the reliability of such transition probability calculations, we recently developed and applied a computational model [10] which is based on systematically enlarged MCDF wave functions [9]. Several complex spectra have been investigated by means of this method, including the low-lying spectra of Cu II [11] and Xe I [12] for which accurate predictions are known to be still a challenge of present-day atomic structure theory. The advantage of our computational procedure is that all dominant effects of relativity, electron correlations, and of the relaxation of the electron density is treated within the same framework. In the present contribution, we apply this method for studying the electric-dipole allowed $2p^53s\ ^{1,3}P_1$–$2p^6\ ^1S_0$ as well as *forbidden* $2p^53s\ ^3P_2$–$2p^6\ ^1S_0$ (M2) and $2p^53s\ ^3P_0$–$2p^53s\ ^3P_1$ (M1) transitions. For the lifetimes of the two metastable $^3P_{0,2}$ levels and the $2p^53s\ ^{1,3}P_1$–$2p^6\ ^1S_0$ transition probabilities, thereby, a good agreement with recently evaluated data from the NIST Spectroscopic Database [13] is found only, if all single, double, and triple excitations from $2s$, $2p$ into the $3s$, $3p$, and $3d$ shells as well as singles and doubles into the $4l$ and $5l$ layer are taken into account properly.

2. Method of computation

The relativistic MCDF method and the programs which were used for calculating the wave functions and transition probabilities have been described in detail elsewhere [14, 15]. Therefore, we will only give a brief account here on the MCDF model and the treatment of relaxation effects for many-electron atoms.

2.1. CALCULATION OF THE LEVEL ENERGIES

In relativistic atomic theory, one often starts from the *no-pair* Dirac–Coulomb Hamiltonian for an atom or ion with N electrons and nuclear charge Z (here written in atomic units)

$$\widehat{H}_{\mathrm{DC}} = \sum_{i=1}^{N} \{c\hat{\alpha} \cdot \hat{\mathbf{p}}_i + (\beta - 1)c^2 + V_{\mathrm{nuc}}(\hat{r}_i)\} + \sum_{i<j}^{N} |\hat{\mathbf{r}}_i - \hat{\mathbf{r}}_j|^{-1}, \tag{1}$$

where the first two terms describe the kinetic energy (taking off the rest energy c^2 of the electrons) and the third term represents their interaction with the nucleus. The last term, finally, is the Coulomb repulsion between all pairs of electrons which typically dominates the effort in all practical computations.

To construct a (N-electron) basis of configuration state functions (CSF), in which the Hamiltonian matrix (1) can later be considered, we apply the standard relativistic spin–orbitals

$$\varphi_{n\kappa m}(\mathbf{r}) = \frac{1}{r}\begin{bmatrix} P_{n\kappa}(r)\chi_{\kappa m}(\theta,\phi) \\ i\,Q_{n\kappa}(r)\chi_{-\kappa m}(\theta,\phi) \end{bmatrix}, \tag{2}$$

where κ is the (relativistic) angular momentum quantum number, and $P_{n\kappa}(r)$ and $Q_{n\kappa}(r)$ are called the large and small components. The functions $\chi_{\pm\kappa m}(\theta,\phi)$ are *spinor spherical harmonics* which behave under rotations in space similarly to the well-known spherical harmonics $Y_{lm}(\vartheta,\varphi)$ in the non-relativistic case, if the orbital angular momentum operator $\mathbf{l}$ is replaced by the (one-electron) total angular momentum $\mathbf{j} = \mathbf{l} + \mathbf{s}$.

In the multiconfiguration Dirac–Fock model, symmetry-adapted CSF with well-defined parity P and angular momentum J, M are typically applied. They can be represented also in terms of Slater determinants from the same *electron configuration*

$$|\gamma_r PJM\rangle = \sum_p B_{rp}|\Psi_p\rangle, \tag{3}$$

but are usually constructed from the coupling of (pre-antisymmetrized) subshell states $|n\kappa^q v J\rangle$ of proper symmetry. In the MCDF model, moreover, the atomic state wavefunctions are taken as linear combination of CSF of the same symmetry (PJM)

$$\big|\psi_\alpha(PJM)\big\rangle = \sum_{r=1}^{n_c} c_r(\alpha)|\gamma_r PJM\rangle, \tag{4}$$

where n_c is the number of CSF and $c_r(\alpha)$ are the mixing coefficients in that basis. For all practical computations, the (proper) choice of the many-electron CSF basis finally determines to which extent electron–electron correlations are taken into account. A systematically enlarged basis of such CSF are often used to distinguish between different *computational models* within the MCDF framework [11].

The energies of the atomic levels with symmetry J^P are obtained by diagonalizing the Hamiltonian matrix (1) in the given basis. Further relativistic corrections to the level structure can be considered by utilizing the Dirac–Coulomb–Breit matrix or by adding some suitable (one-electron) estimates from QED. For neutral neon, however, these relativistic corrections only result in small shift of a few tens cm^{-1} and, thus, can be neglected when compared with missing correlations.

2.2. CALCULATION OF THE TRANSITION PROBABILITIES

Following Fermi's *Golden rule*, the transition probability for a radiative decay from initial state i to final state f is given by

$$A_{fi} = \frac{2\pi}{2j_i + 1} \sum_{M_i, M_f} \sum_{L} |M_{fi}^{(L)}|^2, \tag{5}$$

where one has to average over the initial states of level i and to sum over the final states of f. An additional summation runs, in principle, over all the *multipole* components L of the electro-magnetic field but, since higher-order components of the field are strongly suppressed, only the lowest order $L_{\min}$, which is allowed, need usually to be taken into account. The transition matrix elements

$$\begin{aligned} M_{fi}^{(L)} &\equiv \big\langle \psi_f(P_f J_f M_f) \big| O^{(L)} \big| \psi_i(P_i J_i M_i) \big\rangle \\ &= \sum_{r,s} c_r(f) c_s(i) \langle \gamma_r P_f J_f M_f | O^{(L)} | \gamma_s P_i J_i M_i \rangle \end{aligned} \tag{6}$$

are reduced to the according transition matrix within the CSF basis. Different computational procedures have been developed to calculate this matrix for symmetry-adapted functions which are built from two incomplete orthogonal orbital sets. In our programs, we make use of the expansion (3) to represent the CSF in terms of Slater determinants as well as of Lowdin's expressions [16]

$$\langle \gamma_r P_f J_f M_f | O^{(L)} | \gamma_s P_i J_i M_i \rangle = \sum_{p,q} \sum_{k,l} B_{rp} B_{sq} \langle \varphi_k | O^{(L)} | \varphi_l \rangle D_{pq}(kl) \tag{7}$$

for the *computation* of the matrix elements. In this representation, $D_{pq}(kl) = det\{d_{pq}(kl)\}$ denotes the determinant of the (one-electron) *overlap* matrix elements $d_{kl} = \langle \varphi_k | \varphi_l \rangle$ from which the pth row and qth column has been deleted. The one-particle matrix elements for the interaction with the radiation field

$$\langle \varphi_k | O^{(L)} | \varphi_l \rangle = \left(\frac{(2j_l + 1)\omega}{\pi c} \right)^{1/2} (-1)^{j_l - 1/2} \begin{pmatrix} j_k & L & j_l \\ \frac{1}{2} & 0 & -\frac{1}{2} \end{pmatrix} \overline{M}_{kl} \tag{8}$$

can be reduced further by the Wigner–Eckart theorem and written in terms of the radial integrals $\overline{M}_{kl}$ which need to be distinguished for the different multipoles (E1, M1, E2, ...) and gauges of the radiation field. Usually two different gauges, namely the Babushkin and Coulomb gauges are considered which, in the non-relativistic limit, correspond to the length and velocity gauges. They may help obtain insight into the accuracy of the calculation, although the gauge invariance of the results is itself a *necessary but not sufficient condition* to draw conclusions about the (physical) convergence of the results.

An efficient program for the computation of *relaxed-orbital* transition probabilities has been developed recently by us within the framework of the RATIP package [22] which now supports wave functions expansion of several hundred thousand determinants even at standard PC's.

2.3. RELAXATION AND CORRELATION EFFECTS

Relaxation effects on the transition probabilities arises for inner-shell processes but also optical transitions. While these effects are rather straightforward to take into account in the wave function generation (just by carrying out a separate optimization for each of the levels), transition probability calculations require some more care as outlined in the previous subsection. These requirements also determine how the computations ought to be organized: They are often feasible only, if the levels are arranged in several groups according to their symmetry, i.e. their total angular momentum and parity. In the present study, we divided the levels into four groups, one for the $2s^22p^6$ $J = 0$ ground state and three other groups for the $J = 0$, 1, 2 levels of the first excited $2s^22p^53s$ configuration, respectively. For these groups, separate computations were carried out through all the steps in our serious of approximations (see below), using the (extented) optimal level model of the GRASP92 package [9]. To display, for instance, the influence of the relaxation effects on the low-lying levels of atomic neon explicitly, Table I shows the *mean orbital radius* $\langle r \rangle$ of the spectroscopically occupied orbitals for the $2s^22p^6\ {}^1S_0$ ground state level and the two $2p^53s\ {}^{1,3}P_1$ excited levels. As seen from this table, the mean orbital radii for the ${}^{1,3}P_1$ levels are always smaller, when compared with the ground-state orbitals, and the differences become enhanced for the $2p_{1/2,3/2}$ subshells. For these two (valence) subshells, the differences are remarkable 9.5% and 9.8%, compared to the corresponding ground-state orbitals. Thus, it is obvious that these relaxation effects in the electron density cannot be neglected, in particular, if the decay probabilities are considered for weak lines.

A serious of steps have been carried out to analyze the influence of electron–electron correlations, using the active space method. In this method, virtual excitations of electrons from the spectroscopically occupied into unoccupied (sub-) shells are taken into account systematically. Typically, different *classes* of excitations are considered for different (layers of) shells. In the present computations, we incorporated all virtual single (S), double (D) and triple (T) excitations from

Table I. Mean orbital radius (in atomic unit) of the occupied orbitals of the $^1S_0$ ground-state and the $^{1,3}P_1$ excited levels of atomic neon

Orbitals	Ground state $2p^6\ ^1S_0$	Excited states $2p^53s\ ^{1,3}P_1$
$1s$	0.1583	0.1553
$2s$	0.8882	0.8580
$2p_{1/2}$	0.9686	0.8761
$2p_{3/2}$	0.9755	0.8793

Table II. Numbers of CSFs in the various correlation models

Reference configuration	Correlation model				
	J^P	SC	3SDT	3SDT4SD	3SDT5SD
$2s^22p^6 + 2s^22p^43p^2$	0^+	10	2390	7539	20493
$2s^22p^53s$	2^-	1	2521	4977	11117
	1^-	2	2006	3916	8371
	0^-	1	766	1487	3116

the $2s$, $2p$, and $3s$ shells into the (unoccupied) $3l$ shells as well as all single and double excitations into the $4l$ and $5l$ shells. Below, we refer to these models by the short notation 3SDT, 3SDT4SD and 3SDT5SD, respectively. The helium-like $1s^2$ core has been kept *frozen* in all computation, i.e. *core–core* correlations are not included. Table II list the number of CSF in the wave function expansion (4) of the various groups. Even though these expansions could be easily increased to incorporate excitations also into the $6l$ layer, we will show below that the present approximation includes the most important correlations and, overall, leads to a good agreement with evaluated data.

3. Results and discussions

In the study of transition probabilities and lifetimes, of course, we must first analyze the theoretical $2p^6$–$2p^53s$ excitation energies and those correlation contributions which (might) affect them. In Table III, the excitation energies are given for the four levels of the $2p^53s\ ^1P_1$ and $2p^53s\ ^3P_{0,1,2}$. All of them are relative to the ground state level $2p^6\ ^1S_0$. Results are listed for all four correlation models as explained above. Any single-configuration (SC) calculation for these levels only gives a rough estimate of the energies with deviations from experiment of $\sim$1 eV,

Table III. Excitation energies (in cm^{-1}) of the $2p^5 3s\ ^1P_1$ and $2p^5 3s\ ^3P_{0,1,2}$ levels in different correlation models

Correlation model	$^1P_1$	$^3P_0$	$^3P_1$	$^3P_2$
SC	142 485.20	141 526.02	141 174.12	140 710.45
3SDT	139 953.27	138 936.66	138 524.27	138 140.41
3SDT4SD	135 575.19	134 520.39	134 160.67	133 693.63
3SDT5SD	135 968.92	134 939.59	134 567.14	134 109.75
Experiment [13]	135 888.71	134 818.64	134 459.28	134 041.84

which is (typically) of no help for the experimental analysis. These deviations are clearly reduced, however, if more and more (virtual) excitations are taken into account. In the 3SDT5SD approximation, we obtain excitation energies in good agreement with experiment; here, the deviations are about 0.08% or even less. From the comparison of the 3SDT, 3SDT4SD and 3SDT5SD approximations, it is seen that at least the (virtual) SD excitations into the $4l$ layer are not negligible if accurate excitation energies are to be obtained.

In Table IV, we list the lifetimes $\tau = (1/A_{fi})$ of the $^1P_1$ and $^3P_1$ excited levels, as obtained in the different computational models, and compare them with previous calculations and experiments. These lifetimes are shown in length and velocity gauges to support further and more extended investigations in the future. When compared with the excitation energies, Table IV shows a quite different pattern for the lifetime results of the $^1P_1$ and $^3P_1$ levels in the four models. In going from the 3SDT to the 3SDT5SD, approximations, the lifetime of the $^1P_1$ level increases if more correlation contributions are included, and finally is very close to the data from NIST [13]. Compared with the previous computations (in a variety of different theoretical methods), our results also agree much better for the two gauge forms. For the $^3P_1$ level with longer lifetime, on the other hand, its lifetime decreases if more correlation contributions are included in going from the 3SDT to the 3SDT5SD approximations. For these lifetimes, the excitations into the $5l$ layer are still rather important even though their contribution to the level energy can – almost – be neglected. Compared with the previous computations, our result agrees very good with the latest evaluated data from NIST [13] and the experiment of Zhong *et al.* [6]. Of course, due to a high sensitivity of this lifetime to correlation effect, the present calculation for this level is not so good as the result for the $^1P_1$ level.

Table V, finally, lists the calculated lifetimes of the metastable $^3P_2$ and $^3P_0$ levels which are determined by the (forbidden) $2p^5 3s\ ^3P_2$–$2p^6\ ^1S_0$ (M2) and $2p^5 3s\ ^3P_0$–$2p^5 3s\ ^3P_1$ (M1) transitions, respectively. It also shows the results from earlier computations and experiment. In particular for the $^3P_2$ level, there are still large deviations between (previous) experimental and theoretical results. A lower limit

Table IV. Lifetimes ($1/A_{fi}$) of the $^1P_1$ and $^3P_1$ excited levels (in 10^{-12} sec), compared with previous computations and experiments

Authors and methods	$^1P_1$		$^3P_1$	
	Length	Velocity	Length	Velocity
Present calculations				
SC	1792.404	2029.868	20413.601	22441.780
3SDT	1639.635	1537.885	21194.293	19245.804
3SDT4SD	1681.451	1629.204	20107.484	19182.830
3SDT5SD	1686.935	1641.956	19135.870	18361.548
Previous calculations				
Avgoustoglou *et al.* [4]*	1528.654	1674.240	15422.739	16114.786
Hibbert *el al.* [20]*	1515.653		20225.333	
Aymar *et al.* [21]*	1512.828	1873.580	20559.636	24877.159
Experiments				
NIST Database Data [13]	1636.6		21008.4	
Zhong *et al.* [6]*	1561 ± 90		20062 ± 6148	
Curtis *et al.* [5]	1470 ± 100		29600 ± 1000	
Lawrence *et al.* [7]	1870 ± 180		31700 ± 1600	
Kernahan *et al.* [8]	1300 ± 100		29800 ± 2000	

*As obtained from the oscillator strength in [4, 6, 20, 21].

of this lifetime was obtained by van Dyck *et al.* [19] three decades ago by using beam-of-flight techniques. They obtained a value of 0.8 second for neutral neon. Later, a calculation by Stedman and Setser [2] for the $^3P_2$ level was carried out, but their result was larger than 2 times of van Dyck's lower limit. The further calculation of Small-Warren and Chiu [3] for the $^3P_2$ and $^3P_0$ levels had given even ten times of Stedman and Setser's results [2]; Our present calculations are (reasonable) close to the values of Small-Warren and Chiu and are also found less sensitive from correlation effects. For the lifetime of the $^3P_0$ level, there is also a competitive E2 transition to the $^3P_2$ level, but when compared with the dominant M1 lines, this decay channel is suppressed by about 5 orders of magnitude for neutral neon.

4. Conclusion

In summary, the energies of the $2p^6\ ^1S_0$ and $2p^53s\ ^{1,3}P_J$ levels and the probabilities of the $^{1,3}P_1$–$^1S_0$ (E1), $^3P_2$–$^1S_0$ (M2), $^3P_0$–$^3P_1$ (M1) and $^3P_0$–$^3P_2$ (E2) transitions of neutral neon have been calculated systematically using the Multi-configuration Dirac–Fock (MCDF) method. As a consequence, the lifetimes of

Table V. Comparison of lifetimes (in sec) of the metastable $^3P_2$ and $^3P_0$ levels with previous results

Authors	$^3P_2$	$^3P_0$
These calculations		
SC	20.1	432.4
3SDT	20.2	428.8
3SDT4SD	19.4	431.4
3SDT5SD	18.9	433.3
Previous calculations		
Small-Warren and Chiu [3]	24.4	430.0
Stedman and Setser [2]	2.0	
Previous experiment		
Dyck *et al.* [19]	0.8	

all the four low-excited $^{1,3}P_J$ levels are presented. Comparison is made for our theoretical results with previous computations and measurements; they are found to be in good agreement with recently evaluated data.

In this study, emphasis has been paid also for the effects of relaxation and correlations on the level energies and lifetimes of the low-excited $^{1,3}P_J$ levels. Our study shows that (1) a serious rearrangement (relaxation) of the electron density occurs when neon undergoes a transition from the stable $2p^6$ ground state to a state $2p^53s$ with two open shells. As a result, it is necessary to consider this effect in calculation of the transition probability. For calculation of the excited energies, moreover, (2) strong correlation contributions arise for the 3SDT and 3SDT4SD approximations; for weak lines such as the $^3P_1$–$^1S_0$ intercombination line even excitations into the $5l$ layer 3SDT5SD need to be taken into account.

Acknowledgement

This work has been supported by the National Natural Science Foundation of China (Grant No. 10274062), the Excellent Young Teachers Program of MOE, P.R. China, and Foundation of the Center of Theoretical Nuclear Physics, National Laboratory of Heavy Ion Accelerator, Lanzhou.

References

1. Burkhalter, P. G., Nagel, D. J. and Cowan, R. D., *Phys. Rev. A* **11** (1975), 782.
2. Stedman, D. H. and Setser, D. W., *Prog. React. Kinet.* **6** (1971), 193.
3. Small-Warren, N. E. and Chiu, L. C., *Phys. Rev. A* **11** (1975), 1777.
4. Avgoustoglou, E. N. and Beck, D. R., *Phys. Rev. A* **57** (1998), 4286.

5. Curtis, L. J., Maniak, S. T., Christ, R. W. *et al.*, *Phys. Rev. A* **51** (1995), 4575.
6. Zhong, Z. P., Wu, S. L., Feng, R. F. *et al.*, *Phys. Rev. A* **55** (1997), 3388.
7. Lawrence, G. M. and Liszt, H. S., *Phys. Rev.* **178** (1969), 122.
8. Kernahan, J. A., Denis, A. and Drouin, R., *Phys. Scr.* **4** (1971), 49.
9. Parpia, F. A., Froses Fischer, C. and Grant, I. P., *Compt. Phys. Commun.* **94** (1996), 247.
10. Fritzsche, S., Froses Fischer, C. and Dong, C. Z., *Compt. Phys. Commun.* **124** (2000), 340.
11. Dong, C. Z., Fritzsche, S. and Fricke, B., *J. Electr. Spec. Rel. Phenon.* **114–116** (2001), 157.
12. Dong, C. Z. *et al.*, in preparation.
13. NIST Atomic Spectra Database Data, http://physlab2.nist.gov/cgi-bin/AtData, 2001.
14. Grant, I. P. and Quiney, H. M., *Adv. At. Mol. Phys.* **23** (1987), 37; Grant, I. P., Mckenzie, B. J. and Norrington, P. H., *Compt. Phys. Commun.* **21** (1980), 207; Grant, I. P., In: G. W. F. Drake (ed.), *Atomic, Molecular and Optical Physics, Reference book*, American Institute of Physics, New York, 1996, p. 258.
15. Fricke, B., *Phys. Scr. T* **8** (1984), 129.
16. Löwdin, P. O., *Phys. Rev.* **97** (1995), 1474.
17. Grant, I. P., *J. Phys. B* **7** (1974), 1458.
18. Fritzsche, S. and Froses Fischer, C., *Compt. Phys. Commun.* **99** (1997), 323.
19. Van Dyck, Jr., R. S., Johnson, C. E. and Shugart, H. A., *Phys. Rev. A* **5** (1972), 991.
20. Hibbert, A., Le Dourneuf, M. and Mohan, M., *At. Data Nucl. Data Tables* **53** (1993), 23.
21. Aymar, M., Feneuille, S. and Klapisch, M., *Nucl. Instrum. Methods* **90** (1970), 137.
22. Fritzsche, S., *J. Electr. Spec. Rel. Phenon.* **114–116** (2001), 1155.

Hyperfine Interactions **146/147:** 171–176, 2003.

Threshold Laws for Four-Particle Fragmentation: Mass Effects

V. N. OSTROVSKY
V. A. Fock Institute of Physics, The University of St Petersburg, 198904 St Petersburg, Russia; e-mail: Valentin.Ostrovsky@pobox.spbu.ru

Abstract. The threshold laws for break-up of an atomic particle into N charged fragments reflect dynamics of long-range Coulomb correlation. Progress in experimental technique makes accessible a broader variety of essential parameters: fragment charges and masses. Some sets of fragments for $N = 4$ case are discussed with particular emphasis on mass-dependence of the threshold behaviour. New sets are found with low values of threshold indices suitable for experimental observation.

Key words: threshold laws, break-up, fragmentation, Wannier law, Coulomb correlation.

The processes of a particle break-up into several fragments are of interest in atomic, molecular and nuclear physics. The threshold behaviour of cross sections σ_{fr} for such a process has very different physical status depending on whether the fragments are neutral or charged. In case of N neutral fragments, when interaction is of short range, the threshold behaviour is deducted from the phase-space, i.e. *statistical* arguments as

$$\sigma_{\text{fr}} \sim E^{(3/2)(N-1)-1}, \tag{1}$$

where E is excess energy above the fragmentation threshold.

If $(N-1)$ fragment ('electrons') are *attracted* by Coulomb forces to one fragment ('core'), but the interaction between the electrons is neglected, then the threshold law reads

$$\sigma_{\text{fr}} \sim E^{N-2}. \tag{2}$$

This result also could be obtained by phase-space arguments. In case of Coulomb *repulsion* the fragmentation cross section is exponentially small

$$\sigma \sim \exp(-\alpha/\sqrt{E}), \tag{3}$$

with some constant α. However the threshold laws (2), (3) are incorrect for $N \geqslant 3$ since in fact *all* pairwise interactions between charged fragments are to be taken into account on the same footing and none of them could be neglected in the near-threshold domain. The reason lies, first, in the long-range nature of Coulomb interaction, and, second, in the fact that in the near-threshold regime *all* the fragments necessarily recede slowly, sharing small excess energy E. Therefore they have enough time to be crucially influenced by the long range forces. These circumstances make threshold laws in case of charged fragments* issue of *dynamics* but not statistics, as in case (1). The functional form of threshold laws bears signature of *strong (non-perturbative) dynamic correlations* that makes analysis of these laws particularly interesting.

The breakthrough theoretical study of the subject is due to Wannier [1] (1953) who considered $N = 3$ case. Wannier established his famous power threshold law**

$$\sigma_{\mathrm{fr}} \sim E^{\mu} \tag{4}$$

with non-trivial threshold index

$$\mu = \frac{1}{4}\sqrt{\frac{(100Z_0 - 9Z)m_0 + 128Z_0 m}{(4Z_0 - Z)m_0}} - \frac{1}{4}. \tag{5}$$

In fact the latter expression presents generalization of Wannier's result due to Klar [2]: two identical fragments with masses m and charges Z recede from the 'core' with charge Z_0 and mass m_0. Originally Wannier considered two electrons ($m = 1$, $Z = 1$) and infinitely massive core, $m_0 = \infty$; atomic units are used throughout the paper. Note that an analytical expression for μ is not available for the case when all three fragments are different and have finite masses.

Two years later Wannier published another, very short paper [3] (less than one journal page) where he discussed multiple ionization (i.e. $N > 3$ case) and indicated that crudely the threshold law (2) holds. Further he noticed that

> One can reason by analogy to single ionization that the true exponent is probably somewhat larger than $(N - 2)$

(cast in the present notations). The quantitative measure of Wannier's *somewhat* is given by

$$\delta_W = \mu - (N - 2). \tag{6}$$

This parameter encloses basic information about long-range correlation in multiparticle Coulomb systems being subject of a number of subsequent research, including the present one.

The first case study was carried out by Klar and Schlecht [4] and Grujić [5] who considered escape of three electrons from the charged core ($N = 4$). The comprehensive bibliography could be found in [6] and the most recent papers are

* The presence of neutral fragment results in extra statistical factor $E^{3/2}$ in the formula expressing threshold behaviour, in agreement with Equation (1).

** As usual, the theory of threshold laws establishes functional dependence of cross section on energy. Evaluation of the proportionality coefficient in (4) as well as estimates of the energy range where the law holds is beyond its scope, being considered as separate, and usually quite difficult problems.

cited in [7]. As discussed in detail in these papers, the progress of theory was not smooth and some approaches were questioned. Below we base on the general framework developed by Kuchiev and Ostrovsky [6] applicable to arbitrary number of fragments. Briefly, at first one has to find the *scaling configuration* (SC) in which classical trajectories of fragments correspond plainly to time-dependent blow-up of initial configuration. SCs are always unstable under a small perturbation. The characteristics of instability are related to the index μ in the threshold law of power form (4). The scheme allows us to calculate the threshold index μ for given N and particles masses m_j and charges Z_j, although an analytical expression generally cannot be provided. Applicability to any set of m_j and charges Z_j is essential, since progress in experimental technique broadens spectrum of accessible processes. Note also that the magnitude of threshold index depends only on the charges and masses of fragments in the final state, but not on the initial collision partners. Nevertheless sometime we discuss also incident channels in order to demonstrate that the fragments under consideration could be created in realistic reactions.

For a long time atomic physics studies two 'traditional' fragmentation processes, multiple photoionization

$$\gamma + \mathrm{A} \rightarrow \mathrm{A}^{+n} + ne^- \tag{7}$$

and multiple ionization by electron impact

$$e^- + \mathrm{A} \rightarrow \mathrm{A}^{+n} + (n+1)e^-. \tag{8}$$

Later ionization by positron impact also became subject of active research

$$e^+ + \mathrm{A} \rightarrow \mathrm{A}^{+n} + ne^- + e^+. \tag{9}$$

New development of accelerator facilities make realistic experiments with particles 'exotic' to atomic physics:

$$\mu^-(p^-, \ldots) + \mathrm{A} \rightarrow \mathrm{A}^{+n} + ne^- + \mu^-(p^-, \ldots). \tag{10}$$

From the point of view of threshold indexes evaluation these processes has some specifics: there are two heavy fragments and a number of light ones, whereas the processes (7), (8), (9) produce single heavy fragment. Other processes with two or more heavy fragments involve molecules in incident channel, for instance,

$$\mathrm{C}^+ + \mathrm{AB} \rightarrow \mathrm{C}^+ + \mathrm{A}^+ + \mathrm{B}^+ + 2e^-, \tag{11.1}$$

$$\mathrm{C}^+ + \mathrm{AB} \rightarrow \mathrm{C}^+ + \mathrm{A}^+ + \mathrm{B}^-, \tag{11.2}$$

$$\mathrm{C} + \mathrm{AB} \rightarrow \mathrm{C}^+ + \mathrm{A}^+ + \mathrm{B}^- + e^-, \quad \text{etc.} \tag{11.3}$$

Efficient fragmentation of molecules by photon or electron impact, for example,

$$\gamma + \mathrm{AB} \rightarrow \mathrm{A}^+ + \mathrm{B}^+ + 2e^-, \tag{12.1}$$

$$e + \mathrm{AB} \rightarrow \mathrm{A}^+ + \mathrm{B}^+ + 3e^- \tag{12.2}$$

is usually achieved in two steps. At first a molecule is doubly ionized via 'vertical transitions', i.e. without chaging positions of atomic nuclei. After that the doubly charged molecular ion dissociates. The threshold for such a mechanism ('adiabatic threshold') lies much higher than the energetic threshold considered in the present paper. For energies between energetic and adiabatic threshold efficiency of fragmentation is very low and experimental observation of break up is difficult. For heavy particle collisions (11) observation of threshold behaviour looks more realistic.

We indicate also a number of processes which lead to four fragments of equal masses, with two pairs of positive and negative charges

$$\mathrm{Ps} + \mathrm{Ps} \rightarrow 2e^- + 2e^+, \tag{13.1}$$

$$e^+ + \mathrm{Ps}^- \rightarrow 2e^- + 2e^+, \tag{13.2}$$

$$\gamma + \mathrm{Ps}_2 \rightarrow 2e^- + 2e^+, \tag{13.3}$$

or with one positively and three negatively charged particles:

$$e^- + \mathrm{Ps}^- \rightarrow 3e^- + e^+. \tag{14}$$

It is important to stress that the threshold indexes do not change when all fragment masses are scaled or all fragment charges are scaled. Thus only *ratios* of masses and ratios of charges are important as exemplified by formula (5) for a particular $N = 3$ case. The situation was well illustrated by Klar [2] who in his table cited identical value $\mu = 1.886001$ for three sets of fragments: $2e^- + e^+$, $2e^+ + e^-$ and $2\mathrm{H}^+ + \mathrm{H}^-$. The scaling property means, in particular, that the process

$$\mathrm{A}_2 + \mathrm{A}_2 \rightarrow 2\mathrm{A}^+ + 2\mathrm{A}^- \tag{15}$$

has essentially the same threshold index (within m_e/m_A precision) as the processes (13).

In the systems $\{^\infty\mathrm{A}^{+Z} + (N-1)e^-\}$ (where $^\infty\mathrm{A}^{+Z}$ is infinitely massive core with the charge Z) the threshold index usually is quite close to $N-2$, i.e. Wannier's [3] *somewhat* (6) is small. For instance, one has $\mu = 1.127$ for $^\infty\mathrm{A}^+ + 2e^-$ fragments [1], $\mu = 2.270$ for $^\infty\mathrm{A}^{++} + 3e^-$ fragments [4, 5] $\mu = 3.419$ for $^\infty\mathrm{A}^{+3} + 4e^-$ fragments [8]. The difference δ_W (6) becomes anomalously large with decreasing core *charge* Z, but these sets of fragments could not be produced in realistic reactions [6].

Now we turn to discussion of the *mass* effects. When the '*core*' mass is diminished, the difference δ_W might become large. For $N = 3$ case this was demonstrated by Klar [2]: $\mu = 1.1274$ for $\mathrm{H}^+ + 2e^-$ fragments, $\mu = 1.8860$ for $e^+ + 2e^-$ fragments, $\mu = 69.735$ for $e^- + 2\mathrm{H}^+$ fragments. In the latter case the electron plays a role of the *core*, with two protons receding from it; the mass effects are responsible for huge value of μ. For $N = 4$ a similar situation is observed [7]. The effect of conventional isotopic substitution is small: compare $\mu = 2.8274$ for $\mathrm{H}^+ + 3e^-$ fragments with $\mu = 2.8268$ for $\mathrm{D}^+ + 3e^-$ fragments. However,

Table I. Parameters of scaling configuration (SC) and threshold indices for four-particle systems with a pair of identical particles. Between two other particles, one has a positive and the other a negative charge

Fragments	SC parameters		μ_1	μ_2	μ
	α	β			
$e^+ + 2H^+ + e^-$	29.53°	102.87°	72.15	27.05	99.20
$\mu^+ + 2H^+ + e^-$	30.90°	107.22°	70.27	4.147	74.16
$\pi^+ + 2H^+ + e^-$	31.06°	108.21°	69.70	4.497	74.20
$3H^+ + e^-$ [7]	30°	120°	56.93	56.93	113.86
$e^- + 2H^- + \mu^+$	29.33°	103.20°	19.99	5.196	25.18
$e^- + 2H^- + \pi^+$	29.28°	103.30°	19.89	4.567	24.45
$e^- + 2H^- + H^+$	28.14°	105.33°	18.79	2.045	20.83
$e^- + 2H^- + He^+$	26.15°	109.0°	17.29	1.468	18.76
$e^- + 2H^- + He^{++}$	10.78°	92.40°	1.311	1.265	2.575
$e^- + 2H^- + Li^{++}$	10.63°	93.03°	1.301	1.18	2.481
$e^- + 2H^- + Li^{+3}$	6.42°	90.92°	1.153	1.016	2.169

Table II. Parameters of scaling configuration (SC) and threshold indices for four-particle systems with a pair of identical particles. Two other particles have charges with the sign opposite to the charges of identical particles

Fragments	SC parameters		μ_1	μ_2	μ
	α	β			
$2e^- + 2H^+$ [6]	59.995°	59.995°	50.33	37.46	87.79
$\mu^- + 2H^+ + e^-$	53.00°	65.57°	50.22	2.417	52.63
$\pi^- + 2H^+ + e^-$	48.45°	69.12°	53.80	1.846	55.64

larger changes in the mass ratio lead to drastic increase of the threshold index: $\mu = 4.2218$ for the process (14) (essentially the same threshold index applies to $H^- + 3H^+$ fragments), and $\mu = 113.86$ for $e^- + 3H^+$ fragments.

Some examples of systems with *two or more heavy particles* were considered previously [6, 7], our Tables I and II present new results. Besides the threshold index μ we cite also the partial indices μ_i ($\mu = \sum_i \mu_i$) and SC parameters α and β (or exact definitions and detail see [6, 7]).

The cases of small threshold indices are particularly interesting being easier for experimental observation. There are two such examples in the bottom of Table I. For these processes SC parameter α is small whereas β approaches 90° from above. The meaning of this trend becomes clear if one recalls the definitions of angles α

and β [7]: three heavy particles tend to lie almost on the same line, whereas the light particle (electron) is situated far from them. The partial index μ_1 reflects interaction of light particle with complex of three heavy particles. If the net charge of the complex is negative, then the interaction is repulsive and the threshold index μ_1 is large mimicking the exponential threshold law (3). If the net charge is positive, then the interaction is attractive and μ is small, being somewhat higher than 1 that mimics the threshold law (2).

Large values of μ mean that the fragmentation cross section is very small near threshold that may prohibit experimental test. However even large threshold indices are not of purely academic interest, since they could be useful for interpolation formulae [9].

Note that the break up into $e^-+2H^-+Li^{++}$ fragments is characterized by essentially the same (within m_e/m_p ratio) threshold index as collision with antiproton: $p^-+LiH \rightarrow e^-+p^-+H^-+Li^{++}$. In principle such a heavy particle as antiproton could be replaced by negative ion H^- essentially without changing the threshold index. However, the negative ions are fragile. Therefore one can anticipate that production of H^- in the collisions is suppressed. In terms of threshold law (4) this means small value of constant prefactor in the right hand side. This complicates experimental observation.

As a summary, threshold laws for break up of atomic particle into several charged fragments has form of power dependence (4) of cross section on the excess energy E. The threshold indices μ assume very different values depending on the fragment charges and masses. The theory presents challenges for experimental verification.

Author's participation in APAC2001 was made possible by a grant from the European Union's INCO program.

References

1. Wannier, G. H., *Phys. Rev.* **90** (1953), 817.
2. Klar, H., *Z. Phys. A* **307** (1982), 75.
3. Wannier, G. H., *Phys. Rev.* **100** (1955), 1180.
4. Klar, H. and Schlecht, W., *J. Phys. B* **10** (1976), 1699.
5. Grujić, P., *J. Phys. B* **16** (1983), 2567.
6. Kuchiev, M. Yu. and Ostrovsky, V. N., *Phys. Rev. A* **58** (1998), 321; *J. Phys. IV* **9**(P6) (1999), 51.
7. Ostrovsky, V. N., *Phys. Rev. A* **64**(1–5) (2001), 022715; *Few-Body Systems* **31** (2002), 113.
8. Grujić, P., *Phys. Lett. A* **96** (1983), 233; **122** (1987), 491.
9. Rost, J. M. and Pattard, T., *Phys. Rev. A* **55** (1997), R5; *J. Phys. B* **32** (1999), L457.

Hyperfine Interactions **146/147:** 177–181, 2003.

Subshell Resolved Measurements of Single Electron Capture in Slow Ne^{7+}–Helium Collisions

D. FISCHER[1], B. FEUERSTEIN[1], R. MOSHAMMER[1], J. R. CRESPO LÓPEZ-URRUTIA[1], I. DRAGANIC[1], H. LÖRCH[1], A. N. PERUMAL[1], J. ULLRICH[1] and R. D. DUBOIS[2]

[1] *Max-Planck-Institut für Kernphysik, Saupfercheckweg 1, D-69117 Heidelberg, Germany*
[2] *University of Missouri, Rolla, USA*

Abstract. Single electron capture in collisions of 9 keV/q Ne^{7+}-ions with He has been studied using cold-target recoil-ion momentum spectroscopy (COLTRIMS). With an improved apparatus a longitudinal momentum resolution of 0.07 a.u. has been achieved. This momentum component corresponds to the binding energy of the active electron in the final state. For the first time state-resolved differential cross sections have been determined with respect to the main quantum number, subshell level and spin state of the captured electron.

Single electron capture is the most dominant reaction channel when highly charged ions collide with neutral atoms or molecules at low or intermediate energies ($E \approx 1$ keV/amu). Over the decades detailed investigations of capture reactions contributed significantly to our understanding of ion–atom interactions with considerable relevance for several research fields like, e.g., plasma and astrophysics as well as for the development of novel particle acceleration and trapping technologies. The recently achieved and still ongoing decisive progress concerning both, the production and acceleration of highly charged ions (HCI) and the analysis of reaction products emerging from ion–atom collisions resulted in kinematically complete experiments with low-energy ion-beams at charge states up to 40+ and higher. With EBIT (electron-beam-ion-trap) sources and with the so-called HITRAP project at GSI it is anticipated to fully ionize even the heaviest elements and to produce low-energy beams with unprecedented intensity and quality for collision experiments. Using cold target recoil-ion momentum spectroscopy (COLTRIMS) in combination with highly efficient electron spectrometers (so called REACTION MICROSCOPES) it is possible to measure in kinematically complete experiments the momentum vectors of all emerging particles from a single collision event with high resolution [1]. Another very recent development concerns the usage of ultra-cold atoms in a magneto-optical-trap (MOT) as a target for collisions with low-charged projectiles resulting in unprecedented momentum resolutions for the recoiling target ions [2].

The COLTRIMS method is based on the precise determination of the recoil momentum vector of charged target ions emerging from the capture reaction. For small projectile scattering angles the longitudinal momentum transfer, i.e. the recoil momentum along the projectile beam axis, depends only on the difference between the binding energies of the active electron in the initial and final state, i.e. the Q-value of the reaction $P_{\|} = Q/V_{\mathrm{p}} - V_{\mathrm{p}}/2$ [1], where V_{p} is the projectile velocity. Thus, the longitudinal momentum transfer allows to separate different shells populated in the capture reaction. The transverse recoil-ion momentum, on the other hand, reflects the projectile scattering angle $P_{\perp} = \vartheta \cdot P_0$. The incoming projectile momentum is P_0. Hence, the full information about the collision dynamics and about populated electronic states is accessed via the precise measurement of the recoil-ion momentum vector.

In this report we present results obtained with Ne^{7+}-ions extracted from the former Freiburg EBIT [3] at an energy of 9 keV/q directed into a COLTRIMS apparatus. After passing a well localized and cold beam of He-atoms provided by a three-stage supersonic gas jet those projectiles, which have captured an electron, are guided through a charge-state resolving cylindrical-plate spectrometer and detected by a position-sensitive channelplate. The COLTRIMS spectrometer used is very similar to those, which have been described earlier [4]. The recoil ions are extracted along the projectile beam direction. They are accelerated over a distance of 11 cm by a weak electric field (1–5 V/cm) generated between two ceramic plates covered with resistive layers. To fulfil the time-focusing condition the recoil ions drift after acceleration over 22 cm through a field free region and are measured by a position sensitive detector, which is placed just beneath the projectile beam axis. The He^{1+} momentum vector is deduced from the position on the recoil detector and the time-of-flight measured in a coincidence between the recoil ion and the corresponding charge changed projectile.

The finally reachable momentum resolution is limited mainly by two factors. The major contribution comes from the thermal spread of the target, which amounts to $\Delta p = 0.26$ a.u. in the direction along the gasjet axis and $\Delta p = 0.07$ a.u. perpendicular to it. The spatial extension of the reaction volume, which is determined by the overlap between ion beam and jet, reduces the resolution due to the time-focusing geometry only in the direction transverse to the projectile beam. This is the main reason why longitudinal extraction is most suitable for the study of capture reactions because it allows a very high resolution for the determination of the Q-value. With the present setup at an extraction field of 3.6 V/cm a resolution of $\Delta p_{\perp} = 0.3$ a.u. and $\Delta p_{\|} = 0.07$ a.u. was reached. This corresponds to a Q value resolution of $\Delta Q_{\mathrm{FWHM}} = 0.7$ eV.

In order to reach this high resolution any deviation from ideal extraction conditions like, e.g., electric fringe fields or non-homogeneous electric fields for acceleration of the recoil ions should be avoided. But, at large extraction voltages, which are desirable with respect to a large transverse momentum acceptance, a weak vertical component of the electric field is required to push the ions onto the

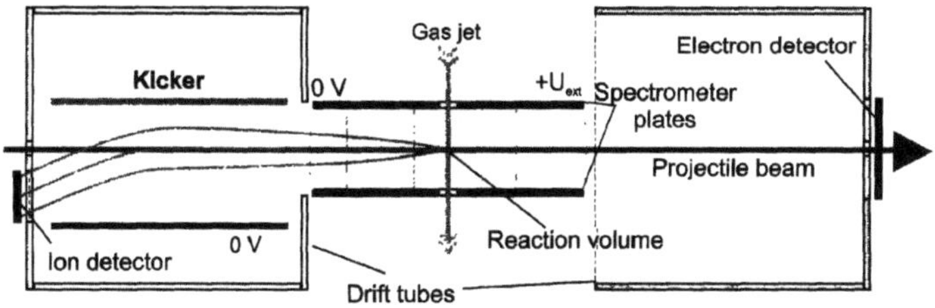

Figure 1. Schematic drawing of the REACTION MICROSCOPE formed by combining a COLTRIMS spectrometer (left part) with an electron analyzer (right part).

off-axis recoil detector. This asymmetric potential gradient is easily generated by applying slightly different voltages to the upper and lower spectrometer plates [4]. However, this technique has various disadvantages because the field is not perfectly oriented along the longitudinal direction resulting in a mixing of transverse and longitudinal momentum components and in a non-acceptable reduction of the momentum resolution.

To circumvent problems arising from asymmetric extraction fields and to achieve the ultimate resolution, the recoil-ion drift tube was supplemented with an electrostatic kicker (Figure 1). This consists basically of two metallic plates where a transverse electric field can be generated by setting the plates on different voltages. Whenever a recoil ion is extracted along the incoming ion beam and resides between the two plates, it gets a kick in the transverse direction towards the recoil detector by applying a short electric pulse. The timing of the kicker pulse is synchronized with the signal from the projectile analyzer, i.e. it is applied in coincidence to a capture reaction. It further ensures that each recoil ion is situated at the same position between the kicker plates when the pulse is starting. In the experiment Gaussian shaped 50 V pulses were used at a pulse width of 1 μs (the total time-of-flight was 11 μs). We like to mention that arbitrarily shaped pulses can be used as long as their duration is considerably shorter than the recoil-ions transit time through the kicker.

With the setup described above the best recoil ion momentum resolution using a supersonic gasjet target has been achieved. Moreover, for the first time finally populated projectile states could be resolved not only with respect to their main quantum number but in addition to different subshell levels and spin states. In Figure 2 the two-dimensional spectrum (scattering angle ϑ versus Q-value) for single electron capture in Ne^{7+}–He collisions is shown. Single capture into Ne^{6+} leads to Be-like final states which give rise to a rich structure in the Q-value spectrum (Figure 3). It consists basically of three groups of lines corresponding to the $2s4l$ and $2s3l$ levels of Ne^{6+}, and to the doubly excited $2p3l$ states. This projectile excitation accompanying the capture reaction has been observed in a recent experiment [5] where different subshells could not be resolved due to limited resolution. Here, in total 22 lines are observed (Figure 3) and in the case of overlapping lines a least-squares fitting procedure has been used to identify the line positions. The

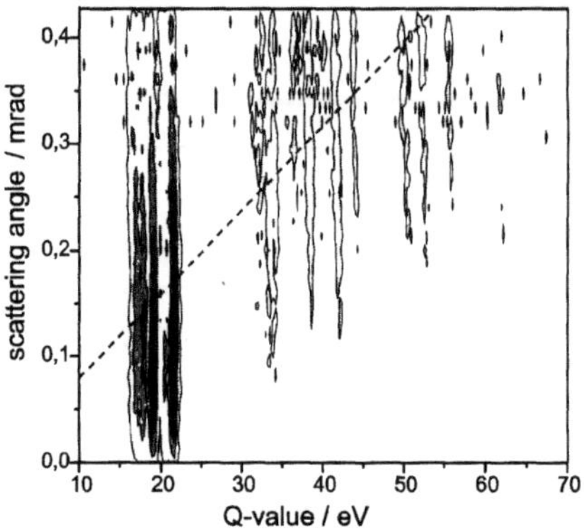

Figure 2. Two-dimensional representation of the projectile scattering angle versus Q-value of the reaction for single electron capture in Ne^{7+}–He collisions. The projectile velocity is 0.355 a.u. ($E = 63$ keV). Dashed line: half-Coulomb angle as function of the Q-value (see text).

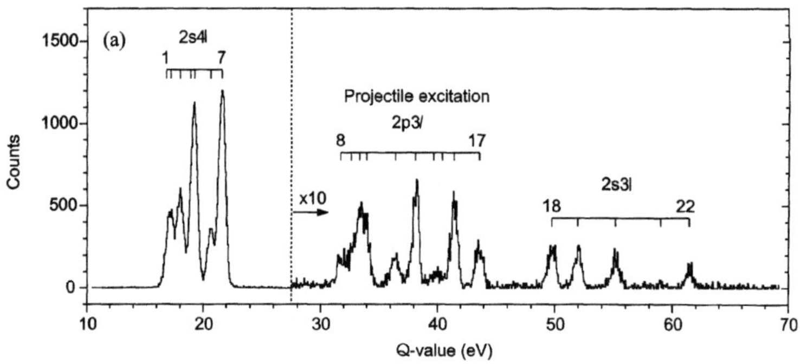

Figure 3. Q-value spectrum for single electron capture into Ne^{7+} from He at 0.355 a.u. projectile velocity. The marks indicate the corresponding level energies of MCHF calculations [8].

calibration of Q-values was performed by means of a linear fit of the experimental raw data (TOF values) to atomic level data from Bashkin and Stoner [6] for the most intense $2s3l$ and $2s4l$ lines. The finally observed level energies, which are precise on a level of 20 meV, are in excellent agreement with the theoretical results except for the $2p3p\ ^{1,3}S$ and the $2p3s\ ^{1}P$ states. For $2p3s\ ^{1,3}S$ states the theoretical values deviate from the present results by more than 0.5 eV. This coincides with differences among several theoretical models [7, 8] which are large in particular for these states.

Since for the present experiment the acceptance of the spectrometer does not cover the full range of scattering angles, relative line intensities are not shown (see Figure 2). Qualitatively, the most dominant channel is capture into the $n = 4$ shell (Figure 3). This is consistent with the curve-crossing model which predicts much larger intermediate crossing radii for the $2s4l$ states than for $2p3l$ and $2s3l$ states, respectively. The decreasing crossing radii lead to an increasing scattering angle as a function of the Q-value (dashed line in Figure 2).

Future experiments with highly charged ions extracted from the EBIT are planned where a REACTION MICROSCOPE will be used to study multi electron exchange accompanied with auto-ionization of dominantly populated highly excited projectile states. With pre-cooling of the gasjet a further increase in resolution by more than a factor of 4 can be anticipated resulting in spectroscopic information on energy levels in highly charged heavy ions not accessible by any other method.

Acknowledgement

This work was supported by the Leibniz Program of the Deutsche Forschungsgemeinschaft.

References

1. Ullrich, J., Moshammer, R., Dörner, R., Jagutzki, O., Mergel, V., Schmidt-Böcking, H. and Spielberger, L., *J. Phys. B* **30** (1997), 2917.
2. Flechard, X., Nguyen, H., Wells, E., Ben-Itzhak, I. and DePaola, B. D., *Phys. Rev. Lett.* **87** (2001), 123203-1.
3. Crespo López-Urrutia, J. R., Bapat, B., Draganic, I., Werdich, A. and Ullrich, J., *Physica Scripta T* **92** (2001), 110.
4. Moshammer, R., Unverzagt, M., Schmitt, W., Ullrich, J. and Schmidt-Böcking, H., *Nucl. Instr. Methods B* **108** (1996), 425.
5. Kamber, E. Y., Abdallah, M. A., Cocke, C. L., Stöckli, M., Wang, J. and Hansen, J. P., *J. Phys. B* **33** (2000), L171.
6. Bashkin, S. and Stoner, J. O., *Atomic Energy-Levels and Grotrian Diagrams*, North-Holland, Amsterdam, 1978.
7. Murakami, I., Safronova, U. I. and Kato, T., *J. Phys. B* **32** (1999), 5331.
8. Buchet-Poulizak, M. C., Bogdanovich, P. O. and Knystautas, É. J., *J. Phys. B* **34** (2001), 233.

Hyperfine Interactions **146/147:** 183–187, 2003.

Vertical Blow-up in a Low-Current, Stored, Laser-Cooled Ion Beam

NIELS MADSEN[1], PAUL BOWE[2], LARS E. SIEGFRIED[2], JEFF S. HANGST[2] and JØRGEN S. NIELSEN[3]
[1] *CERN, Geneva, Switzerland; e-mail: Niels.Madsen@cern.ch*
[2] *Institute of Physics and Astronomy, University of Aarhus, 8000 Aarhus C, Denmark*
[3] *ISA, University of Aarhus, 8000 Aarhus C, Denmark*

Abstract. Using a novel technique for real-time transverse beam profile diagnostics of a stored ion beam, we have observed the transverse size of a stored, laser-cooled ion beam. Earlier we observed that the density of the beam is independent of the beam current. At very low currents we observe an abrupt change in this behavior: the vertical beam size increases suddenly by about an order of magnitude. This observation implies a sudden change in the indirect vertical cooling mediated by intrabeam scattering. Our results have serious implications for the ultimate beam quality attainable by laser-cooling.

Key words: laser-cooling, ion beam, storage ring.

Laser-cooling in a storage ring [1, 2] has been demonstrated to be a powerful tool for creating ultra-cold and dense stored ion beams [8]. Cold and dense beams are of interest for many storage ring applications, and the ultimate state for such beams is the attainment of ion beam crystallization [10].

A single high-speed CCD camera with an image intensifier is used to continuously monitor the transverse beam profiles. We observe one dimension at a time. A simple lens system images the light from the laser-excited ion beam onto the image intensifier. The spatial resolution of our system is approximately 20 μm. In order to produce a spatially flat laser light distribution, the focused (FWHM $\sim$ 1.0 mm) laser beam is actively swept in the desired plane, at a frequency of $\sim$100 Hz. A spatially flat distribution means that all ions are illuminated evenly, thus the charge distribution on the CCD reflects the ion beam density distribution directly [8].

We have used the ASTRID storage ring [9] in Aarhus to store a beam of 100 keV $^{24}Mg^+$ ions. Laser cooling is performed in one straight section, with co- and counter-propagating laser beams, having radii of $\sim$3 mm, overlapping the ion beam. The overlap between the laser beams and the ion beam is carefully optimized for optimum longitunal and transverse cooling [7, 8].

In Figure 1 the transverse beam sizes, extracted from Gaussian fits to the measured distributions are shown as a function of the time after injection. The longitu-

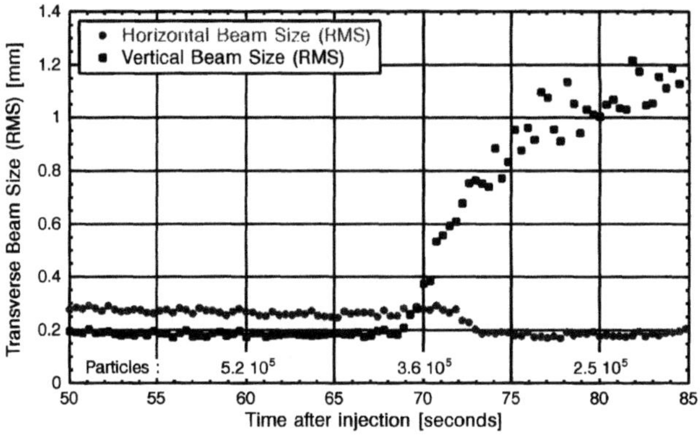

Figure 1. Transverse beam sizes as a function of the time after injection. About 5.0×10^6 particles were injected. Cooling laser powers ~9 mW. Longitudinal temperature ~1 K.

dinal temperature is <1 K. We observe that at 68 s after injection the vertical beam size increases dramatically. The beam lifetime was 26.6 ± 1 s. $(5.0 \pm 0.3) \times 10^6$ particles were injected. There were thus $(3.8 \pm 0.2) \times 10^5$ circulating particles at the start of the blowup. One also observes that the beam size does not schrink below about 0.18 mm. Calculations of the orbit response based on the stability of the dipole-corrector power supplies of ASTRID ($\pm 3 \times 10^{-4}$) indicates that this limitation is due to fluctuations in the beam position in ASTRID.

In considering possible heating sources, we performed a simple IBS calculation based on binary collisions using the program INTRABTC [3] with the ASTRID magnetic lattice and the measured beam parameters right before the blowup. This calculation indicates that the vertical degree of freedom should be cooled with a characteristic time of 94 ms; thus IBS is not the heating source. The horizontal degree of freedom on the other hand should be heated with a characteristic time of 27 ms (the IBS calculation does not include the direct horizontal cooling through dispersion).

A possibly important and generally overlooked source is the heating due to the spontaneous emission of photons during laser-cooling. In order to investigate whether the laser induced diffusion is the source of the heating, we can exploit the fact that the spontaneous emission from the decay of the $(3^2S_{1/2}) \leftrightarrow (3^2P_{3/2})$ transition is spatially non-isotropic. It turns out that the intensity of light emitted perpendicular to the polarization of the cooling laser is a factor 2.5 higher than that emitted parallel to the polarization. Thus by investigating the dependence of the blow-up on the polarization of the cooling lasers we may clarify the source of the heating.

Figure 2 shows two sets of measurements of the transverse beam sizes as a function of the time after injection. The top plot (a) is for both cooling lasers polarized horizontally and the lower plot (b) is for both polarized vertically. We observe that the blowup occurs significantly earlier for the horizontal polarization,

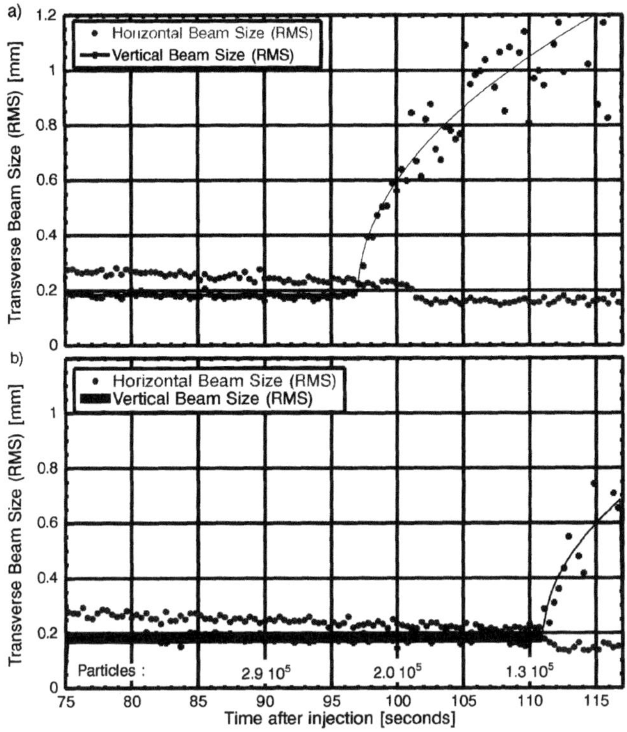

Figure 2. Transverse beam sizes as a function of the time after injection. About 8.4×10^6 particles were injected. Cooling laser powers ~7 mW. (a) Both cooling lasers were horizontally polarized, (b) both cooling lasers vertically polarized.

i.e. at a higher current. Horizontally polarized cooling lasers will cause more spontaneous photons to be emitted vertically and thus more heating vertically. This observation supports the hypothesis that the spontaneous emission is a significant heating source.

With constant diffusion the square of the velocity spread increases linearly with time [12]. The measured diffusion coefficient can therefore be extracted from a square root fit to the blowup, as indicated by the lines in Figure 2. We measure diffusion coefficients in the range $(2.5\pm0.9)\times10^3$ m^2/s^3 to $(3.9\pm0.9)\times10^3$ m^2/s^3.

The diffusion coefficients were measured with a longitudinal temperature of 1 K, which corresponds to a detuning of about 70 MHz. Estimating from measured beam/laser parameters we calculate a diffusion coefficient of 3.6×10^3 m^2/s^3 in fair agreement with the measurements. The determination of the absolute detuning of the laser with respect to the transition is limited by the beam energy spread at injection, and is not better than about 50 MHz.

The calculation above shows that the diffusion rate depends strongly on the laser detuning (for small detunings). We therefore expect the time of the blowup to be very sensitive to the laser detuning. Measurements carried out with changes of the detuning in steps of 20 MHz confirmed this hypothesis. It was observed that for more blue-detuned lasers the blowup occured earlier indicating more heating in the transverse degrees of freedom. Unfortunately, the uncertainty in the measured diffusion coefficients was too large for this information to be quantitatively decisive. We have also observed a tendency of the blowup rate to scale with the cooling laser-power. More detailed quantitative investigations of the blow-up are unfortunately hampered by the long term frequency stability of the laser system.

The experimental evidence presented supports the hypothesis that the heating source which drives the vertical blow-up is the laser-induced diffusion. However, the binary IBS calculation indicates that the vertical motion should be cooled at a rate much faster than the measured blow-up. This contradiction implies that some feature of the dynamics is overlooked by the binary IBS model.

For a cold beam, correlations between particles can be important. These correlations are not included in a binary scattering model.

An important effect, observed in electron cooled beams [6, 11] is a type of quasi-ordering in which the particles no longer drift past one another longitudinally. One signature of this effect is that the longitudinal beam temperature drops suddenly at the onset of ordering. This drop in longitudinal temperature of electron cooled beams has been interpreted as a kind of de-coupling between the transverse and longitudinal degrees of freedom at the onset of the quasi-ordering. If this decoupling happens in a laser-cooled ion beam, the transverse sympathetic cooling of the beam would be strongly suppressed, and the beam heating by laser-induced diffusion could become dominant. The loss of transverse cooling would be more critical vertically, since the horizontal motion would still be cooled by dispersive cooling [7]. The observed decrease of the horizontal beam size after the blow-up starts supports this hyphothesis as the IBS calculation indicated heating. The size reduction can thus be interepreted as a reduction in the IBS mediated heating compared to the dispersive cooling. This hypothesis was further strengthened in a measurement carried out with a setup in which the vertical and horizontal tunes were identical ($Q = 2.277$). In this measurement the blow-up was not observed. Thus, a possible interpretation of our results is that the beam has become quasi-ordered longitudinally, leading to an undesired loss of sympathetic transverse cooling.

A parameter of interest in this regard is the linear charge density of the beam. For the sake of comparison between different machines, the linear charge density can be expressed in dimensionless form by using the Wigner–Seitz radius as a length scale. The Wigner–Seitz radius is the average interparticle spacing for a zero-temperature beam. The dimensionless linear charge density is then defined to be $\lambda = a_{WS}/d$, where d is the average longitudinal distance between ions in the ring. For ASTRID, the Wigner–Seitz radius is about 41 μm, and the circumference of the machine is 40 m. Thus the λ parameter at the onset of blow-up (Figure 1) is about 0.4. The transition between one and two-dimensional ordered structures occurs for $\lambda = 0.709$ [5], so we are well into the linear regime.

If the above interpretation is correct, laser cooling as implemented in ASTRID may not be suitable for obtaining true one-dimensional order in a storage ring. Schottky diagnostics of the beam could help to clarify the experimental situation, but our earlier experience indicates that the longitudinal Schottky signals are strongly distorted by the effect of the laser [4]. Whether this behavior persists in the regime of one-dimensional beams remains to be determined.

This work was supported by the Aarhus Center for Atomic Physics.

References

1. Schröder, S. *et al.*, *Phys. Rev. Lett.* **64** (1990), 2901.
2. Hangst, J. S. *et al.*, *Phys. Rev. Lett.* **67** (1991), 1238.
3. Giannini, R. and Möhl, D., Report no. PS/AR/Note 96-19, CERN, Geneva.
4. Hangst, J. S., *Phys. Rev. Lett.* **74** (1995), 86.
5. Hasse, R. W. and Schiffer, J. P., *Ann. Phys.* **203** (1990), 419.
6. Hasse, R. W., *Phys. Rev. Lett.* **83** (1999), 3430.
7. Lauer, I. *et al.*, *Phys. Rev. Lett.* **81** (1998), 2052.
8. Madsen, N. *et al.*, *Phys. Rev. Lett.* **83** (1999), 4301.
9. Møller, S. P., In: L. Lizama and J. E. Chew (eds), *Proc. 1991 IEEE Particle Accelerator Conference,* San Francisco, USA, 1991, p. 2811.
10. Rahman, A. and Schiffer, J. P., *Phys. Rev. Lett.* **57** (1986), 1133.
11. Steck, M. *et al.*, *Phys. Rev. Lett.* **77** (1996), 3803.
12. Stenhofm, S., *Rev. Mod. Phys.* **58** (1986), 699.

Hyperfine Interactions **146/147:** 189–195, 2003.

Laser-Cooled Ions and Atoms in a Storage Ring

J. KLEINERT, S. HANNEMANN, B. EIKE, U. EISENBARTH, M. GRIESER, R. GRIMM⋆, G. GWINNER, S. KARPUK⋆⋆, G. SAATHOFF, U. SCHRAMM⋆⋆⋆, D. SCHWALM and M. WEIDEMÜLLER
Max-Planck-Insitut für Kernphysik, Postfach 103980, 69029 Heidelberg, Germany; e-mail: m.weidemueller@mpi-hd.mpg.de

Abstract. We review recent experiments at the Heidelberg Test Storage Ring which apply advanced laser cooling techniques to stored ion beams. Very high phase-space densities are achieved by three-dimensional laser cooling of a coasting $^{9}Be^{+}$ beam at 7.3 MeV. Laser-cooled, trapped Cs atoms are used as an ultracold precision target for the study of ion–atom interactions with a 74 MeV beam of $^{12}C^{6+}$ ions.

1. Introduction

The application of advanced laser cooling techniques at heavy-ion storage rings opens unique possibilities. On the one hand, unprecedented phase-space densities can be achieved by direct laser cooling of stored ion beams. On the other hand, laser-cooled atoms can be used as ultracold precision targets for the study of ion-atom interactions at MeV center-of-mass energies. In this article, we report on recent advances into these two directions at the Heidelberg Test Storage Ring TSR which is schematically depicted in Figure 1.

Laser cooling has allowed one to produce beams of singly-charged ions at unprecedented phase-space densities [1–9]. The holy grail of these efforts is the achievement of a regime where space-charge effects dominate the beam dynamics and Coulomb-ordered structures are formed [10]. To cool the ions by the light-pressure force, two laser beams are merged with the ion beam in one of the straight sections of the TSR over a length of about 5 m (TSR circumference: 55.4 m). Two efficient laser cooling techniques have been established so far: cooling of coasting ion beams by the radiation pressure produced by two counterpropagating laser beams (optical molasses) [1–3, 8, 9], and cooling of bunched beams using a single laser beam in combination with longitudinal confinement by radio-frequency fields [4, 5, 7].

⋆ Permanent address: Institut für Experimentalphysik, Universität Innsbruck, 6020 Innsbruck, Austria.
⋆⋆ Permanent address: Institut für Physik, Universität Mainz, 55099 Mainz, Germany.
⋆⋆⋆ Permanent address: Sektion Physik, Universität München, 85748 München, Germany.

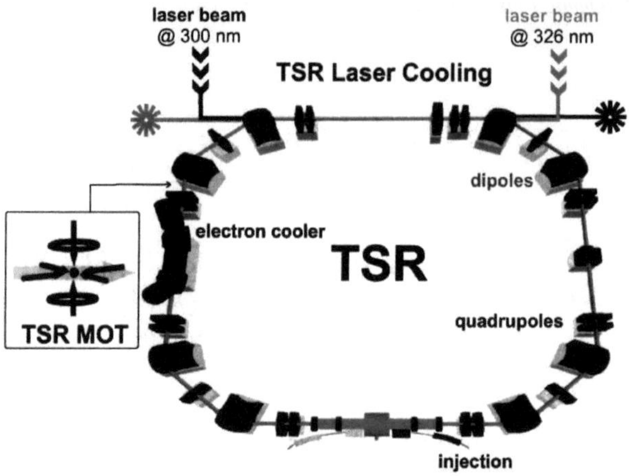

Figure 1. Heidelberg Test Storage Ring TSR. Laser cooling is applied in one of the straight sections of the ring (TSR, Laser Cooling), while the laser-cooled cesium target is placed behind the electron cooler (TSR-MOT).

A target based on laser-cooled atoms confined in an optical trap [11] can either serve as a high-precision diagnostic tool for weak ion beams or may be used to study ion–atom collisions at negligible inital energy of the target particles [12]. The laser trap for cold target atoms at TSR is situated about 6 m behind the electron cooler between a quadrupole and a dipole magnet. Cesium atoms are trapped in the intersection of three mutually orthogonal standing light waves, which are slightly detuned below the atomic resonance, and a magnetic quadrupole field generated by a pair of coils with counterpropagating currents. The cloud of cold atoms can be freely positioned by additional homogeneous magnetic fields in order to maximize the overlap with the ion beam or to measure the transverse beam profile.

2. Laser cooling of fast stored ion beams

The longitudinal ion motion of fast stored ion beams can be cooled by the light-pressure force [13] in a near-resonant laser field. The light force can be extended to the transverse motion by coupling the different degrees of freedom. Two coupling mechanisms have been identified for efficient three-dimensional laser cooling of fast stored ion beams: the density-dependent coupling through intrabeam scattering [5] and the density-independent coupling by storage ring dispersion in combination with linear betatron coupling [7].

2.1. LASER COOLING AT TSR

The laser cooling activities at TSR in the last years have concentrated on bunched beams of $^{9}Be^{+}$ ions at 7.3 MeV [14]. For beam currents around 1 μA, longitudinal temperatures below 1 K are achieved corresponding to a relative momentum spread $\Delta p/p \sim 10^{-7}$. Transverse emittances at these longitudinal temperatures are in the range below 10^{-2} π mm mrad mainly due to dispersive laser cooling [7]. Novel bunch forms have been applied to create homogeneous ion distributions inside the bucket [15]. Comparison with elaborate models including the effects of intrabeam scattering (IBS) has led us to a deeper understanding of the dynamics of laser-cooled bunched ion beams [14].

Recent measurements at TSR revealed an unexpected and still unexplained effect in a bunched, laser-cooled ion beam [16]: When the beam has decayed to about 10^6 stored ions, corresponding to a mean particle distance to 50 μm, one observes an abrupt transition of the beam into a state in which IBS has disappeared and the longitudinal beam temperature is dramatically decreased. The observations are reminiscent to the signature of Coulomb ordering as reported from experiments with highly-charged ions at GSI in Darmstadt where the transition occurred at mean ion distances around 10 cm [17, 18]. To further approach this intriguing effect, we have recently realized laser cooling of a coasting $^9Be^+$ beam, which provides additional diagnostic capabilities, in particular via analysis of Schottky signals. In the following, we present first results on three-dimensional laser cooling of a coasting beam at TSR.

2.2. LASER COOLING OF COASTING BEAMS

The alkali-like $^9Be^+$ can be laser cooled on a strong dipole-allowed transition line in the near-UV spectral range ($D2$ line $2^2S_{1/2} \rightarrow 2^2P_{3/2}$ at a rest frame wavelength of 313.13 nm). As shown in Figure 1, two counterpropagating laser beams are merged with the ion beam in one of the straight sections of the storage ring. In the laboratory frame, the resonance wavelengths is Doppler-shifted to 300.35 nm (copropagating laser beam) and 325.91 nm (counterpropagating laser beam) at the ion beam energy of 7.3 MeV which corresponds to a velocity of $0.042 \times c$. The light at 300.35 nm is provided by a fixed-frequency argon ion laser, while the radiation at 325.91 nm is created by frequency-doubling the output of a dye laser in a lithium iodate crystal [14]. Due to the hyperfine structure of the $D2$ line, two laser frequencies separated by 1.3 GHz and 1.2 GHz, respectively, are required to realize a closed excitation cycle. The resulting ring-averaged radiation-pressure force amounts to approx. 60 meV/m at 300 nm and 20 meV/m at 326 nm.

Figure 2(a) shows the longitudinal temperature of the ion beam as a function of relative detuning between the two laser fields. The temperature decreases with decreasing detuning due to the increasing friction coefficient of the light pressure force. Since the capture range of the optical molasses decreases with decreasing detuning, the number of atoms which are actually cooled out of the initial momentum distribution decreases as well [see inset in Figure 2(a)]. Yet, the achievable temperatures are far above the limit given by the random character of the spontaneous emission (Doppler limit) which, in the case of Be ions, lies at 325 μK. The major

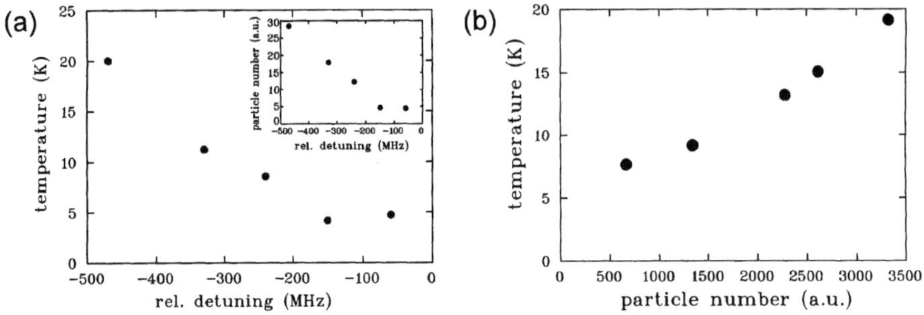

Figure 2. Longitudinal temperature of a coasting, laser-cooled $^9Be^+$ beam. (a) Temperature as a function of relative detuning between the two laser fields. The inset shows the corresponding number of cooled ions. (b) Temperature as a function of the number of stored particles for a relative detuning of −300 MHz.

mechanism counteracting laser cooling of fast stored ion beams is heating through IBS. The role of IBS is elucidated in Figure 2(b) which depicts the temperature versus the number of stored ions. As the number of stored ions decreases, the rate of Coulomb collisions decreases as well leading to lower final temperatures. Currently, we are studying the role of IBS in more detail employing our model simulations.

Besides longitudinal cooling, we have also found clear indications of transverse cooling by dispersive coupling. If the horizontal position of the 300 nm laser beam is shifted towards the ring center, while the 326 nm laser beam is slightly shifted outwards, one observes cooling of the horizontal motion. Through linear betatron coupling the vertical motion is also cooled at an equal rate. Thus, efficient three-dimensional laser cooling of MeV ion beams has become possible at TSR opening a new route to Coulomb ordering effects.

3. Laser-cooled atoms as a precision target for stored ion beams

A magneto-optical trap (MOT) for ultracold cesium atoms has recently been implemented inside the TSR (TSR-MOT). 10^4–10^5 atoms are stored with a temperature below 100 μK, i.e. at thermal energies in the neV range. The cloud of cold atoms has a diameter of 100 μm resulting in a density of the order of 10^{10} atoms/cm^3. The tiny cloud of cold atoms inside a storage ring for heavy ion beams serves for two purposes: as a cold target for collision experiments and as a high-precision monitor for stored ion beams. Details of the set-up before it was installed into the TSR are described in [12]. Here we will therefore concentrate on the operation of the MOT inside the storage ring and present recent experiments demonstrating the application of laser-cooled atoms as a target and as a beam-profile monitor.

3.1. LASER-COOLED ATOMS AS A TARGET

To demonstrate the application of the TSR-MOT as a target, we have investigated the interaction of the trapped Cs atoms with a $^{12}C^{6+}$ beam at an energy of 74 MeV. Collisions of the high-energetic ion beam with cesium atoms lead to a loss of particles from the MOT if the energy transfer by the collision is larger than the kinetic energy needed for a cesium atom to escape from the trap ($E_{esc}/k_B \approx 1$ K). The rate of decay the number of trapped particles is proportional to the ion flux at the position of the atomic cloud. The decay rate is measured by monitoring the fluorescence of the cesium atoms in the MOT after blocking the loading flux.

The decay rate of the MOT for various ion beam intensities is depicted in Figure 3. The measurement proceeds along the following scheme: First the MOT is loaded at a position displaced by about 1 cm from the closed orbit of the ion beam. Then, the ion beam is injected and cooled for 10 seconds with the electron cooler at TSR, which reduces the beam diameter from initially >30 mm to 0.6 mm. The loading flux into the MOT is interrupted and the cloud of cold atoms is moved into the center of the ion beam by changing the current through the control field coils. A CCD camera monitors the fluorescence decay of the MOT and allows for a precise determination of the position of the atom cloud. The ion flux is deduced via a calibrated beam profile monitor at TSR.

As shown in Figure 3, the atom loss rate scales linearly with the ion flux. The finite loss rate at vanishing ion flux is due to collisions with thermal particles from the background gas. This decay rate ultimately limits the sensitivity of the TSR-MOT. Due to the low background pressure at TSR, the effect of collisions with the ion beam can be identified down to ion currents of only 10 nA. The slope of the linear curve in Figure 3 yields the total collision cross section $\sigma = 3(1) \times 10^{-15}$ cm^2. First estimations based on the LOSS code developed by V.S. Shevelko *et al.* [19] indicate that the major contribution of the cross section stems from

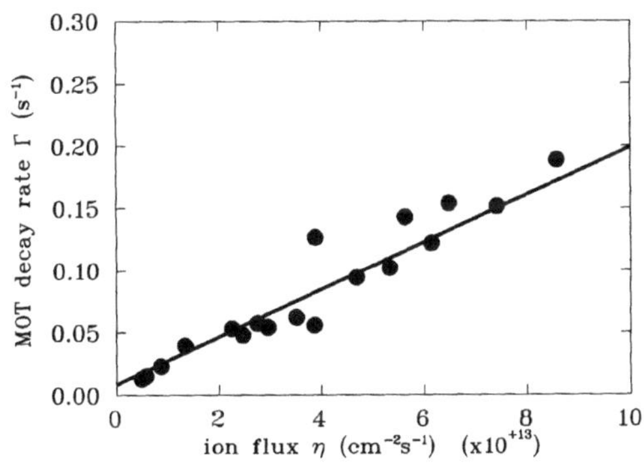

Figure 3. Decay rate of cesium atoms in the magneto-optical trap as a function of ion flux for a $^{12}C^{6+}$ beam at 74 MeV.

collisions ionizing the target atom. Detailed studies of the ion–atom interaction are currently underway.

3.2. LASER-COOLED ATOMS AS A BEAM-PROFILE MONITOR

The cloud of cold cesium atoms can be moved by adding a homogeneous magnetic field to the MOT quadrupole field. By measuring the decay rate of the MOT as a function of position one can map the transverse profile of the ion beam. The result of a beam profile measurement is shown as the dots in Figure 4. The measurement procedure is similar to the one described in the previous section. Each data point corresponds to a new injection of ions into the ring. The measured decay rate is corrected for fluctuations in the ion flux which differs by up to 50% for different injections. The solid line in Figure 4 gives the profile of the ion beam as measured with the beam profile monitor at TSR. Since the beam profile monitor is placed at a different position at TSR, the profile is scaled to the position of the MOT using the known local values of the betatron function for TSR. Both independent measurements agree well with each other.

The estimated uncertainty of the MOT position is indicated by the error bar in Figure 4. The MOT position was not been measured directly but was derived from the magnetic control field. Although the zero of the magnetic field moves linearly with the strength of the control field, the MOT position can slightly deviate from magnetic zero due to uncompensated light forces in the interference pattern of the six overlapping laser beams. In the near future, we will implement a second CCD camera in order to take stereoscopic pictures of the MOT and, hence, its three-dimensional position with an accuracy given by the resolution of the optical imaging system. This will allow us to reach the intrinsic resolution limit for the TSR-MOT given by the minimum spatial extension of the atom cloud ($<100\ \mu m$).

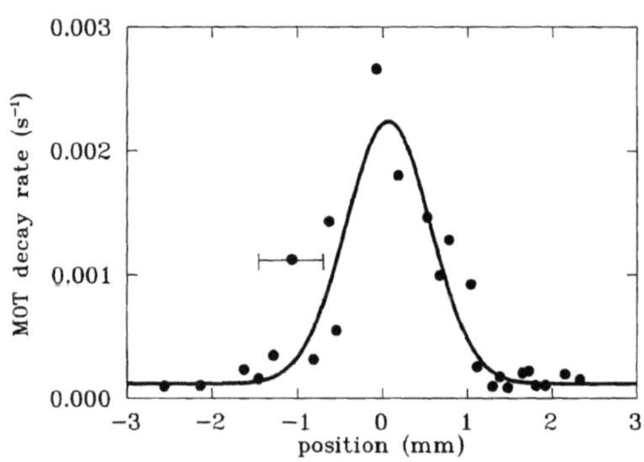

Figure 4. Measurement of the transverse beam profile for a $^{12}C^{6+}$ ion beam. The data points give the decay rate of the laser-cooled target as a function of position. For comparison, the solid line shows the transverse density distribution as measured by the TSR beam profile monitor.

4. Conclusions

The question whether Coulomb ordering or even beam crystallization can be achieved by laser cooling of fast stored ion beams still remains open. With the demonstration of three-dimensional laser cooling of a coasting $^9Be^+$ beam we have opened a new approach to achieve very high phase-space densities close to the onset of Coulomb ordering. In this way, we hope to get further insights into the ultimate limits of laser cooling of ion beams.

With respect to the transverse diagnostics of dilute ion beams, and ultimately Coulomb ordered structures, the laser-cooled cesium target inside TSR offers unique possibilities through its high sensitivity even at low ion currents. Besides the use as a beam profile monitor, we will employ the TSR-MOT to study collisions with atomic and molecular ion beams with the final goal to spectroscopically resolve the recoil-ion momentum by techniques such as COLTRIMS [20]. This technique, recently demonstrated with laser-cooled targets in keV ion beams [11], opens new ways into the deeper understanding of ion-atom collision at high center-of-mass energies.

Acknowledgements

We thank H. Krieger for technical assistance in maintaining our laser systems and during beam times. Discussions with the other members of the Laser Cooling Group at the MPI für Kernphysik are gratefully acknowledged.

References

1. Schröder, S. *et al.*, *Phys. Rev. Lett.* **64** (1990), 2901.
2. Hangst, J. S. *et al.*, *Phys. Rev. Lett.* **67** (1991), 1238.
3. Petrich, W. *et al.*, *Phys. Rev. A* **48** (1993), 2127.
4. Hangst, J. S. *et al.*, *Phys. Rev. Lett.* **74** (1995), 4432.
5. Miesner, H.-J. *et al.*, *Phys. Rev. Lett.* **77** (1996), 623.
6. Wanner, B. *et al.*, *Phys. Rev. A* **58** (1998), 2242.
7. Lauer, I. *et al.*, *Phys. Rev. Lett.* **81** (1998), 2052.
8. Madsen, N. *et al.*, *Phys. Rev. Lett.* **83** (1999), 4301.
9. Schätz, T., Schramm, U. and Habs, D., *Nature* **412** (2001), 717.
10. Habs, D. and Grimm, R., *Annu. Rev. Nucl. Part. Sci.* **45** (1995), 391.
11. van der Poel, M. *et al.*, *Phys. Rev. Lett.* **87** (2001), 123201; Turkstra, J. W. *et al.*, *ibid.*, 123202; Flechard, X. *et al.*, *ibid.*, 123203.
12. Eike, B. *et al.*, *Nucl. Instrum. Methods Phys. Res. A* **441** (2000), 81.
13. Metcalf, H. and van der Straten, P., *Laser Cooling and Trapping*, Springer, New York, 1999.
14. Eisenbarth, U., Dissertation, Universität Heidelberg, 2001.
15. Eisenbarth, U. *et al.*, *Nucl. Instr. Methods Phys. Res. A* **441** (2000), 209.
16. Eisenbarth, U. *et al.*, *Hyp. Interact.* **127** (2000), 223.
17. Steck, M. *et al.*, *Phys. Rev. Lett.* **77** (1996), 3803.
18. Hasse, R. W., *Phys. Rev. Lett.* **83** (1999), 3430.
19. Shevelko, V. S., private communication.
20. Dörner, R. *et al.*, *Phys. Rep.* **330** (2000), 95.

Hyperfine Interactions **146/147:** 197–201, 2003.

Electron Cooling of Positrons in LEPTA

I. N. MESHKOV, A. O. SIDORIN, A. V. SMIRNOV and G. V. TRUBNIKOV
Joint Institute for Nuclear Research, Dubna, Russia

Abstract. Electron cooling of positrons is the essential peculiarity of the method of antihydrogen and positronium fluxes generation, which is based on low-energy positron storage ring. Such a ring, the Low Energy Positron Toroidal Accumulator (LEPTA), is under construction in JINR. The calculation algorithm of electron cooling of positrons is discussed. Electron cooling of positrons leads to generation of the positronium monochromatic fluxes with small angular spread. With positronium fluxes in vacuum one can perform new original setting up of the experiments without the distortion caused by medium when positronium is generated in a target.

Key words: positron, electron cooling, storage ring, recombination.

1. Physical program

Positronium (Ps) – the bound state of an electron and its antiparticle the positron is an ideal test object for bound state QED: QCD effects and the weak interaction play no role at the present state of the accuracy. With positronium fluxes in vacuum one can perform new original setting up the experiments (so-called positronium-in-flight set-ups) [1]. Among experiments of this type are the following:

(1) comparison of electron (e^-) and positron (e^+) electric charges;
(2) the measurement of the orthopositronium lifetime and the parapositronium lifetime;
(3) the positronium spectrum;
(4) the search for exotic and rare decay channels of p-Ps $\to n\gamma, n > 2$;
(5) search for a light, neutral, short-lived boson, via which o-Ps annihilation can occur:

$$\text{o-Ps} \to b + \gamma \to 2\gamma.$$

Perhaps the most intriguing problem in positronium physics is the search for the "mirror Universe". The accuracy of the measurement of the positronium life time, the probability of decays with momentum conservation and charge invariant infringement (CPT violation), fine structure of the positronium spectrum, Lamb shift measurements can be much higher than in traditional methods.

The new approach to the experimental studies of Ps physics relates to the idea of antihydrogen generation [2]. Key point of this idea is an application of electron cooling of positrons, which immediately leads to generation of Ps in-flight. A de-

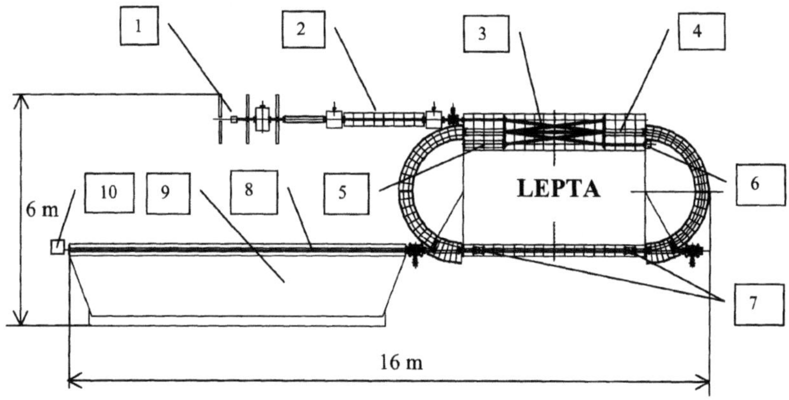

Figure 1. The LEPTA installation: 1 – positron source, 2 – positron trap, 3 – septum coils, 4 – kicker; 5 – electron gun, 6 – electron collector, 7 – pick-up stations, 8 – decay channel, 9 – dipole magnet, 10 – Ps detector.

velopment of this idea [3, 4] would allow the production of intense pencil beams of orthopositronium (30–1.7 × 10^4 sec^{-1}).

Nowadays similar experiments with spectroscopy of antihydrogen atoms, stored in magnetic traps, are under development by several groups [5]. One should underline that these experiments and the proposed ones in [1] are complimentary, and information to be obtained in both types of experiment has independent value.

2. LEPTA ring

The LEPTA ring (Figure 1, Table I) is filled with positrons from the injector which provides the positron beam of the radius of 0.5 cm at the energy of 10 keV, at intensity of 10^8–10^9 particles per pulse and of the pulse duration of 300 nsec or less. The injection periodicity is about 100 sec (this value corresponds to the positron beam life time in the ring). At the first stage of the LEPTA operation we plan to use a positron source on the base of β^+-radioactive isotope ^{22}Na (Figure 1, pos. 1). The positron beam dynamics in LEPTA is described in [2–4].

3. Electron cooling of positron and investigation of friction force

First experiments on LEPTA will be connected with friction force measurements and study of recombination process. Essential peculiarity of the electron cooling of positrons is magnetisation of both interacting particles. In this case an impact parameter of the collision can be determined as a distance between magnetic field lines crossing the centers of positron and electron Larmour circles (Figure 2).

Numerical calculation of the friction was performed in the model of binary collisions, the calculation algorithm includes calculation of the positron momentum losses $\Delta p(V_\perp, V_\parallel)$ for given impact parameter and electron velocity compo-

Table I. General parameters of the LEPTA

Circumference	m	18.12
Positron energy	keV	10.0
Longitudinal magnetic field	G	400
Major radius of the toroids	m	1.45
Gradient of the spiral quadrupole field	G/cm	10.0
Positron beam radius	cm	0.5
Number of positrons	–	$1 \cdot 10^9$
Vacuum	Torr	$1 \cdot 10^{-10}$
Positron beam life time	sec	100
Electron cooling system		
Cooling section length	m	4.53
Beam current	A	0.5
Beam radius	cm	1
Characteristic cooling time	msec	100
Orthopositronium beam parameters		
Intensity	$\sec^{-1}$	$1 \cdot 10^4$
Angular spread	mrad	1
Energy spread	–	$1 \cdot 10^{-3}$
Flux diameter at the ring exit	cm	1.1
Decay length	m	8.52

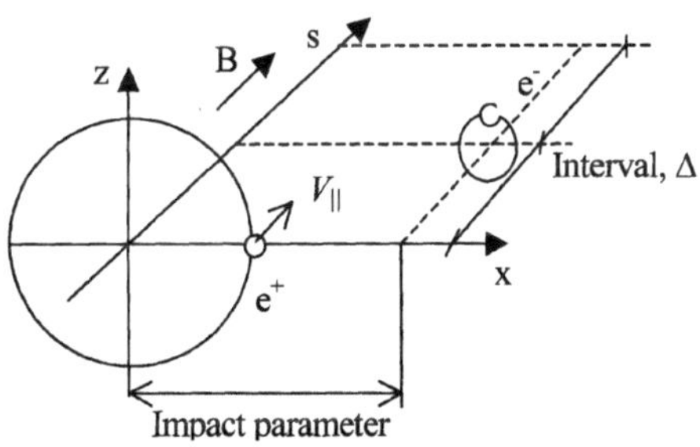

Figure 2. Binary collision schematics.

nents and then the averaging of the $\Delta p(V_\perp, V_{||})$ value over the electron velocity distribution function $f(V)$:

$$f(v)\, \mathrm{d}^3v = \left(\frac{m}{2\pi}\right)^{3/2} \frac{1}{T_\perp\sqrt{T_{||}}} \mathrm{e}^{-mv_\perp^2/2T_\perp - mv_{||}^2/2T_{||}} 2\pi v_\perp\, \mathrm{d}v_\perp\, \mathrm{d}v_{||}. \tag{1}$$

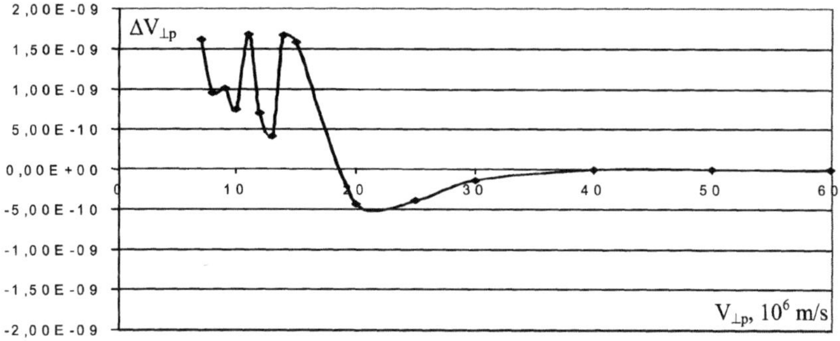

Figure 3. Positron velocity variation as function of their transverse velocities. Transverse electron temperature = 0.1 eV, longitudinal = 0.001 eV.

Integration of the momentum losses over the impact parameter ρ gives the force value:

$$F_{\perp} = n_e v_{||} 2\pi \int_0^{\rho_{\max}} \langle \Delta p \rangle \rho \, d\rho, \tag{2}$$

where $\rho_{\max} = \min(R_D, a_e)$, R_D and a_e are radii of Debye shielding and electron beam correspondingly.

Longitudinal component of the friction force has to be calculated taking into account a finite length of cooling section (interval Δ in Figure 2). For calculation of transverse friction force was elaborated special program code, which realizes the described algorithm.

Values of the electron beam parameters were chosen in accordance with LEPTA design parameters. Changing of the positron velocity during a binary collision when interval of integration on time is increased comes to saturation if this interval is equal to 10–20 periods of particle Larmour rotation. The dependence of positrons velocity change on maximum impact parameter falls into saturation when value of maximum impact parameter equal to 3–5 radii of Larmour rotation of positron that differs this process with electron cooling of ions. The example of numerical calculation of positron velocity variation after crossing the cooling section is shown in Figure 3. This value is proportional to friction force. Zero value of the friction force corresponds to the equality of positron velocity and the electron thermovelocity $V_{\perp} = \sqrt{2 \cdot T_{\perp}/m}$. Maximum of the friction force lies near the value of $V_{p\perp} = 1.22 \cdot V_{\perp}$.

Expected temperature of thc transverse degree of freedom of the positron beam after termination of cooling process lies in the interval of $1\text{–}1.5T_{\perp}$. Oscillating character of the dependency of positron velocity variation at small velocities related to insufficient accuracy of the calculations. The preliminary analysis of dependence in the range of larger velocities points that obtained dependence differs from the similar dependence of electron cooling of ions. In further calculations the accuracy will be improved both in the range of small, and in the range of larger velocities.

4. Electron–positron recombination

The experimental study of positronium formation in recombination of positrons with cooling electrons can be performed in very beginning of the LEPTA ring operation when electron cooling process is to be tuned to get its optimal regime. This experiment is of great interest for theory because formation of positronium was never observed at pure vacuum conditions.

The crucial point is the control of positron and electron temperature $T_\perp$. One can obtain the value of the positron transverse temperature $T_\perp$, measuring the dependence of the recombination rate R on the average velocity difference of electrons and positrons and evaluating the recombination coefficient from the formula

$$R = \frac{\mathrm{d}N}{\mathrm{d}t} = \alpha_r n_e n_p V, \tag{3}$$

where α_r is the recombination coefficient, $n_{e,p}$ are the electron and positron beam densities, $V = \ell_{\mathrm{rec}}\pi a^2$, ℓ_{rec} – the length of the recombination section, a – minimal of the electron and positron beam radii.

The experimental dependence $\alpha_r(v_{\mathrm{shift}})$ can be fitted by the theoretical function, where $T_\perp$ is used as fitting parameter in the least square method.

5. Status

At present the significant part of the ring elements is manufactured: straight section solenoids, toroidal solenoids, helical quadrupole. The vacuum chamber is manufactured and tested. Homogeneity of the magnetic fields inside the cooling section was measured and corrected up to the level better than $\Delta B/B \leqslant 10^{-3}$. The measurement and correction of the field in other magnetic elements (toroids and quadrupole) is in progress. The test of the ring with circulating electron beam is scheduled to be performed next year. At the same time the positron source and trap have to be constructed.

References

1. Meshkov, I. N., *Fiz. Elem. Chastits At. Yadra* **28** (1997), 495, translated into English in *Phys. Part. Nucl.* **28**(2) (1997).
2. Meshkov, I. N. and Skrinsky, A. N., *NIM A* **391** (1997), 205.
3. Meshkov, I. N. and Sidorin, A. O., *NIM A* **391** (1997), 216.
4. Meshkov, I., Sidorin, A., Smirnov, A. and Syresin, E., In: *6th EPAC*, Stockholm, 1998, p. 1067.
5. Charlton, M. *et al.*, *Phys. Rep.* **241**(2) (1994), 65.
6. Meshkov, I., Sidorin, A., Smirnov, A., Syresin, E., Trubnikov, G. and Zeinalova, O., In: *7th EPAC*, Vienna, 2000, p. 1408.

Hyperfine Interactions **146/147:** 203–207, 2003.

Stability of Crystalline Ion Beams

T. SCHÄTZ, U. SCHRAMM and D. HABS
Sektion Physik, LMU München, D-85748 Garching, Germany;
e-mail: ulrich.schramm@physik.uni-muenchen.de

Abstract. In this paper, the conditions for which three-dimensional crystalline ion beams were attained in the rf quadrupole storage ring PALLAS at a velocity of around 2800 m/s are presented. The stability of these crystalline beams when subjected to discontinuous cooling is discussed for beams of different linear density and thus of different crystal structure. The survival time seems to only weakly depend on the linear density, in contrast to the heating of the beam after melting.

Key words: crystalline ion beams, strong coupling, storage rings.

1. Introduction – the rf quadrupole storage ring PALLAS

For the attainment of Coulomb-ordered crystalline ion beams [1–3], a high symmetry and multiplicity of a storage ring lattice has been postulated to be essential [4]. The technique of combined focusing and bending of laser-cooled ion beams in the rf quadrupole storage ring PALLAS incorporates these requirements and recently allowed the realization of crystalline beams at a velocity around $v = 2800$ m/s (beam energy 1 eV) [5, 6]. The table-top ring PALLAS, sketched in Figure 1, is constructed similar to the ring-shaped Paul trap, where the first stationary cylindrical ion crystals [7] were observed [8]. The transverse confinement of the ion beam is characterized by the number of 'focusing sections' or rf cycles per revolution,

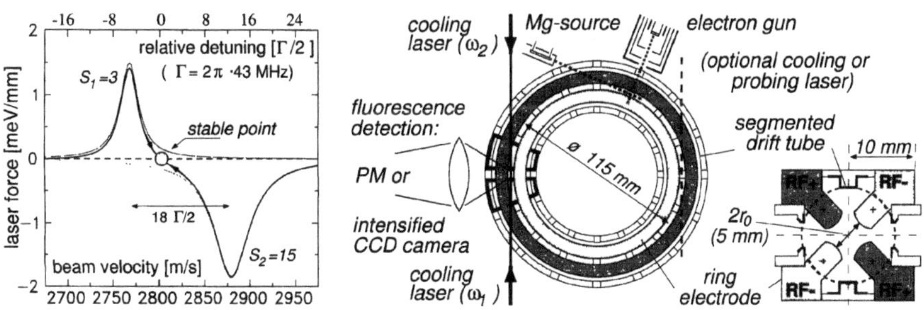

Figure 1. Axial and radial cut through the rf quadrupole storage ring PALLAS with a circumference of 0.36 m. $^{24}Mg^+$ ions are generated inside the storage ring and accelerated by the light pressure of the continuously tuned laser beam ($\omega_1(t)$). At a velocity of $v = 2800$ m/s, the ion beam is laser-cooled on a fraction of only 1% of the orbit by means of the velocity dependent force of two counter-propagating laser beams. A typical stationary situation is illustrated in the left graph.

the periodicity $P = \Omega/\omega_{rev} \approx 800$, and by the corresponding number of transverse secular or betatron oscillations, the tune $Q = \omega_{sec}/\omega_{rev} \approx 60$ (for details, see refs. [5, 9]).

2. Focusing conditions for crystalline beams

The feasibility of altering the focusing conditions of an rf quadrupole storage ring ($Q \propto U_{rf}/v$) allows the systematic mapping of the parameter range where crystalline beams are attainable in PALLAS. In Figure 2, the rf amplitude U_{rf} is plotted for ion beams of different dimension-less linear densities λ (for details, see [7, 5, 9]). Crystalline beams were observed to occur in a narrow band [9] which markedly follows the curvature of the dashed line. This dependency ($\sqrt{\lambda} \cdot U_{rf} =$ const.) is based on the argument that the mean energy of the periodic transverse motion of particles in the time-varying confining potential equalizes the melting temperature of the crystal [4, 9]. The dotted lines express the influence of bending shear ($v\sqrt{\lambda}/U_{rf} = 41$ [m/Vs] for PALLAS), which increases with rising linear density and velocity of a crystalline ion beam cooled to constant linear velocity [4, 9]. For the largest crystalline beams ($\lambda \approx 5$), which were at random taken at a lower velocity of 1900 m/s, shear could have already affected the crystal, although the model does not include any shear elasticity of the beam [4]. For the lower dimensional crystalline beams, a higher rf amplitude is required. This effect can be attributed to the reduced number of neighboring ions in the beam, which, at the upper border, reduces the amplitude and the coupling of rf motion into thermal motion. At the lower border, sympathetic cooling of the transverse motion

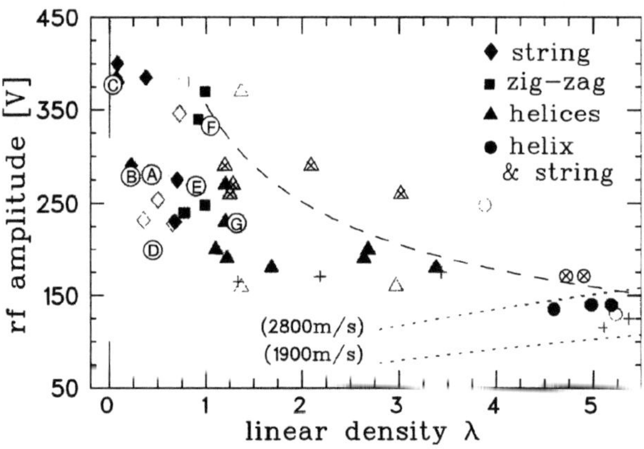

Figure 2. Correlation of the dimensionless linear density λ of crystalline (filled symbols) and non-crystalline (open and crossed symbols, + mark beam losses) ion beams and the amplitude of the confining rf potential. The shaded area indicates the parameter space for possible crystalline beams. The circles (A)–(G) themselves represent measurements with discontinuous cooling, presented in Figures 3 and 4. For a beam velocity of $v = 2800$ m/s, an rf amplitude of $U_{rf} = 250$ V corresponds to a tune of $Q = 58$.

is reduced. However, more efficient longitudinal and especially direct transverse cooling is expected [10] to lower this border which is important for the realization of crystalline beams in 'high-energy' rings.

3. Stability of crystalline beams with discontinuous cooling

For a quantitative study of the heating and melting properties of crystalline beams both cooling lasers were simultaneously blocked for a variable period of time t_b. With the lasers un-blocked, the remaining relative fluorescence rate Δ_r and its further development was recorded, as exemplarily presented in Figure 3. The instantaneous recovery ($\Delta_r = 100\%$) after $t_b = 40$ ms (Figure 3b) indicates the persistence of the string of ions, marked (A) in Figure 2. After a blocking period of $t_b = 400$ ms (Figure 3c), Δ_r amounts to about 50%. The maximum fluorescence rate of a non-crystalline beam of identical ion number amounts to $\lesssim 30\%$ [5] and thus a less dense crystalline beam or a variation of the state of the beam (two phase regime) along the orbit of 36 cm is assumed. Still, the beam is restored to full order in about $t_r = 30$ ms. This behaviour only weakly depends on the dimensionality of the crystalline beam, as demonstrated in Figure 4 for the beams marked A, E, and G. Only for dilute crystalline beams of $\lambda = 0.06$ the full fluorescence yield is conserved over blocking periods of up to $t_b \gtrsim 1$ s and maintained on a constant level of $\Delta_r \approx 60\%$ for $t_b \gtrsim 40$ s. Once a crystalline beam melts, the heating rate and therefore the slope of $\Delta_r(t_b)$ strongly depends on the linear density. A related behaviour was observed for a crystalline beam close to the lower border of stability (beam (B) in Figure 2). After the blocking, a reduction of the number of constituent ions shifts the crystalline beam out of the region of stability. The beam melts and

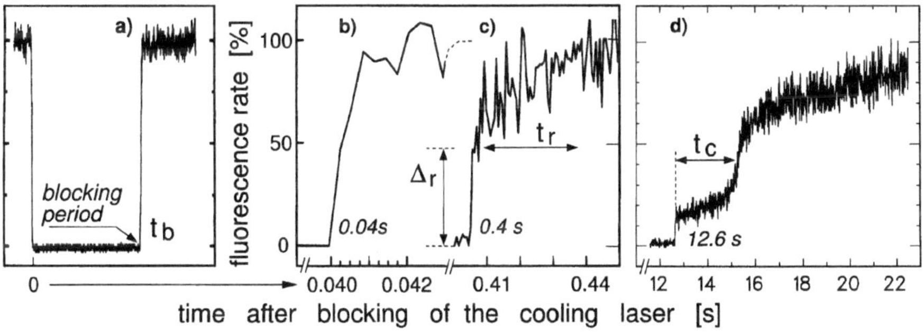

Figure 3. Fluorescence signal of the 1D crystalline ion beams (A) and (B) in Figure 2 while blocking and unblocking of the cooling lasers, as indicated in (a). After a blocking period of $t_b = 40$ ms (b), the signal of beam (A) immediately recovers to its original strength while the time resolution of 250 μs excludes the possibility of melting and subsequent re-crystallization of the crystalline beam. After $t_b = 400$ ms (c), the signal reappears at $\Delta_r \approx 50\%$, which still lies above the characteristic 30% of a non-crystalline beam [5] and fully recovers within $t_r \approx 30$ ms. In (d), the crystalline beam (B) melts during the blocking period of $t_b = 12.6$ s and requires a cooling time of $t_c \approx 2.7$ s to reach the temperature needed for the re-crystallization.

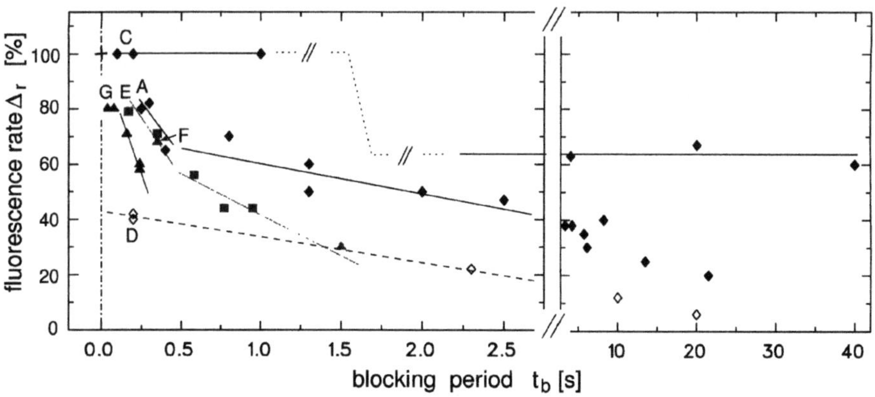

Figure 4. Remaining relative fluorescence yield (Δ_r) of crystalline beams of different linear density λ after a blocking period t_b. The persistence of the crystalline beams A, C, and E–G (marked as in Figure 2) only weakly depends on the linear density λ, except for the very dilute beam C. After the first fast decline of $\Delta_r(t_b)$, the behaviour is similar to the one of the non-crystalline beam D and the slope of $\Delta_r(t_b)$ decreases with λ. The lines are solely ment to guide the eye.

it takes a comparatively long time t_c to restore the conditions where the beam re-crystallizes, as shown in Figure 3d.

The main source of heating of crystalline ion beams in PALLAS seems to be the coupling of the collective periodic motion of the ions in the confining rf potential into random thermal motion. This coupling is reduced by several orders of magnitude compared to the intra-beam scattering rate of a dense non-crystalline beam [10–13]. The reason for this reduction lies in the nature of long range Coulomb ordering, where close collisions of particles are suppressed [2, 4]. Without ordering, no suppression of the heating takes effect and the remaining rate after blocking and un-blocking the laser beams, Δ_r, immediately ($t_b < 20$ ms) drops to $\lesssim$40% of the maximum rate of the non-crystalline beam (Figure 4D). The remaining fluorescence of this strongly diluted beam behaves in a similar way as that of the melted crystalline beams.

MD simulations of a crystalline beam (single shell) in a model storage ring (10 FODO cells, $P/Q \approx 4$) revealed a minimum relative heating rate of about $\Delta T/T \approx 10^{-6}$ per focusing period [13], which faster than linearly increases with the temperature T. Thus, it would take of the order of 10^6 rf cycles for a (3D) crystalline beam with an initial temperature of $T \approx 4$ mK to melt in PALLAS which is consistent with survival times of crystalline beams in the simulations [11]. For the high periodicity of PALLAS ($P \approx 800$), this value corresponds to 1,500 round trips or 150 ms. It surprisingly well matches the measured survival times of the order of 100 ms ($\Delta_r \approx 100\%$ criterion) for dense ion strings. At the heavy ion storage ring ESR (GSI), survival times of the order of 0.5 s were observed for extremely dilute ($\lambda \approx 0.0004$) weakly ordered beams of highly charged heavy ions [14]. Since these times also correspond to about $3 \cdot 10^5$ focusing periods, one could imagine the order of magnitude of the survival time to be, at least for the 1D beams, rather independent from the storage ring lattice. However, this coincidence should be interpreted carefully, since for 3D beams the amplitude of the transverse motion is quadratically reduced with the number of focusing elements [4], and later simulations for the TARN II lattice [10] revealed an increased heating rate for 3D beams.

Summarizing the latter, crystalline beams were observed to survive for several 10^6 focusing periods in PALLAS without cooling. Up to a single-shell structure, the heating of the crystalline beam was found to only weakly depend on the linear density. To answer the question whether a high multiplicity reduces the 'real' survival time, dedicated MD simulations with realistic cooling conditions would be desirable.

References

1. Schiffer, J. P. and Kienle, P., *Z. Phys. A* **321** (1985), 181.
2. Schiffer, J. P. and Rahman, A., *Z. Phys. A* **331** (1988), 71.
3. Maletic, D. M. and Ruggiero, A. G. (eds), *Crystalline Beams and Related Issues*, World Scientific, Singapore, 1996.
4. Schiffer, J. P. in [3], p. 217.
5. Schätz, T., Schramm, U. and Habs, D., *Nature* **412** (2001), 717, and Schramm, U. *et al.*, In: F. Anderegg, L. Schweikhard, C. F. Driscoll (eds), *Proc. Non-Neutral Plasma Physics IV*, SanDiego, CA, 2001, *AIP Conf. Proc.* **606** (2002), 235.
6. Schramm, U., Schätz, T. and Habs, D., *Phys. Rev. Lett.* **87** (2001), 184801.
7. Hasse, R. W. and Schiffer, J. P., *Annals of Phys.* **203** (1990), 419.
8. Birkl, G., Kassner, S. and Walther, H., *Nature* **357** (1992), 310.
9. Schramm, U., Schätz, T. and Habs, D., *Phys. Rev. E* **66** (2002), 036501.
10. Wei, J., Okamoto, H. and Sessler, A. M., *Phys. Rev. Lett.* **80** (1998), 2606.
11. Wei, J., Li, X.-P. and Sessler, A., *Phys. Rev. Lett.* **73** (1994), 3089.
12. Spreiter, Q., Seurer, M. and Toepffer, C., *Nucl. Instr. Meth. A* **364** (1995), 239; Seurer, M., Spreiter, Q. and Töpfer, C., *Hyp. Interact.* **99** (1996), 253.
13. Wei, J., Dräsecke, A., Li, X.-P. and Sessler, A., in [3], p. 229.
14. Steck, M. and Hasse, R., In: *Proc. EPAC 2000*, Vienna, 2000, p. 274; and Steck, M., In: *Proc. PAC 2001*, Chicago, 2001, PaperID TOAA002.

Hyperfine Interactions **146/147:** 209–213, 2003.

Observations of Ordered Beams in CRYRING

A. SIMONSSON, H. DANARED, A. KÄLLBERG and K.-G. RENSFELT
Manne Siegbahn Laboratory, Frescativägen 24, SE-104 05 Stockholm, Sweden

Abstract. In the CRYRING storage ring weak beams of electron-cooled highly charged ions have been studied. When the intensity of the ion beam decays below a certain limit the momentum spread of the beam show a sudden transition to a state with much lower momentum spread. Similar observations have earlier been made at GSI, and have been interpreted as a one-dimensional ordering of the beam. For a beam of 7.4 MeV/u Xe^{36+} we have varied the electron density and studied how that affects the particle number at transition, the momentum spread of the ion beam, and the Schottky power. The transition occurs at particle distances between 1 and 35 cm, depending on the cooling rate.

Key words: ordered beam, phase transition, intrabeam scattering, Schottky noise, Schottky power.

1. Introduction

Some years ago it was discovered at the ESR ring at GSI in Darmstadt that the momentum spread of electron-cooled beams of highly charged heavy ions dropped abruptly to very small values when the ion beam current decayed below a certain threshold [1]. The drop has been seen for fully stripped ions in charge states from 18 to 92. The measurements have been interpreted as a transition to a state where the ions line up after each other [2]. Later the same phenomenon has been seen at the SIS ring, also at GSI.

Similar observations have now been made at CRYRING using beams of $^{58}Ni^{17+}$ at 8.1 MeV/u, $^{129}Xe^{36+}$ at 7.4 MeV/u and 3.1 MeV/u, and Pb^{53-55+} at 4.2 MeV/u. Because of more available beam time and suitable beam lifetimes the data with best quality was obtained with 7.4 MeV/u Xe^{36+} ions, and the following discussion will only concern these measurements.

The beam lifetime was 190 s without electron cooling, mainly determined by electron capture from residual gas molecules. With an electron density of 2.9×10^{13} m^{-3} the lifetime decreased to 29 s.

Before the transition the momentum spread and the transverse size of the beam is given by the balance between intrabeam scattering growth and electron cooling. Intrabeam scattering is Coulomb collisions between ions in the beam, that on average transfer energy from the huge amount of longitudinal kinetic energy to incoherent transverse or longitudinal motion.

Ions in an ordered beam no longer make any collisions and the intrabeam scattering disappears and the ion temperature can be much lower. The lower limit is given by the electron beam temperature.

2. Observations

The momentum spread of the ion beam was determined from the Schottky spectra. Figure 1 shows a sequence of Schottky spectra with relatively strong electron cooling. Each spectrum is an average of 224 individual spectra recorded by a spectrum analyzer, and centered to compensate for a slow frequency drift. The frequency scale is relative the tenth harmonic of the revolution frequency (10×727 kHz).

At $t = 0$ around 50000 ions were stored and cooled with the electron cooler. The electron density was 73×10^{12} m^{-3}, and due to collective effects the spectrum has double peaks, but they will not be discussed further here. After 125 s the beam intensity has decayed to one tenth, and then the frequency spread suddenly shrinks one order of magnitude.

2.1. PARTICLE NUMBER AT THE TRANSITION

The particle number at transition is too small to measure directly, so we have used a combination of a DC current integrator, a pick-up signal, and the rate from a residual gas beam profile monitor.

First the signal from an electrostatic pick-up was calibrated against a DC current transformer in a separate measurement. This gave an uncertainty of $\pm 10\%$, which dominates the uncertainty of the particle numbers. The pick-up signal gives the

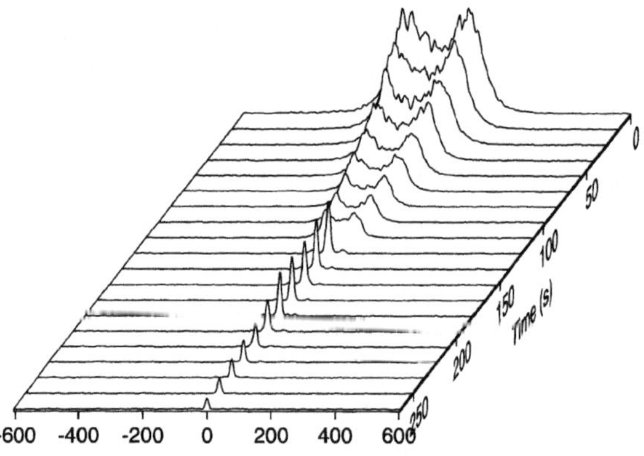

Figure 1. A sequence of Schottky spectra (measurements of the momentum spread in the beam) for a beam of $^{129}Xe^{36+}$ at 7.4 MeV/u. Around 50000 ions were stored and cooled, and the electron density was 7.3×10^{13} m^{-3}. The transition to low momentum spread takes place after 125 s.

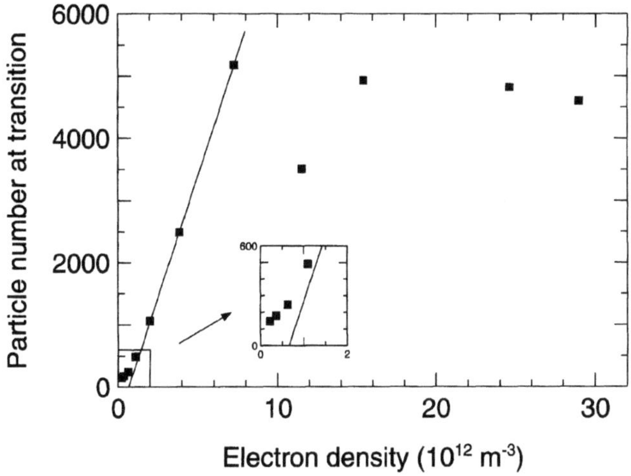

Figure 2. Number of particles in the ring at the transition to low momentum spread as a function of the electron density in the cooler.

particle number for each new injection. Since the RF system is turned off after acceleration the pick-up signal can only be used to get the initial value.

The lifetime of the beam was measured through the count rate of ionized rest gas molecules created by the ion beam. The decay of the ion beam can be followed down to 10–20 ions. The count rate cannot be calibrated directly to the beam transformer since it is sensitive to variations of the residual gas pressure, so the pick-up signal is used as a link.

Sequences like the one in Figure 1 were recorded for a number of different electron densities, and Figure 2 shows the particle number at the transition as a function of the electron density. A straight line is fitted to the data points for electron densities between 2×10^{12} m^{-3} and 8×10^{12} m^{-3}, and the data points for electron densities above 1×10^{13} m^{-3} have relatively constant particle numbers around 5000 ions. At electron densities below 1×10^{12} m^{-3} the number of ions at the transition deviates slightly from a straight line.

As shown in Figure 2 ordering of the ion beam appeared at electron densities down to 2×10^{11} m^{-3} and 145 particles in the ring. This means that before the transition intrabeam scattering limits the beam quality even when the mean ion–ion distance is as large as 35 cm.

2.2. MOMENTUM SPREAD AT TRANSITION

From Schottky sequences as the one in Figure 1 the momentum spread of the beam can be extracted as functions of particle number and electron density. This is shown in Figure 3 for a range of electron densities corresponding to the linear part of Figure 2. Although the individual data points scatter somewhat, the particle number

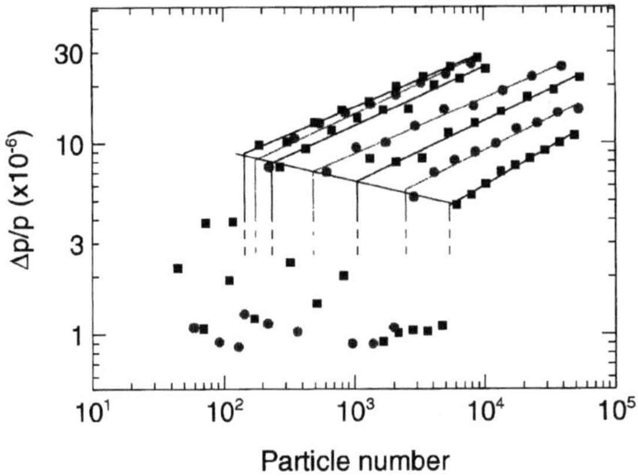

Figure 3. Momentum spread for the electron densities in the linear part of Figure 2 as a function of particle number. This is the width of the peaks in sequences of Schottky spectra like in Figure 1. The transition points are calculated separately for each sequence, and then the horizontal line is fitted to these points.

at transition can be determined quite accurately according to the discussion above. The momentum spread at the transition is thus also accurate, and a straight line is drawn through these points. The direct measurement of momentum spread is somewhat distorted by the double-peak effect but this is corrected in the data in Figure 3 where the momentum spread is extracted from a fit.

From the average slope of the fitted lines it is found that the momentum spread before the transition scales with particle number as $\Delta p/p \sim N^{0.32\pm0.03}$. This can be compared to the power 0.3 stated in [1].

The width of the Schottky peaks after the transition indicate a relative momentum spread in the order of 10^{-6}, but this is only an upper limit since ripple on the dipole power and the electron energy should contribute to at least that amount of widening.

2.3. SCHOTTKY POWER

The Schottky power at the tenth harmonic of the revolution frequency is plotted in Figure 4 for one high and one low electron density. Open symbols represent the power before the transition and filled symbols represent the ordered beam. With low electron density 2.0×10^{12} m^{-3}, the Schottky power is almost exactly proportional to the particle number both before and after the transition. When the electron density is higher, 7.3×10^{12} m^{-3}, the Schottky power is reduced. The reduction before the transition is caused by the coherent oscillations in the beam that also can be seen in the double-peak structure.

For a perfectly ordered beam the Schottky power becomes zero, except at harmonics of the revolution frequency times the number of particles (for us 3.6 GHz, which we cannot measure). After the transition the Schottky power decreases more

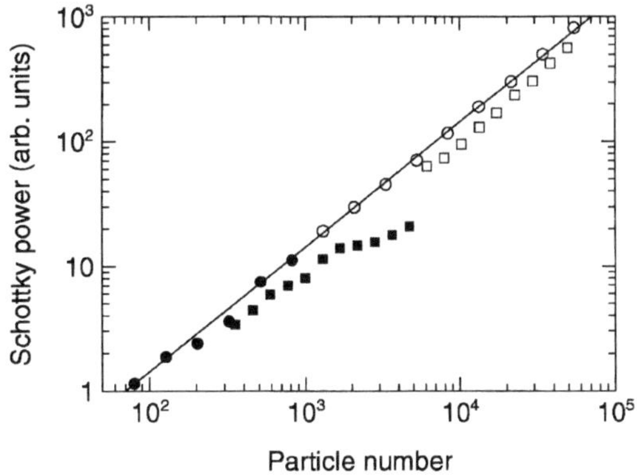

Figure 4. Schottky power as a function of particle number for two electron densities, 2.0×10^{12} m^{-3} (circles) and 7.3×10^{12} m^{-3} (squares). Open symbols – before the transition, filled symbols – ordered beam.

than a factor of two for the case with the highest electron density, and that is an indication of some kind of ordering of the positions in the beam.

3. Interpretations

The observations at GSI were interpreted as a transition to a liquid-like state. Plasma theory has been used to predict ordering in beams, and according this theory liquid order can be expected when the plasma parameter Γ is in the range 1–170. Γ is defined as the ratio between potential energy and kinetic energy, i.e. for a string of charges

$$\Gamma = E_{\text{pot}}/E_{\text{kin}} = Z^2 e^2/(4\pi\varepsilon_0 d)/kT.$$

With 5000 ions d equals 1 cm and the mean potential energy becomes 0.2 meV. The kinetic energy kT is in the order of 100 meV before the transition. After the transition the upper limit from the measurements is 2 meV longitudinally. The lower limit is given by the temperature of the electron beam, which are approximately 0.1 meV longitudinally and 1 meV transversely [3].

Thus Γ is much smaller than one before transition, but may be in the order of one for an ordered beam. On the other hand we see ordering also when the particle number and thus also Γ is down a factor 100, so clearly a large Γ is not necessary for this kind of beam ordering.

References

1. Steck, M. *et al.*, *Phys. Rev. Lett.* **77** (1996), 3803.
2. Hasse, R. W., *Phys. Rev. Lett.* **83** (1999), 3430.
3. Danared, H., *Nucl. Inst. Meth. A* **441** (2000), 123.

Hyperfine Interactions **146/147:** 215–218, 2003.

Cold Electrons from Cryogenic GaAs Photocathodes: Energetic and Angular Distributions

D. A. ORLOV[1], U. WEIGEL[1], M. HOPPE[1], D. SCHWALM[1], A. S. JAROSHEVICH[2], A. S. TEREKHOV[2] and A. WOLF[1]
[1]*Max-Planck-Institut für Kernphysik, 69029 Heidelberg, Germany*
[2]*Institute of Semiconductor Physics, 630090 Novosibirsk, Russia*

Abstract. For the investigation of electron–ion collisions at the Test Storage Ring with a much enhanced energy resolution a cold electron source with cryogenic GaAs(Cs,O) is being developed. To ensure optimal performance of GaAs photocathodes the two-dimensional energy distributions of emitted electrons are studied at room and low (90 K) temperature. The photocathode performance data promise electron beams of several 10 mA/cm^2 with an energy spread $\leqslant$10 meV.

Key words: cold electron beams, NEA-photocathodes, energy and angular distributions.

1. Introduction

Intense electron beams with low energy spreads, especially in the transverse degree of freedom, are required to study electron–ion interactions at low relative energies at the Heidelberg Test Storage Ring [1]. A promising candidate for the electron source is GaAs, activated by cesium and oxygen to the state of Negative Electron Affinity (NEA), when the vacuum level (E_{vac}) lies below the bottom of the conduction band (E_c) in the bulk of the semiconductor [2]. Inside the bulk an intense electron ensemble with an energy spread as low as the crystal temperature can be obtained by photoexcitation of electrons near the conduction band minimum. In addition, for epitaxial layers of GaAs with the electron diffusion length exceeding the optical absorption length, almost all photoelectrons reach the surface and due to the NEA condition can escape to vacuum. The quantum efficiency (QE) as well as the broadening of the energy and angular distributions of photoemitted electrons are controlled mainly by elastic and inelastic electron processes near the interface. To study the photoelectron escape process we developed a method [3] to measure the complete energy distribution $N(E_{||}, E_{\perp})$ of photoelectrons in vacuum as a function of longitudinal and transverse energies.

In this paper we present $N(E_{||}, E_{\perp})$ distributions for GaAs(Cs,O) measured at room temperature and under cryogenic conditions. It was shown that elastic as well as inelastic electron scattering processes are of crucial importance in the electron transfer through the interface. The possibility to obtain intense cold electron beams is demonstrated.

2. Experimental

We used GaAs transmission mode photocathodes bonded to a transparent glass carrier. The emitting 1.5 μm thick p^+-GaAs(100) layer was doped with Zn to a concentration of about 5×10^{18} cm^{-3}. The final heat cleaning and subsequent activation of the photocathode with (Cs,O) to QE of about 25% (in the reflection mode at 670 nm) were performed in an ultra-high vacuum chamber with a base pressure below 10^{-12} mbar. The details of the photocathode preparation setup are described in [4]. After the activation the photocathode was transferred into the electron gun of the measurement setup, where it could be cooled to 90 K. The photocathode was illuminated in the reflection mode with a 800 nm diode laser and distributions of photoelectrons in vacuum were studied.

The measurements of two-dimensional electron distributions as a function of longitudinal and transverse energies were performed using a new method [3] based on "marking" electrons with fixed longitudinal energies $E_{||}^b$ directly after the photoemission. Applying an adiabatically changing magnetic field, a well-known portion of the transverse energy was transferred to the longitudinal one [5]. The transverse energy distribution of marked electrons $N_\perp(E_{||}^b, E_\perp)$ was then obtained by measuring the longitudinal energies of these electrons by a retarding field analyzer. A series of measurements for different $E_{||}^b$ samples directly the two dimensional distribution $N(E_{||}, E_\perp)$. The longitudinal energy resolution of the analyzer, defined as the standard deviation of a Gaussian, was 17 meV.

3. Results and discussion

Figure 1 shows two-dimensional energy distributions of electrons for the same photocathode measured at room temperature and at 90 K. The positions of the

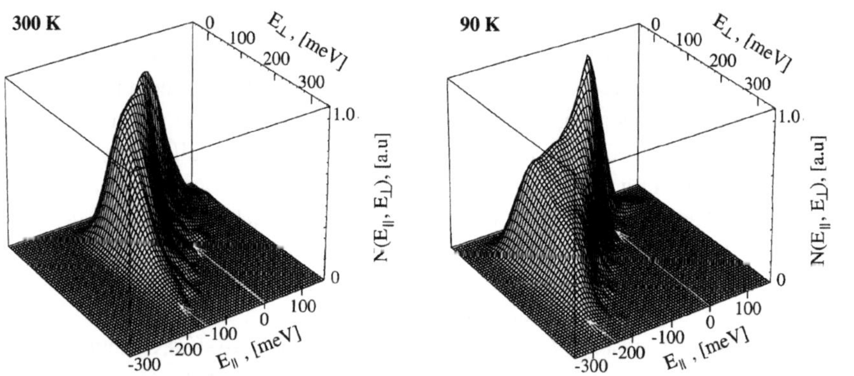

Figure 1. Two-dimensional energy distribution of electrons as a function of longitudinal ($E_{||}$) and transverse ($E_\perp$) energies measured at two temperatures. The positions of the conduction band minimum in the bulk (at $E_{||} = 0$) and of the vacuum level are shown by arrows.

conduction and vacuum levels are shown in the figure. The value of |NEA| is 155 meV and 245 meV at room and low temperature, respectively. It is seen that instead of the narrow distribution in the bulk, where photoelectrons are located at E_c with an energy spread of about kT, in the vacuum the electron distributions are strongly broadened. The $N(E_{||}, E_{\perp})$ are dominated by a peak slightly below E_c, i.e., a considerable part of electrons leaves to vacuum without strong energy and momentum relaxation. A shift of the peak to lower energies might be explained [4, 6] by an efficient capture of electrons into the upper two-dimensional quantized subband in the near-surface potential well, from where they can escape to the vacuum by tunneling through the thin potential barrier. The high energy side of the peak ($E_{||} \geqslant E_c$), which originates from the Boltzmann distribution, drops exponentially with $1/e$-intervals of about 29 meV (at 300 K) and 12 meV (at 90 K), i.e., of about kT. At low temperature the low energy side, after an initial steep decrease, gradually evolves into a plateau extending down to the vacuum level with a width continuously growing with the decrease of $E_{||}$. At room temperature the plateau is not so strongly pronounced due to the smaller |NEA| and the kT-broadening. Looking down from the peak, two-dimensional distributions are found to be limited by the line of zero transverse energy to the right, and by the line of constant total energy $E_{\perp} + E_{||} \approx E_c$ to the left. At the right-hand limit the electrons lost a part of their total energy but kept a low transverse energy. At the left-hand limit the photoelectrons passed through the band bending region without energy loss, but were scattered elastically and emitted at a large angle to the surface normal. The plateau itself corresponds to electrons which were scattered and suffered energy loss as well. Although the mechanisms of the electron elastic and inelastic scattering are still controversial, the observation [6] of phonon replicas in some longitudinal energy distributions proves that phonon emission is at least one of the relevant mechanisms of electron energy loss near the interface. Moreover, scattering of electrons on local fluctuations of the potential near the surface can give a contribution to the momentum broadening.

The obtained data demonstrate that the transverse and longitudinal energy spreads from NEA photocathodes are limited mainly by electron scattering near the surface. At ultra-high vacuum conditions, when a degradation of the QE due to the cryoadsorption effect is suppressed, decreasing the temperature leads mainly to a rise of the |NEA| ($\approx$85 meV in our measurements) due to the increase of the GaAs band gap, in agreement with earlier observations [7]. This rise of the |NEA| opens additional momentum space which can be occupied by electrons which are scattered during the escape process. All together this leads to an increase of the QE (from 23% to 29%) and to a broadening of the energy distribution: we found that the mean transverse energy increased from 59 meV to 66 meV and the longitudinal one changed from -70 meV to -132 meV (with $E_{||}$ measured above E_c). Thus, photocathodes behave opposite to thermocathodes: lowering the temperature causes larger energy spreads. On the other hand, it is seen (Figure 1) that by extracting electrons above a suitable longitudinal energy barrier ($\approx E_c$), an electron beam with longitudinal and transverse energy spreads of about the bulk temperature can indeed be obtained.

The selection of these "cold" electrons can be achieved by an external potential barrier, by extracting in the space charge mode, or/and by preparing the surface with about zero NEA. It was found that for optimally activated photocathodes, cutting the electrons with longitudinal energies below a carefully chosen barrier, we can prepare an electron beam with transverse and longitudinal energy spreads of about 25 meV at 300 K (effective QE of about 2.5–3.0%), and less than 10 meV at 90 K (effective QE of about one per cent). This means that in order to obtain several mA of dc current a laser power as high as 1 W has to be applied in a spot with a diameter of about 3 mm. For our present setup this power level still causes a strong temperature increase and short life times of the photocathodes [8]. To improve the photocathode performance we studied the electron escape process at different surface conditions. We found that with thin (Cs,O) activation layers the effective QE could be increased by a factor of 1.3–1.5 [9]. We also developed a new cooling system where: (a) the number of critical interfaces is minimized, (b) instead of glass substrate a sapphire carrier is used, which has much higher thermoconductivity at low temperatures, and (c) a high force of 200–300 N is applied to the crucial sapphire-cooler boundary (with a contact area of $\approx$2.5 cm^2) in order to decrease the thermal resistance of the interface. First measurements in a test chamber showed only a small increase of the temperature of a sample (few degrees) at the incident light power of 1 W.

In conclusion, the two-dimensional energy distributions of electrons emitted from GaAs(Cs,O) surfaces were studied. It was found that elastic as well as inelastic electron scattering strongly influences the photoelectron transfer through the GaAs(Cs,O)-vacuum interface. Based on these measurements we are confident that cold electron beams of several 10 mA/cm^2 with longitudinal and transverse energy spreads $\leqslant$10 meV can be produced.

References

1. Pastuszka, S., Schramm, U., Grieser, M., Broude, C., Grimm, R., Habs, D., Kenntner, J., Miesner, H.-J., Schüssler, T., Schwalm, D. and Wolf, A., *Nucl. Instr. Meth. A* **369** (1996), 11.
2. Scheer, J. J. and van Laar, J., *Solid State Commun.* **3** (1965), 189.
3. Orlov, D. A., Hoppe, M., Weigel, U., Schwalm, D., Terekhov A. S. and Wolf, A., *Appl. Phys. Lett.* **78** (2001), 2721.
4. Pastuszka, S., Hoppe, M., Kratzmann, D., Schwalm, D., Wolf, A., Jaroshevich, A. S., Kosolobov, S. N., Orlov, D. A. and Terekhov, A. S., *J. Appl. Phys.* **88** (2000), 6788.
5. O'Neil, T. M. and Hjorth, P. G., *Phys. Fluids* **28** (1985), 3241.
6. Orlov, D. A., Andreev, V. E. and Terekhov, A. S., *JETP Lett.* **71** (2000), 151.
7. Terekhov, A. S. and Orlov, D. A., *JETP Lett.* **59** (1994), 864.
8. Weigel, U., Diploma thesis, Univ. Heidelberg, 2000.
9. Orlov, D. A., Weigel, U., Schwalm, D., Terekhov, A. S. and Wolf, A., in preparation.

Hyperfine Interactions **146/147:** 219–224, 2003.

Study of Cooling and Storage Properties of a Gas-Filled Linear Paul Trap Coupled to a Time-of-Flight Mass Spectrometer for Mass Measurements of Exotic Nuclei

M. WEIDENMÜLLER[1,*], S. A. ELISSEEV[1], W. R. PLASS[1,**], U. CZOK[1], M. WINKLER[1], H. WOLLNIK[1], H. GEISSEL[1,2], C. SCHEIDENBERGER[2], H. WEICK[2], Z. ZHOU[1,2], A. F. DODONOV[3] and V. KOZLOVSKI[3]
[1]*II.Physikalisches Institut, Justus-Liebig-Universität Gießen, Heinrich-Buff-Ring 14-16, D-35392 Gießen, Germany; e-mail: Wolfgang.R.Plass@exp2.physik.uni-giessen.de*
[2]*GSI, Planckstraße 1, D-64291 Darmstadt, Germany*
[3]*Institute of Chemical Physics, Chernogolovka, Russia*

Abstract. A gas-filled radio-frequency quadrupole has been developed, which acts as an interface between a time-of-flight mass spectrometer and an atmospheric-pressure ion source or a gas-filled ion beam stopping cell. It delivers cooled continuous or bunched beams to the mass analyzer. The radio-frequency quadrupole has been characterized and the longitudinal energy distributions, bunching efficiencies and beam losses as a function of storage time, and the duty cycle improvement have been measured. Using the bunch mode, the mass-resolving power of the time-of-flight mass spectrometer could be increased by a factor of about two and now amounts to 24 000 (full width at half maximum).

Key words: time-of-flight mass spectrometer, exotic nuclei, collisional cooling, radio-frequency quadrupole, linear Paul trap.

1. Introduction

Buffer-gas cooling is widely used in ion traps and ion-guide systems for phase-space reduction of ion beams. Here, buffer-gas cooling is applied in order to reduce the velocity spread in a radio-frequency quadrupole (RFQ) ion-guide and cooler system, which couples an atmospheric-pressure ion source to a time-of-flight mass spectrometer (TOF-MS) with orthogonal extraction [1]. In such an energy-isochronous TOF-MS, phase-space reduction of the injected ions is essential to achieve a high mass-resolving power, because the mass-resolving power is usually limited by the transverse velocity spread before ion extraction.

For the present experiments, the RFQ was operated as ion guide in the conventional continuous mode and as a linear Paul trap [2] in bunch mode. It is the

* Part of doctoral thesis, Justus-Liebig-Universität Gießen, 2002.

** Corresponding author.

goal of the present paper to study the cooling and storage properties of the system, to investigate the properties of the extracted ion beams, and to compare the two operation modes regarding duty cycle and mass resolution.

2. Experimental setup

A schematic view of the experimental setup is shown in Figure 1. The atmospheric pressure electrospray ion source [3] can continuously deliver, among others, singly charged sodium, potassium and cesium ions. The first RFQ is used to pre-cool the ions and to transmit them through a differential pumping stage at a pressure of approximately 1 mbar. The ions enter the second RFQ with kinetic energies of a few electron-volts. The buffer-gas pressure in this region amounts to approximately $1 \cdot 10^{-2}$ mbar. The second RFQ has a length of 170 mm and an inner diameter of 7 mm, the rods have a diameter of 8 mm. A special arrangement [4] of four additional rectangular electrodes, which are placed between neighboring rods parallel to them and with increasing distance from the quadrupole axis, allows to superimpose an axial field, which drags the ions through the system. The RFQ is operated in RF-only mode at a frequency of 1.8 MHz and at the Mathieu parameter $q = 0.35$ for the ions of interest, corresponding to amplitudes ranging from 300 to 700 V. Behind the quadrupole, there is a trap electrode and the exit skimmer electrode. Depending on the voltages applied to these electrodes the RFQ can be operated in two different modes: (i) In continuous mode, the trap electrode is set to the same potential as the exit skimmer electrode and the ions pass through the RFQ continuously. (ii) In bunch mode, the trap electrode is used to form a linear Paul trap [2] with a longitudinal potential well depth of 1 to 2 V, in which the ions are accumulated. In order to extract the ions, the potential of the exit skimmer electrode is lowered and the ions leave the RFQ in a bunch. After extraction, the ions pass through a drift region and hit a micro-channel-plate detector. A grid system [5] in front of the detector, to which retarding potentials can be applied, allows to measure the longitudinal kinetic energy distribution of the ions.

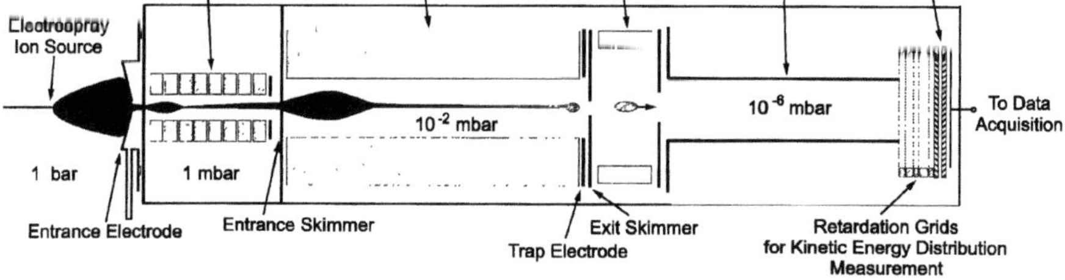

Figure 1. Schematic view of the setup used for the present measurements.

3. Results and discussion

The cooling properties of the second RFQ were studied by simulations with the ion optics program SIMION 7 [6] employing a Monte-Carlo-type user program, which implements a hard-sphere collision model. In the limit of low ion density, where the ion temperature is determined by RF heating, the calculations show that in both continuous mode and bunch mode the ions can be cooled inside the RFQ to temperatures 2 to 3 times larger than the buffer gas temperature. The ion temperature corresponds to a kinetic energy of about 80 meV. A radial spread in the RFQ of only a few hundred micrometers can be achieved. These results are in good agreement with the results of an analytical treatment [7].

Experimentally, the longitudinal kinetic energy distribution in continuous mode was determined. As a representative example, the measured retardation curve of a $^{133}Cs^{+}$ beam is shown in Figure 2. From the derivative of the measured intensity distribution one obtains the longitudinal kinetic energy distribution of the ions, which is characterized by a mean kinetic energy of 14.5 eV and a width as small as 260 meV. The width is larger than theoretically expected (see above). This could be due to several factors: RF fringe fields in axial direction at the exit of the quadrupole may influence the longitudinal energy distribution, collisions with the buffer gas during or after extraction may induce energy straggling, and the limited energy resolution of the retardation measurement technique due to field penetration through the retardation grids could be responsible for the observed width of the energy distribution.

In the next step the cooler was operated in the bunch mode. Singly-charged sodium, potassium and cesium ions were cooled and stored in air and in helium. The measured widths of the longitudinal kinetic energy distribution are one order of magnitude larger than in the continuous-extraction mode and almost independent from the ion species and the buffer gas type. They increase with extraction field

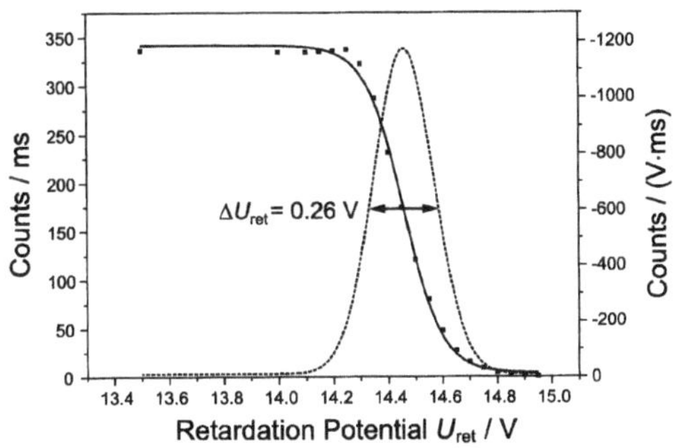

Figure 2. Measured retardation curve for $^{133}Cs^{+}$ ions (full square symbols), showing the number of ions arriving at the detector as a function of the applied retardation voltage. A Boltzmann fit is applied (solid line) and its derivative is depicted with the dashed line (right ordinate). The latter reflects the longitudinal energy distribution which is approximated by a Gaussian fit.

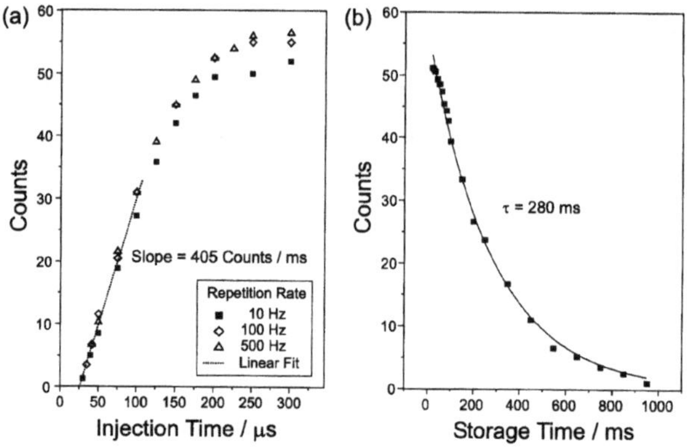

Figure 3. Accumulation and storage properties of the trap studied for cesium ions cooled with dried air as buffer gas at $1 \cdot 10^{-2}$ mbar pressure. Number of detected ions as a function of (a) the injection time for different cycle repetition rates (b) the storage time in the trap.

strength and thus do not reflect the ion temperature in the trap but the different potential energies of the ions located at different positions during extraction from the trap.

For the study of the capture and storage properties of the linear Paul trap, the injection time and the storage time was varied and the number of ejected ions was measured (Figure 3a). For injection times below 120 μs the number of ions ejected from the trap increases linearly with the injection time, as expected. The slope of a linear fit to the data yields a rate of 405 detected ions per millisecond. This accumulation rate is in agreement with the measured rate of transmitted ions in the continuous mode. Thus the capture efficiency of the trap is close to unity. This result is independent of the repetition rate of the whole cycle. For injection times longer than 150 μs a saturation effect becomes visible and the number of detected ions does not grow further, presumably because the storage capacity of the trap is reached. From ion current measurements the maximum number of stored ions is estimated as approximately 1000. This is a very small number compared to the expected storage capacity in excess of 10^6 ions [8]. The reason for this discrepancy is currently not understood and will be an issue for further investigations. Most likely ion losses occur at the exit skimmer.

A further possible reason could be ion losses in the trap due to neutralization processes arising from charge exchange or other collision processes. Therefore the intensity of the ejected ion bunch was studied as a function of storage time in the trap in the range between 18 ms and 950 ms. The result is shown in Figure 3b. The time constant for ion loss from the trap is derived from an exponential fit to the data and it is found as $\tau = 280$ ms. It is much longer than the injection time of 200 μs required to reach storage saturation. Thus collisional ion losses cannot account for the observed low number of stored ions.

The operation of the RFQ in bunch mode as a linear Paul trap yields several advantages for the performance of the TOF-MS compared to the operation in continuous mode: (i) In continuous mode the duty cycle of the instrument is only about 4% for singly charged cesium ions at an extraction repetition frequency of 3 kHz. In bunch mode the incoming beam can be accumulated and the duty cycle of the system, and hence its sensitivity, is enhanced. It was measured that the number of detected ions increases by a factor of 100. This number includes improvements in duty cycle and probably also in ion transmission. (ii) In bunch mode, the ions are spatially focused in the extraction region of the TOF-MS and suffer less from the unavoidable mechanical misalignment and field imperfections in the injection zone. This leads to a gain in mass resolution (FWHM) by roughly a factor of two from typically 13 000 in continuous mode to 24 000.

4. Summary and outlook

Cooling and storage properties of a gas-filled RFQ have been studied theoretically and experimentally. In the continuous mode, longitudinal energy distributions with a width of 260 meV (FWHM) have been measured at a mean kinetic energy of 14.5 eV. In bunch mode, a high injection efficiency (of the order of unity) into the linear Paul trap is achieved. However, the storage capacity is lower than expected and will be the topic of further investigations. A TOF-MS with orthogonal extraction has been coupled to the linear Paul trap. In bunch mode, the number of detected ions could be increased by a factor of 100 and the improved injection conditions increase the mass resolution by a factor of two.

It is intended to use such a system for direct mass measurements of exotic nuclei with half-lives of the order of 10 ms. These nuclei will soon become available at the SHIPTRAP [9] facility, where fusion-reaction products selected with SHIP will be thermalized in a gas-filled stopping cell and extracted with an RFQ system. Our system is ideally suited for this purpose because it is coupled to an ion source which is operational up to atmospheric pressure and because the cycle length is of the order of 10 ms only. The system will ideally complement the mass measurements planned with the Penning-trap system at SHIPTRAP.

References

1. Dodonov, A. F., Kozlovski, V. I., Soulimenkov, I. V., Raznikov, V. V., Loboda, A. V., Zhen, Z., Horvath, T. and Wollnik, H., *Eur. J. Mass Spectrom.* **6** (2000), 481.
2. Campbell, J. M., Collings, B. A. and Douglas, D. J., *Rapid Commun. Mass Spectrom.* **12** (1998), 1463.
3. Gaskell, S. J., *J. Mass Spectrom.* **32** (1997), 677.
4. Loboda, A., Krutchinsky, A., Loboda, O., McNabb, J., Spicer, V., Ens, W. and Standing, K., *Eur. J. Mass Spectrom.* **6** (2000), 531.
5. Weidenmüller, M., Doctoral thesis, Justus-Liebig Universität, Gießen, 2002.
6. Dahl, D. A., *Int. J. Mass Spectrom.* **200** (2000), 3.

7. Moriwaki, Y., Tachikawa, M., Yashiharu, M. and Shimizu, T., *Jpn. J. Appl. Phys. Part 2* **31** (1992), L 1640.
8. Dehmelt, H. G., *Adv. Atom. Mol. Phys.* **3** (1967), 53.
9. Dilling, J. *et al.*, *Hyp. Interact.* **127** (2000), 491.

Hyperfine Interactions **146/147:** 225–230, 2003.

New Magnetic Storage Rings

MARKUS STECK
GSI, Darmstadt, Germany

Abstract. Magnetic storage rings have been successfully employed for atomic and nuclear physics experiments. Experience gained at existing magnetic storage rings with ions and molecules will be the basis for projects of new magnetic storage rings presently being designed or constructed. These storage rings will be mostly devoted to radioactive beams using many of the available techniques. Some techniques relevant for the new storage rings will be quickly reviewed. An overview of the new storage ring projects will be given.

Key words: storage ring, beam cooling.

1. Introduction

Magnetic storage rings are a powerful tool for atomic and nuclear physics experiments. The existing storage rings [1] are shared by the two physics communities. Three storage rings (CELSIUS, IUCF Cooler, COSY) are mostly used for nuclear physics with proton beams and internal targets. Three storage rings (ASTRID, CRYRING, TSR) provide positively charged light ions, molecules and sometimes also negatively charged ions mainly for atomic physics. The ESR storage ring, finally, can be filled with basically all ion species up to bare heavy ions. Radioactive beams which are produced by projectile fragmentation and separated in a magnetic system can also be injected into the ESR. All these storage ring are preferentially used with stored beams, although in a few cases also beam extraction is available. The stored beam can interact with internal targets consisting of gaseous material, pellets or free electrons. The heating due to the presence of a target is compensated by beam cooling. Beam cooling also improves the quality of the ion beam which provides conditions for experiments with unprecedented precision.

2. Techniques applied to stored ion beams

2.1. INTERNAL TARGETS

The interaction of a particle in a circulating beam which passes about 10^6 times per second through a target limits the maximum target density to values which are orders of magnitude smaller than those of the thinnest solid targets. For nuclear physics mainly hydrogen targets, sometimes polarized, are used in combination with proton beams. Gas-jets, cluster-jet targets and pellet targets with maximum target densities of up to 10^{16} hydrogen atoms/cm^2, are available. Atomic physics experiments require lower densities, because of the larger cross sections, but a larger range of target materials. This requirement can be met by gas jets which provide target densities of some 10^{12} to 10^{14} atoms/cm^2 for gases ranging from hydrogen up to heavy noble gases like xenon [2]. The internal targets are particularly powerful in combination with beam cooling which compensates adverse effects like energy loss and straggling in the target which impair the quality of the circulating beam.

2.2. BEAM COOLING

Three methods of cooling stored ion beams have been demonstrated. Their application is quite different. Stochastic cooling is most powerful for hot beams, secondary beams produced by bombardment of a solid target are the typical application of stochastic cooling. Antiprotons are routinely cooled this way as they have large emittances and energy spread after production in the target. Stochastic cooling of radioactive ions which are produced by projectile fragmentation has been demonstrated just recently in the ESR. It can be employed for accumulation of higher beam intensities in a similar fashion as it has been used in the accumulation of intense antiproton beams for collider rings, but with cooling times below 1 s. For radioactive isotopes with lifetimes of the order of seconds it allows accumulation of intense radioactive beams for experiments with stored ions. For highest precision it has to be complemented by electron cooling.

Electron cooling is the standard method available in all existing storage rings to provide ion beams of superb quality. The cooling of the hot ion beam by Coulomb interaction with a much colder electron beam is usually limited by intrabeam scattering. The Coulomb scattering in the dense ion beam causes a heating rate which is proportional to the intensity and to the phase space density of the ion beam. The beam quality is determined by an equilibrium between this heating process and electron cooling. Due to the q^2/A-dependence of the cooling rate and the q^4/A^2-dependence of the intrabeam scattering rate the equilibrium values are highest for highly charged ions. For low intensity beams a suppression of intrabeam scattering has been observed which occurs as a reduction of the momentum spread by up to one order of magnitude. The momentum spread of order 10^{-7} of the low intensity ion beam facilitates the determination of the mass of radioactive nuclei by measurement of their revolution frequency with correspondingly high accuracy. In Schottky mass spectrometry the masses of the unknown isotopes are determined by comparison of their revolution frequency with that of known isotopes which are stored simultaneously [3]. The cooling times for the hot radioactive beam by electron cooling is on the order 10 s. Even with stochastic precooling the total cooling time will hardly fall below 1 s. For much shorter lifetimes of the radioactive nuclei other methods are required, such as isochronous ring operation.

The third cooling method laser cooling can provide beams of very small momentum spread. Transverse cooling, however, can only be achieved by coupling the transverse and longitudinal degree of freedom and therefore is much weaker. As laser cooling can only be applied to a few incompletely stripped ions with appropriate atomic transitions it is useful only for a few dedicated experiments.

2.3. ISOCHRONOUS OPERATION OF THE STORAGE RING

In synchrotrons where the beam is accelerated by simultaneous ramping of the magnetic field and the frequency of the accelerating rf system the transition energy must be either avoided or crossed rapidly. The transition energy corresponds to an isochronous circulation of beam particles with different momenta. This causes a loss of phase stability in the rf bucket and therefore unavoidable beam loss. For a coasting beam which is not manipulated with the rf system the transition energy is not harmful.

Therefore, if acceleration is not required, a coasting beam can circulate at the transition energy stably. This is a way to have isochronous circulation for ions with different momenta, but fall into the acceptance of the storage ring [4]. As there is basically no frequency spread due to the momentum spread of the beam, the revolution frequency can be determined precisely even without cooling. This mode is preferentially used for very short lived nuclei. Drawbacks are that the ring acceptance is strongly reduced compared to normal operation and that the revolution frequency is more sensitive to variations of the magnetic bending field. For nuclei with lifetimes below about 10 ms Schottky noise analysis is not applicable because of the long measuring times required for the frequency analysis. A solution has been developed by measuring the revolution time with a fast detector which measures the passage of individual particles through a thin foil and detection of secondary particles.

2.4. BEAM ACCUMULATION

Radioactive ion beams after a fragmentation target are, due to the low production cross sections, produced with low intensity. For internal experiments in a storage ring higher beam intensities must be provided by accumulation. Standard accumulation methods are limited by the emittance of the incoming beam and the acceptance of the storage ring. This limitation can be overcome if cooling is introduced into the accumulation process. For the accumulation of antiprotons schemes have been developed which are based on stochastic cooling systems. For low energy ion beams the application of electron cooling to large emittance beams after transverse multiturn injection or longitudinal rf stacking have been demonstrated successfully. For highly charged ions cooling times down to 0.2 s have been achieved. The ion current in this scheme is either limited by the lifetime of the ion beam, mainly due to recombination with the electrons, or by the high phase space density of the cold beam which causes beam loss due to transverse instabilities. In the first case intensity gain factors of several tens can be achieved, whereas the instabilities limit the ion current to about 10 mA.

2.5. BEAM DECELERATION

High beam energies of several hundred MeV/u are required to ionize the ions to high charge states by stripping in a foil as well as to produce radioactive ions by projectile fragmentation. For precision experiments with stored ions low energies can often be beneficial in order to diminish Doppler shift and line broadening originating from the beam velocity. Further deceleration to rest is required for injection into traps. Two storage rings are providing decelerated beams for experiments with slow particles. In the antiproton decelerator AD at CERN antiprotons are routinely decelerated from 2.8 GeV to 5 MeV and in the ESR at GSI virtually all highly charged ions can be decelerated from energies around 300 MeV/u to a variable end energy as low as 10 MeV/u. In both accelerators preparations are under way for further deceleration to almost rest and final injection into traps.

3. Projects for new storage rings

The new storage ring projects use many of the features of the existing storage rings. In contrast to the existing rings they will be operated mainly with radioactive beams produced by projectile fragmentation of a fast heavy ion beam. All three new complexes which are proposed or already under construction use existing heavy ion beam facilities as injectors. A major effort is devoted to intensity upgrades for these injectors in order to achieve maximum output of radioactive ions. The secondary beams will be accumulated employing stochastic cooling, electron cooling or a combination of both.

3.1. CSR

At the Institute of Modern Physics IMP in Lanzhou, China, a synchrotron-storage ring complex is under construction which uses ions from the existing cyclotron [5]. By accumulation of ions in the synchrotron CSRm by a combination of multiturn injection, rf stacking and electron cooling intensities in the range 10^6–10^9 ions per pulse will be available. The heaviest species can be accelerated to a maximum energy of 400 MeV/u. The primary heavy ion beam will be used to bombard a fragmentation target producing a secondary beam of radioactive ions which will be stored in the storage ring CSRe. The stored beam, which can also be the primary beam, will be used for internal experiments. An internal gas jet target (maybe also for polarized target atoms) and an electron cooling system will be available. Mass measurements by Schottky noise detection of the cooled fragment beams or by isochronous operation of the ring are foreseen.

3.2. MUSES

The MUSES complex which is part of the RI beam factory project [6] at RIKEN, Japan, will also use primary heavy ion beams from a cyclotron for the production of radioactive beams by projectile fragmentation. Presently a cascade of new cyclotrons is under construction which allows to accelerate even the heaviest ions up to an energy of 350 MeV/u. The radioactive products will be accumulated in a dedicated collector ring ACR by a combination of stochastic and electron cooling. Cooling times of 10 ms will be required in order to collect radioactive beams of sufficient intensity for internal experiments in the subsequent storage ring. To support fast stochastic cooling the special ion optical split-ring design is considered which gives the best mixing conditions for stochastic cooling. The ion optical structure can also be tuned to transition energy allowing isochronous operation.

The collector ring will provide intense radioactive beams for the subsequent storage ring. This storage ring is designed for collision experiments of the ion beam with a counterpropagating electron beam which circulates in an electron storage ring with an energy up to 1 GeV. This will allow electron scattering on radioactive nuclei with luminosities of order 10^{26} cm^{-2} s^{-1}. The highest electron energies will be used to generate high energy X-rays for interaction with the ion beam.

3.3. THE NEW GSI COMPLEX

A proposal for a new synchrotron-storage ring facility is presently worked out at GSI, Germany. This new facility is based on the new high intensity, fast cycling superconducting synchrotrons SIS 100/200 which will provide heavy ion beams of up to 1 GeV/u for the production of radioactive secondary beams and 29 GeV protons to generate antiprotons [7]. The radioactive beams or the antiprotons can be cooled stochastically in a dedicated collector ring CR. A split-ring ion optical lattice provides optimum conditions for stochastic cooling and can also be used for isochronous mass measurements. Cooling times around 0.5 s for radioactive ions and 5 s for antiprotons will allow fast beam accumulation of the precooled beams in the subsequent storage ring NESR. This new storage ring will have many features of the existing ESR, like electron cooling, stochastic cooling, internal target, beam deceleration, etc., and in addition one straight section will overlap with an electron storage ring. The ions will collide with electrons of energy up to 500 MeV circulating in a storage ring to study charge radii and charge distributions of radioactive nuclei. The new high energy storage HESR is designed for internal experiments with intense antiproton beams of up to 14 GeV, which are available after acceleration of the accumulated antiprotons in the synchrotron. Special features of this storage ring will be a high-density hydrogen target and an electron cooling device for the high energy antiprotons, aiming at high-resolution spectroscopy of highly excited nuclear states.

References

1. Reistad, D., In: *Proc. of the 6th Europ. Part. Acc. Conf.*, Stockholm, 1998, p. 141.
2. Ekström, C., *Nucl. Phys. A* **626** (1997), 405c.
3. Franzke, B., Beckert, K., Eickhoff, H., Nolden, F., Reich, H., Schwinn, A., Steck, M. and Winkler, T., In: *Proc. of the 6th Europ. Part. Acc. Conf.*, Stockholm, 1998, p. 256.
4. Hausmann, M., Attallah, F., Beckert, K., Bosch, F., Dolinskii, A., Eickhoff, H., Falch, M., Franczak, B., Franzke, B., Geissel, H., Kerscher, Th., Klepper, O., Kluge, H.J., Kozhuharov, C., Löbner, K. E. G., Münzenberg, G., Nolden, F., Novikov, Yu., Radon, T., Schatz, H., Scheidenberger, C., Stadlmann, J., Steck, M., Winkler, T. and Wollnik, H., *Nucl. Instr. Meth. Phys. Res. A* **446** (2000), 569.
5. Wei, B. W., Luo, Y. X., Jin, G. M., Zhang, W. L., Xia, J. W., *Nucl. Phys. A* **626** (1997), 561c.
6. Yano Y., Goto, A., Kase, M. and Katayama, T., In: *Proc. of the 2001 Part. Acc. Conf.*, Chicago 2001, p. 509.
7. Nolden F., Beckert, K., Beller, P., Boine-Frankenheim, O., Dolinskii, A., Franzke, B., Steck, M., Weick, H., In: *Proc. of the 2001 Part. Acc. Conf.*, Chicago, 2001, p. 503.

Hyperfine Interactions **146/147:** 231–235, 2003.

Research with Trapped keV Ion Beams

H. B. PEDERSEN, D. STRASSER, B. AMARANT, O. HEBER,
M. L. RAPPAPORT and D. ZAJFMAN*
Department of Particle Physics, Weizmann Institute of Science, Rehovot, 76100, Israel;
e-mail: fndaniel@wicc.weizmann.ac.il

Abstract. The motion of an ion bunch trapped between two electrostatic mirrors in an ion trap resonator has been studied. Under certain conditions the ion motion becomes correlated and either synchronization of the ion motion or enhanced diffusion occurs.

1. Introduction

In the last few years a new class of ion traps has been developed [1–3], where ions oscillate between a pair of electrostatic mirrors much like photons in an optical resonator. The trapped ions have kinetic energies of a few keV in the central part of the trap; inside the mirrors they are slowed down to a few meV near their turning points where they are also subject to strong radial focusing forces. Due to these effects, the density of ions can increase by two to three orders of magnitude inside the mirrors.

We have studied the evolution of bunches of ions during storage in such a trap. We find that the Coulomb repulsion among the ions, in combination with the kinematical properties dictated by the electrostatic fields, can either enhance the diffusion, and hence speed up the transition to a steady state beam, or cause ion motion synchronization and self-ordering phenomena to occur.

2. Experimental setup

A drawing of our electrostatic ion trap is shown in Figure 1 [1, 4]. The ion trap consists of two coaxial electrostatic mirrors each composed of a stack of cylindrical electrodes. The stability of the ion trap can be well understood with analogy to an optical resonator [1, 5]. Injection of an ion beam into the trap is performed by rapidly switching the potentials of the entrance electrodes [1].

The configuration of the trap is characterized by the set of potentials on five electrodes: [V_1, V_2, V_3, V_4, V_z] (see Figure 1), the others being grounded. Here, a particular set of trap configurations is investigated, namely [V_1, 4.875 kV, 3.25 kV, 1.625 kV, 4.06 kV], where $4.0 < V_1 < 6.5$ kV. These configurations have a high

* Corresponding author.

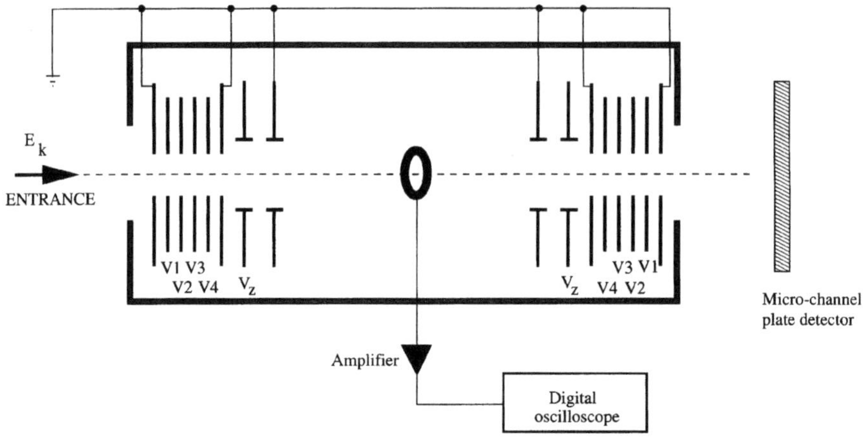

Figure 1. Schematic view of the ion beam trap. The bunch is injected through the left side of the trap. The central ring is the pick-up electrode. The trap length is 400 mm. A microchannel plate detector is located to the right of the exit mirror.

value of V_z which causes the ions to be strongly focused inside the mirror region for all values of V_1 [5].

The experiments were performed with bunches of 4.2 keV Ar^+, produced by an electron impact ion source. The evolution of the bunch was monitored with a pick-up electrode 7 mm in length and 18 mm in diameter, located in the center of the trap. The pick-up was connected to a charge sensitive amplifier and its signal was recorded with a digital oscilloscope.

3. Results

Figure 2(a) shows the signal observed for a bunch of Ar^+ 200 μs after injection with $V_1 = 5.5$ kV. Each (negative) peak corresponds to the passage of a bunch of about 10^6 ions through the pick-up. The initial width of the bunch was set to $W_0 = 170$ ns corresponding to a bunch length of 2.4 cm. The oscillation time was about 3 μs. The bunch evolution was quantified by fitting a Gaussian profile to each measured peak. For all trap configuration for which $V_1 > 4.9$ kV, the bunch was observed to broaden rapidly (on a scale of a few hundred oscillations) until it fills the whole length of the trap. The overall lifetime of the ions in the trap was of the order of a few seconds.

When changing the trap configuration so that $V_1 < 4.9$ kV, a completely different bunch evolution was observed. This is illustrated in Figure 2(b) where the evolution of a bunch of $W_0 = 130$ ns is shown as a function of the number of oscillations when $V_1 = 4.7$ kV. After a brief increase and some jumps, the width of the bunch W_n stabilizes at a value around 120 ns. At the same time, an exponential decay of the number of ions in the bunch is observed with a typical lifetime of

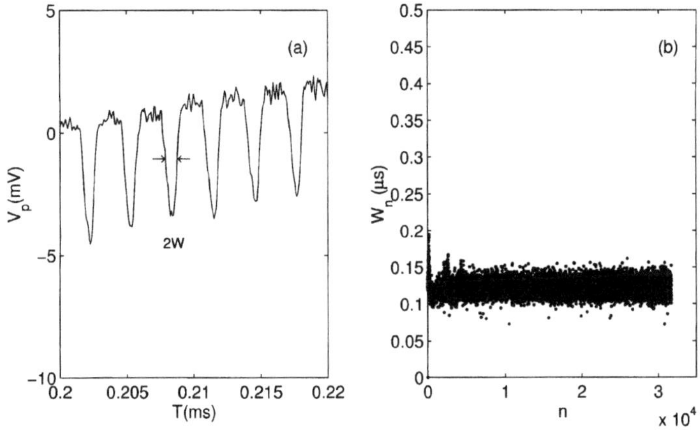

Figure 2. (a) Signal observed on the pick-up electrode 500 μs after injection for $V_1 = 5.5$ kV. (b) Bunch width as a function of the number of oscillations for $V_1 = 4.7$ kV.

240 ms. Clearly, a very strong correlation is created between the ions, as they now stick together.

Larger bunches were also injected into the trap. In these cases, and when $V_1 <$ 4.9 kV, an initially wide bunch ($W_0 \sim 1$ μs) was seen to display very complex behavior for the first 5000–20,000 oscillations, where the bunch breaks into several smaller bunches which merge at later time. In all cases, on long time scales, only one bunch remained. The asymptotic width and decay time was found to be a characteristic of the voltages applied to the electrodes.

4. Discussion

The special bunch evolution illustrated in Figure 2(b) stems from the correlation of the ions due to their mutual interaction [5].

To illustrate that such synchronization can indeed occur and to gain insight into its dynamics we have performed a numerical study of a one-dimensional model system, where ions are trapped in a flat-bottomed potential with a field free region of 200 mm and potential walls of various linear slopes. The kinematics of a single ion trapped in such a potential can be characterized by the logarithmic derivative of the oscillation time T with respect to energy E:

$$\alpha = \frac{1}{T} \cdot \frac{\mathrm{d}T}{\mathrm{d}E}, \tag{1}$$

which is bounded as $-1 < \alpha \cdot 2E < 1$ corresponding to potential slopes between infinity and zero.

A number $N = 30$ of 4.2 keV Ar^+ ions were propagated in this potential. To allow the ions to overtake each other in the one-dimensional model, the Coulomb ion–ion interaction is augmented by a minimum impact parameter of 1 μm. The

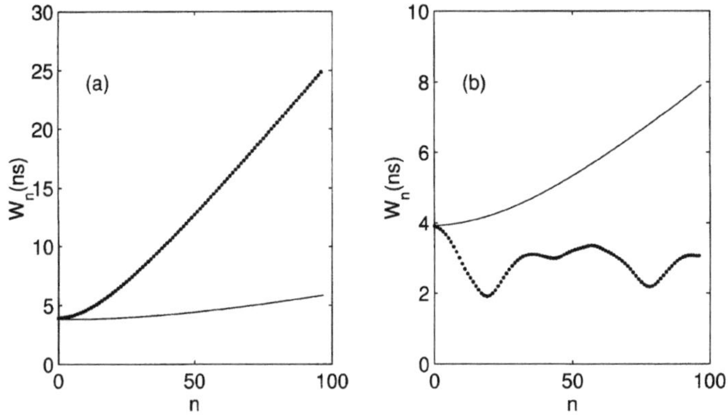

Figure 3. Bunch width as a function of the number of oscillations for 30 Ar^+ ions trapped in a one dimensional flat-bottomed potential with (dots) and without (lines) Coulomb repulsion for (a) $\alpha \cdot 2E = -0.3$, and (b) $\alpha \cdot 2E = 0.3$.

initial positions of the ions were chosen randomly according to a uniform distribution of mean $\langle z \rangle = 0$ and width $\Delta z = 0.6$ mm, while the kinetic energies of the ions were selected from a uniform distribution of $\langle E \rangle = 4200$ eV and $\Delta E = 0.6$ eV.

Figures 3(a) and 3(b) show the width of the bunch as a function of the number of oscillations n with (dots) and without (lines) Coulomb repulsion between the ions for two different values of α. In the first case, where $\alpha \cdot 2E = -0.3$, the width of the bunch increases much faster with interaction than without, i.e., the repulsion enhances the diffusion. On the contrary, in the second case where $\alpha \cdot 2E = 0.3$ the bunch width remains bound with interaction, so that the Coulomb repulsion helps to synchronize the ion motion. Thus, correlated ion motion emerges as an interplay between the kinematical properties (α) of the electrostatic potential and the repulsion between the ions [7]. Specifically, we found that synchronization can occur for $\alpha \geqslant 0$.

We were also able to calculate the parameter α for the trap used in the experiment using three dimensional trajectory simulations, and found that for all the values of V_1 for which synchronization was observed, $\alpha > 0$, while $\alpha < 0$ when diffusion was observed. With a more detailed calculation [7], we have demonstrated that this criterion ($\alpha > 0$) is one of three to be observed to obtain motion synchronization in the trap.

The so called "negative mass instability" [8], which occurs at relativistic velocities in heavy ion storage rings, resembles in some sense the synchronization observed in our trap. However, the fact that synchronization is seen to occur for ions of different masses for which the oscillation frequencies are very different and the fact that the effect depends strongly on the trap configuration indicates that wake fields from resonator type structures along the beam path do not drive the synchronization. The ion motion synchronization observed here is perhaps better compared to synchronization among coupled oscillators [9], where each ion can be regarded as an oscillator coupled to the others by their mutual Coulomb interaction.

References

1. Zajfman, D., Heber, O., Vejby-Christensen, L., Ben-Itzhak, I., Rappaport, M., Fishman, R. and Dahan, M., *Phys. Rev. A* **55** (1997), 1577.
2. Brenner, W. H., *Anal. Chem.* **69** (1997), 4162.
3. Fardi, A., Cederquist, H. and Schmidt, H. T., *Hyp. Interact.* **127** (2000), 247.
4. Dahan, M., Fishman, R., Heber, O., Rappaport, M., Altstein, N., van der Zande, W. J. and Zajfman, D., *Rev. Sci. Instrum.* **69** (1998), 76.
5. Pedersen, H. B., Strasser, D., Heber, O., Rappaport, M. L. and Zajfman, D., *Phys. Rev. A*, in press.
6. Pedersen, H. B., Strasser, D., Ring, S., Heber, O., Rappaport, M. L., Rudich, Y., Sagi, I. and Zajfman, D., *Phys. Rev. Lett.* **87** (2001), 055001-1.
7. Pedersen, H. B., Strasser, D., Heber, O., Rappaport, M. L. and Zajfman, D., submitted to *Phys. Rev. A*.
8. See, e.g., Chao, A., *Physics of Collective Beam Instabilities*, John Wiley & Sons, New York, 1993.
9. See, e.g., Rabinovich, M. I., Ezersky, A. B. and Weidman, P. D., *The Dynamics of Patterns*, World Scientific, Singapore, 2000.

Hyperfine Interactions **146/147:** 237–244, 2003.

First Results of Ion Trapping in the Dresden EBIT II

U. KENTSCH[1], G. ZSCHORNACK[1,*], F. GROSSMANN[2], V. P. OVSYANNIKOV[2] and F. ULLMANN[2]

[1] *Technische Universität Dresden, Institut für Kern- und Teilchenphysik, Mommsenstr. 13, D-01069 Dresden, Germany*
[2] *Leybold Vakuum Dresden GmbH, Zur Wetterwarte 50, D-01109 Dresden, Germany*

Abstract. With the Dresden EBIT it is possible to produce highly charged ions in a compact, economical and long-term stable ion source working at room temperature. The operation principle is based on electron impact ionization of primarily neutral atoms in a high-density electron beam in combination with electrical potentials for axial trapping. Radial trapping is realized by the negative space charge of the electron beam itself. After build-up the first Dresden EBIT in 1999 first results from a second EBIT device, the so-called Dresden EBIT II are presented. By means of X-ray spectroscopy we analyze the influence of different source parameters as ion trapping time, height of wall potentials and pressure on the production and storage of highly charged xenon ions.

Key words: highly charged ions, ion trap, xenon ions, EBIT.

1. Introduction

Highly charged ions play an important role in plasmas which are found in astrophysical X-ray sources, controlled fusion devices or X-ray lasers. X-ray emission spectra from highly charged ions are utilized for diagnostics of plasma properties. Slow-moving HCIs also interact with solid state surfaces, that is a recent branch of investigation, where among others applications in nanotechnology are in focus. Thus, it is desirable to have an apparatus where slow highly charged ions can be provided whether for spectroscopic investigations or for interactions with highly charged ions. Up to now such investigations preferently take place at cryogenic EBIT devices or at ECR ion sources, where highest charge states can only be reached for low Z elements. The Dresden EBIT device presented here has proven its ability to produce highly charged ions for a variety of elements [1–7].

Thus, it is of basic interest to investigate the ability to repeat the reported results for a second device, the so-called Dresden EBIT II. The aim of the present paper is to report first results derived at the Dresden EBIT II. The results presented here show that it is possible to create duplicates of the Dresden EBIT I deriving results similar to those reported from the Dresden EBIT I.

*Corresponding author.

2. Experimental setup

The measurements reported here have been performed at the Dresden electron beam ion trap facility. An electron beam ion trap uses a magnetically compressed electron beam to produce highly charged ions and to hold them captive. The magnetic field compressing the electron beam to high density is produced by two permanent magnet ring structures at room temperature. Ions stripped to high charge states by the electron beam are trapped radially by the negative space charge of the electron beam whereas axial trapping is achieved by voltages of some ten volts applied to the drift tube system. The confined ions undergo successive electron impact ionization leading to ion charge states which are limited by recombination processes as charge exchange and electron capture as well as by ion loss from the trap region due to electron beam heating of the ions.

The Dresden EBIT is equipped with a gas injector. Beside noble gases this injector is used to feed metal ions into the trap by the MIVOC (Metal Ions from VOlatile Compounds) method, where the vapour pressure of the compounds which include the elements of interest is sufficiently high at room temperature. We have thus produced bare nuclei of Ti, Mn, Fe and Ni ions by insertion of titanium tetrachloride ($TiCl_4$), methylcyclopentadienyl manganese tricarbonyl ($(CH)_5CH_3Mn(CO)_3$), ferrocene ($(C_5H_5)_2Fe$) and nickelocene ($(C_5H_5)_2Ni$). Bare ions have been accounted for by X-rays from radiative recombination processes, where an electron from the monoenergetic electron beam is captured into the bare ion [8].

The Dresden EBIT has been used to perform spectroscopic measurements of X-rays from highly charged ions in order to show what charge states of the investigated ions can be accounted for. These investigations have been done by varying the trap parameters, that are the electron beam energy, pressure, trapping time and trapping potential. Thereby the signature of spectral lines from direct excitation and radiative recombination processes was used to determine the reached mean ion charge state (maximum of the charge state distribution) in the trap. A 3D representation of the used spectroscopic setup is shown in Figure 1. As a matter of principle we can detect the emitted X-rays by an energy dispersive solid state detector or wavelength-dispersive by a crystal diffraction spectrometer. Results from measurements with the crystal diffraction spectrometer are reported in [7, 8]. In the present paper we will discuss results from energy dispersive measurements.

3. Results

After starting first investigations of the Dresden EBIT I in 1999 [1] we now have begun to study a second exemplar of the Dresden EBIT, built up with somewhat small modifications. Photographs of the Dresden EBIT I and Dresden EBIT II devices are shown in Figure 2. The construction of both devices is very similar, where only small modifications are applied to the second apparatus.

First studies at the Dresden EBIT II have been directed to get information on quality parameters of the electron beam and the operation of the ion trap. Investi-

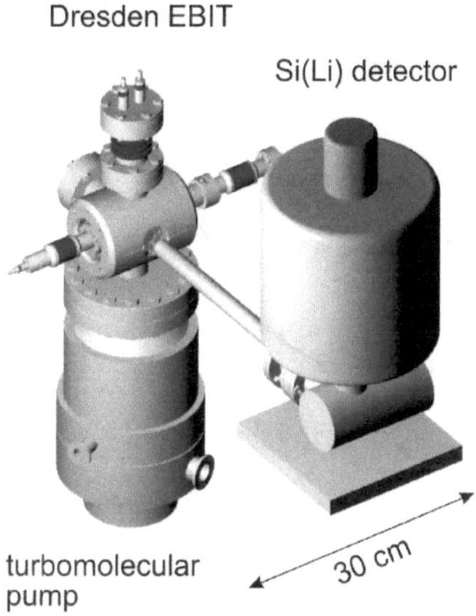

Figure 1. 3D representation of the spectroscopic setup for investigations of the ion charge state distribution in the electron beam of the Dresden EBIT by energy dispersive X-ray spectroscopy.

Figure 2. Photoprints of the Dresden EBIT I (*left*) and Dresden EBIT II (*right*).

gations have been done by X-ray spectroscopy of lines from direct excitation (DE) processes in different xenon ions.

First the integral output of xenon L X-rays from DE processes as a function of the trapping time has been analyzed at an electron energy of 14 keV, an electron beam current of 36.5 mA and a gas pressure of $2.2 \cdot 10^{-9}$ mbar (Figure 3). Here we note, that the gas pressure is measured about 15 cm away from the ion trap region,

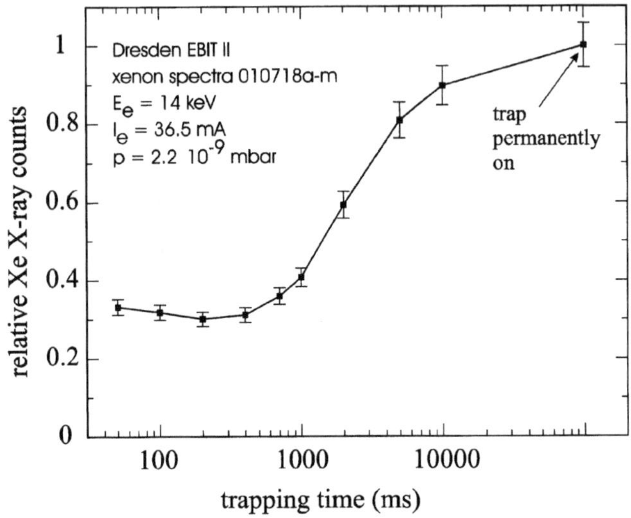

Figure 3. Measured relative xenon L X-ray quanta output as function of the trapping time.

i.e. it is expected that the real pressure around the electron beam in the trap region is somewhat different.

For ion trap times greater than 1 s we observe a clear increase of the quanta output. After a trap time of 10 s a level of about 90% of the output that is typical for the trap permanently switched on is reached. The dependence given here does not directly indicate the number of ions of a certain ionization stage but reflects the mean number of ions stored in the trap. With increasing trap time the mean ionization stage increases (see Figure 4) and by this way the probability of radiative deexcitation channels increases by a simultaneously decreasing probability for non-radiative deexcitation processes.

In Figure 4 xenon L X-ray spectra from DE processes are shown in dependence of the trap time including the cases where the trap is permanently switched on and off, respectively. Here the trapping potential is the same as shown in Figure 5. With increasing trap time the mean ion charge state increases significantly and converges to a final ionization stage for longer trap times. This behaviour reflects the fact, that the ionization factor $j\tau_q$ depends on the ionization cross sections σ^q for ions of the charge state q in the following manner

$$j\tau_q = \sum_{k=1}^{q} \frac{1}{\sigma_k^q}, \tag{1}$$

whereby the product $j = n_e v_e$ describes the electron current density with n_e – electron density and v_e – electron velocity and τ_q is the ion life-time in the trap.

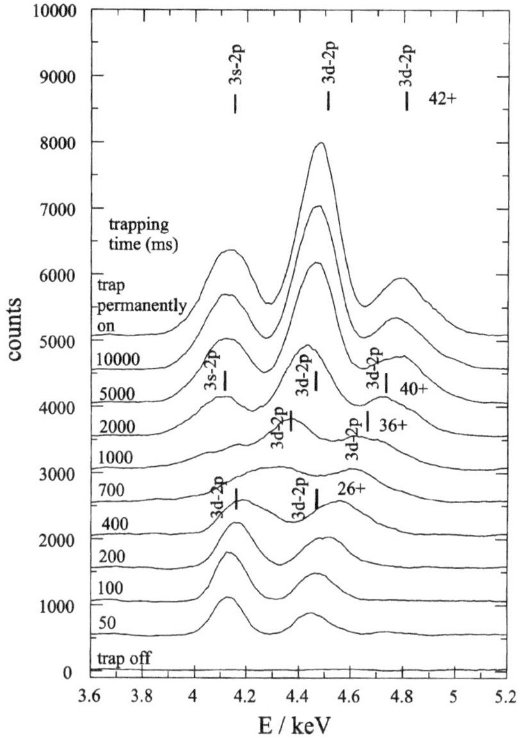

Figure 4. Measured xenon L DE X-rays for different ion trap times (shown as parameter) at $E_e = 14$ keV, $I_e = 36.5$ mA and $p = 2.2 \cdot 10^{-9}$ mbar.

The production of ions of a certain ionization stage then depends on the realization of an appropriate ionization factor $j\tau_q$, i.e. the condition

$$j\tau_q \geqslant \sum_{k=1}^{q} \frac{1}{\sigma_k^q} \tag{2}$$

must be satisfied. Estimations show, that for the highest observed ion charge states corresponding to the spectra shown in Figure 4 an ionization factor of about $1 \cdot 10^{21}$ cm^{-2} is necessary [9]. On the other hand ionization factors of $\approx 1 \cdot 10^{20}$ cm^{-2} are sufficient to generate nickel-like xenon ions. Thus, beside the variation of the electron energy a carefully chosen ion trap time also allows a selection of different ionization factors and hence ion charge state distributions.

Ions are trapped in transversal direction in the electron beam of the Dresden EBIT by the self-charge of the electron beam. For ion storage in axial direction a potential distribution is realized, where electrostatic potential barriers at the ends of the trap control the trap operation. In Figure 5 different xenon L X-ray spectra with transitions from DE processes are shown at varying electrostatic trap potentials for Dresden EBIT operation indicated in the figure. The reached maximum ionization

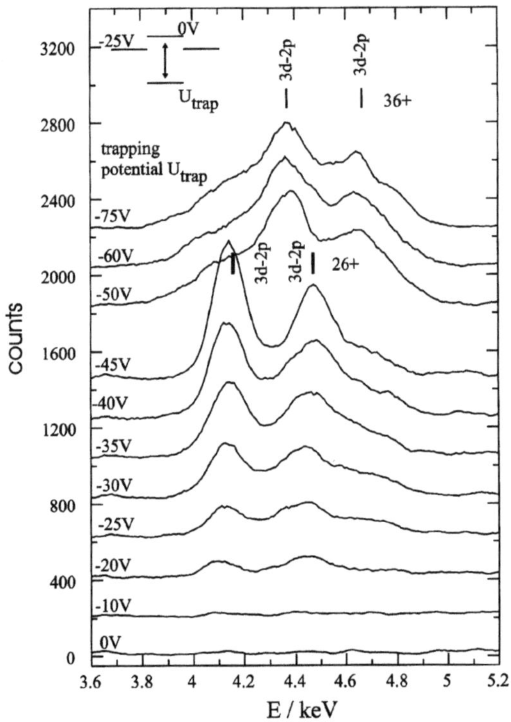

Figure 5. Measured xenon L DE X-rays for different ion trap potentials (shown as parameter) for an ion trap time of 1 s at $E_e = 14$ keV, $I_e = 35$ mA and $p = 2.1 \cdot 10^{-9}$ mbar.

stages correspond to those observed in Figure 4 for a trapping time of 1 s. For trap potentials higher than 20 V ($U_{trap} = -45$ V) a significant change in the X-ray spectra is observed corresponding to the production of higher charge states in the trap. For orientation calculated energetic positions of E1 transitions for DE lines from Xe^{26+} and Xe^{36+} are labelled. The shown behaviour suggests that for an optimal production of highly charged ions in the Dresden EBIT II potential barriers higher than 20 V should be applied. For the case of a switched off ion trap no X-ray lines are detected. This is explained by the fact, that the xenon atoms do not live long enough to be ionized in inner L subhells. Thus, the trap can be cleaned after each ionization cycle almost completely to liberate the trap from admixtures of heavy ions stemming from the cathode material to the benefit of the trap capacity to store working gas ions.

In Figure 6 measured xenon L X-ray spectra from $n = 2$ to $n = 3$ DE processes are shown for three different pressures. As well known from other ion sources lower vacua lead to structures corresponding to ions of higher charge states. As seen from Figure 6 there exists an optimum for the quanta output depending on the gas pressure.

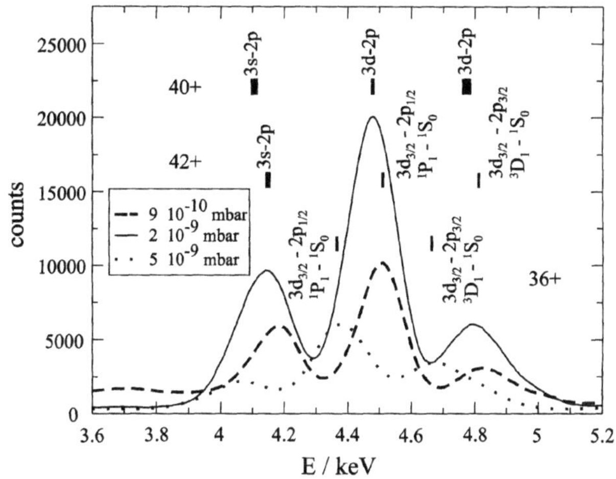

Figure 6. Measured xenon L DE X-rays for different working gas pressures for a permanently switched on ion trap and a trap potential of 75 V at $E_e = 14$ keV and $I_e = 35.1$ mA.

4. Conclusions

With the Dresden EBIT exists the ability to produce highly charged ions in a very small laboratory apparatus that does not need any special equipment beside standard UHV technics and high voltage supplies. The Dresden EBIT has proven that it is most suitable for long running and stable operation. It has been shown that it is especially qualified for any spectroscopic investigations of highly charged ions including processes as direct excitation, radiative recombination as well as dielectronic resonances. It was shown, that the Dresden EBIT device can be rebuilt and is so available for any other applications. The apparatus also has been designed for the extraction of highly charged ions. Appropriate investigations are in progress and extracted highly charged ions can hopefully be accounted for in the near future.

Acknowledgements

The work is supported by the EFRE fund of the EU and by the Freistaat Sachsen (Project 6937/1087) and by DFG (Project Zs 14/7-3). The authors would like to thank Dipl.-Phys. S. Landgraf and E. Steinhorst for their technical support during the experiments.

References

1. Ovsyannikov, V. P. and Zschornack, G., *Rev. Sci. Instr.* **70** (1999), 2646.
2. Ovsyannikov, V. P., Zschornack, G., Grossmann, F., Landgraf, S., Ullmann, F. and Werner, T., *Rev. Sci. Instr.* **71** (2000), 690.
3. Ovsyannikov, V. P., Zschornack, G., Grossmann, F., Kulthachev, O. K., Landgraf, S., Ullmann, F. and Werner, T., *Nucl. Instr. Methods Phys. Res. B* **161** (2000), 1123.

4. Werner, T., Zschornack, G., Grossmann, F., Ovsyannikov, V. P. and Ullmann, F., *Rev. Sci. Instr.* **71** (2000), 2038.
5. http://www.physik.tu-dresden.de/apg and http://www.dresden-ebit.de.
6. Werner, T., Zschornack, G., Grossmann, F., Ovsyannikov, V. P. and Ullmann, F., *Nucl. Instr. Methods B* **178** (2001), 260.
7. Werner, T., Zschornack, G., Grossmann, F., Ovsyannikov, V. P. and Ullmann, F., *Physica Scripta* **T92** (2001), 241.
8. Kentsch, U., Zschornack, G., Grossmann, F., Ovsyannikov, V. P., Ullmann, F., Fritzsche, S. and Surzhykov, A., *Nucl. Instr. Methods Phys. Res. B* **187** (2002), 238.
9. Donets, E. D., *Soviet Journal of Particles & Nuclei* **13** (1982), 387.

Hyperfine Interactions **146/147:** 245–251, 2003.

SHIPTRAP is Trapping: A Capture and Storage Device on Its Way towards a RIB-Facility

G. MARX, J. DILLING, H.-J. KLUGE, M. MUKHERJEE, W. QUINT, S. RAHAMAN, D. RODRIGUEZ, G. SIKLER, M. TARISIEN, C. WEBER and the SHIPTRAP Collaboration
GSI Darmstadt, Postfach 110552, D-64220 Darmstadt, Germany

Abstract. First off-line tests at the ion trap facility SHIPTRAP took place. The facility is being set up to deliver very clean and cooled beams of singly-charged recoil ions (Rare Isotope Beam) produced at the SHIP (Separator for Heavy Ion Production) velocity filter at GSI, Darmstadt. SHIPTRAP consists of a gas cell for stopping and thermalizing high-energy recoil ions from SHIP, an rf ion guide for extraction of the ions from the gas cell, a linear rf trap for accumulation and bunching of the ions, and a Penning trap for isobaric purification. The physics programme of the SHIPTRAP facility comprises mass spectrometry, nuclear spectroscopy, laser spectroscopy and chemistry of transeinsteinium elements. The progress in testing the sub-systems separately and in combinations is reported.

Key words: ion traps, RFQ buncher, Penning traps, buffer gas.

1. Introduction

Important pre-requisites for high-accuracy experiments are the extremely small phase space in which ions reside in a trap, the possible long storage time and the extremely high purity that can be obtained. From this the interest in coupling ion traps to sources of radionuclides arises. The purification is very important since contaminants always plagued the investigation of exotic radioactive species. This separation not only allows for very sensitive and accurate experiments to be performed within the ion trap itself but also allows the collected radionuclides to be extracted from the trap and studied in a well-defined low-emittance ion beam. Together, these possibilities open a wide range of physics applications.

SHIPTRAP will enable very sensitive and accurate experiments, performed within the ion trap itself or with an extracted beam [1]. The half-lives of the most neutron-rich isotopes of the elements up to hassium ($Z = 108$) are longer than one second, which is also predicted by theory for a large number of ions in this region of the chart of nuclei. This is long enough to prepare the ions and to perform precision experiments. The advantage of such facilities is that they enable the rich variety of physics experiments currently performed at ISOL facilities to be extended to isotopes for which target/ion source systems do not exist at ISOL facilities. The particular advantage that SHIPTRAP will have is the ability to extend these experiments to transuranic nuclides. Thus, SHIPTRAP will enable the application of refined trapping techniques for the first time to nuclides with nuclear charge $Z > 92$, extending even to the superheavy elements of which many have a sufficiently long half-life for trap technology.

2. The SHIPTRAP facility

SHIP is a kinematic separator for reaction recoils from thin targets irradiated by beams from the heavy-ion linear accelerator UNILAC at GSI [2]. It is optimized for the separation of heavy elements produced by fusion of projectiles from $A = 40$ to 80 with heavy target nuclei such as lead or bismuth. The primary beam has an energy close to 5 MeV/u and time-averaged intensities of typically $2 \cdot 10^{12}$–$5 \cdot 10^{12}$ ions/s.

The SHIPTRAP facility is set-up at the exit of SHIP. It stops and thermalizes the produced recoil ions in a noble gas from which they are then extracted and collected in a trap. The system is outlined in Figure 1. It consists of a stopping chamber containing the noble gas, an extraction system to bring the stopped ions into a vacuum region, a radiofrequency trapping system to collect the ions in this vacuum region and to cool them into well-defined bunches which are then extracted and injected into a Penning trap for isobaric purification. The Penning trap accumulates the ion bunches, filters out possible contaminants and further cools the ion ensemble to room temperature. This ensemble is then extracted on the demand of downstream experiments.

The noble gas in the stopping chamber, at pressures around 100 mbar, will thermalize recoil ions preferentially in the singly ionized state. An electric field, together with the gas flow, then guides the ions out of the chamber into the extraction system where they are separated from the gas. This system is a quadrupole rod structure with a length of 18 cm that confines the ions to its axis by an rf field while the noble gas is pumped away. An axial electric dc field applied to the

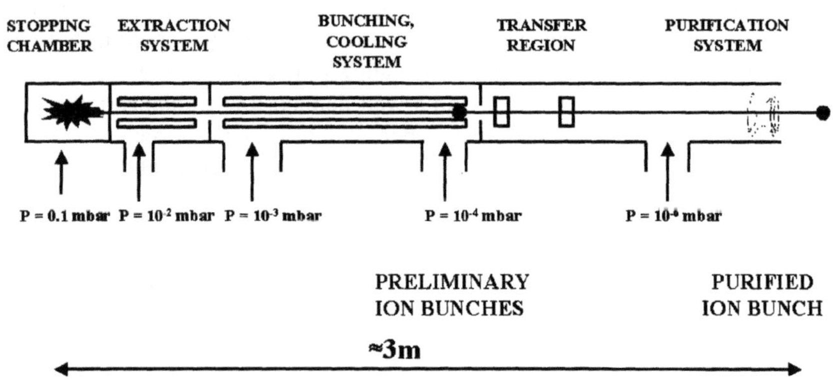

Figure 1. The overall SHIPTRAP configuration.

segmented rods guides the ions along the axis towards the bunching system. The recent development for the SHIPTRAP stopping chamber and the extraction RFQ is described elsewhere [3].

In the ion bunching system, a 1 m quadrupole rod structure immersed in a low-pressure buffer gas, the ions are trapped by a proper choice of longitudinal dc and transverse rf fields and cooled in collisions with the buffer gas. The RFQ-buncher was set up and tested. The purification system into which these preliminary bunches are collected is based on a Penning trap similar to the one used for this purpose at the ISOLTRAP facility at ISOLDE. In such a system the contaminating isotopes are very effectively suppressed due to the high mass resolving power of the cooling process. Depending on the trapping time it can reach M/δ M(FWHM) = 10^5.

The ion bunches delivered by this Penning trap system will have extremely low emittances, thermal energy spreads and pulse durations of typically less than a microsecond, although bunches of longer duration could be extracted if desired. These bunches could be delivered at any energy desired up to about 60 keV. In some cases it may be advantageous to use specifically those ions from SHIP that become neutral atoms in the stopping gas. These can then be re-ionized by a laser beam (resonance ionization spectroscopy, RIS) for delivery to the vacuum system. Such a scheme provides element selectivity, and for heavy elements isotopic selectivity as well, in the stopping chamber itself.

3. The RFQ buncher

The radiofrequency quadrupole is operated by applying to its rods a radiofrequency of amplitude V and frequency ω_{rf} and a dc bias voltage U. An ion with mass m and charge e, which enters the RFQ, performs an oscillation which is described by the solution of the Mathieu equations. In this equation one defines a pair of dimensionless parameters $a = 4eU/(m\omega_{rf}^2 r_0^2)$, and $q = 2eV/(m\omega_{rf}^2 r_0^2)$. This pair of parameters characterizes the working point in the a, q-stability diagram. The motion of the ion is stable if the amplitude of the oscillation remains finite, and unstable if its amplitude rises exponentially. In our case the RFQ was tested in rf-only mode, i.e. without a dc voltage U. Then the operating region lies on the q axis. For a certain rf voltage applied to the rods, all ions pass the filter for which q is less than the stability limit 0.908. The rods of the SHIPTRAP RFQ have a diameter of 9 mm. The distance between two opposite rods is 7.86 mm.

In first tests in the rf-only mode with a calibrated ion source a transmission of 93(5)% was achieved. Figure 2 shows a transmission plot for Ar^+ ions. The driving field frequency was set to $\upsilon_{rf} = 600$ kHz. The asymmetry in this transmission curve is well understood and explained by theory [4]. Trapping times of up to 30 s without significant ion loss have been achieved. Accumulation of ions with a dc-injection as well as a pulsed injection was demonstrated. The ions can be extracted in a pulsed mode in bunches with high phase-space density and can be delivered

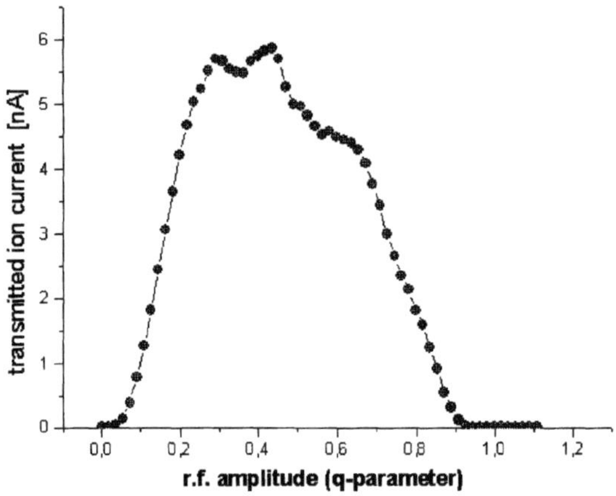

Figure 2. Transmission plot in rf-only mode for Ar^+ ions.

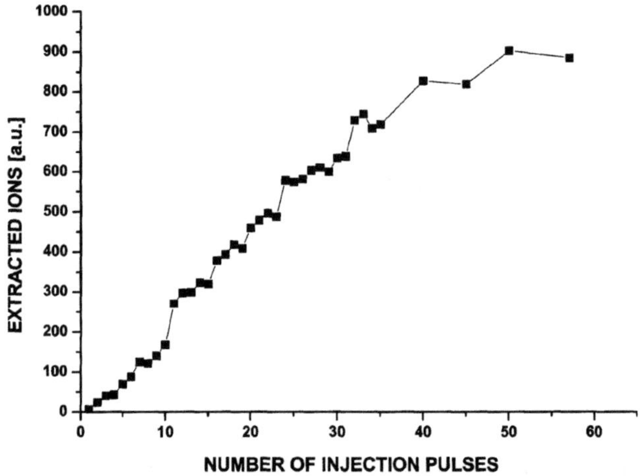

Figure 3. Accumulation of ions in the RFQ. The number of ions, which are extracted from the RFQ buncher and detected by a microchannel plate, is plotted as a function of the number of injection cycles.

to the high-vacuum region. Figure 3 shows an example for accumulation of ions in the RFQ. For this measurement the extraction electrode of the off-line ion source was pulsed, delivering about 10^3 ions per bunch.

4. The Penning traps

A double-trap system is installed within one superconducting magnet at SHIPTRAP. Both traps have cylindrical electrodes with an inner diameter of 30 mm. The purpose of the first trap is to catch ion bunches from the RFQ buncher and to purify the ion bunches. The second trap is for, e.g., mass measurements. The traps

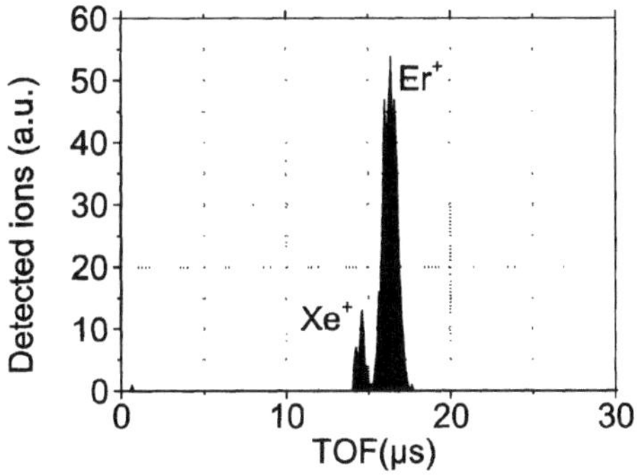

Figure 4. Ion bunch injected from the RFQ buncher, trapped for 200 ms in the purification trap and extracted through the diaphragm and the measurement trap.

(distance between the centers of traps: 20 cm) are separated by a diaphragm for differential pumping.

In a first test the injection into the magnetic field, the transmission between both Penning traps, and the extraction out of the magnet have been studied. Figure 4 shows an ion bunch injected from the RFQ buncher, trapped for 200 ms in the purification trap and extracted through the diaphragm and the measurement trap. The pure transmission efficiency through the Penning traps has been determined to be about 80%. The trapping efficiency, i.e. the ratio of the number of ions captured in the Penning trap compared to the number of ions delivered from the RFQ buncher is about 80%. The tests have been done under UHV conditions at 10^{-8} mbar.

In a second step the motion of stored ions in the Penning trap was excited by an electric rf field. The driving rf field was coupled to a split electrode of the trap for a dipole excitation of the cyclotron motion of the trapped ions. The excitation frequency was scanned and the number of extracted ions was measured. At each frequency step, ions were loaded into the trap, the excitation field was switched on, then the ions were detected on a microsphere plate (MSP). An example of such a cyclotron-frequency measurement with trapped Ar^+ ions is shown in Figure 5. When the driving rf frequency is in resonance with the cyclotron frequency ν_+ (at about 2.7 MHz) of the trapped Ar^+ ions, the size of their cyclotron orbits is increased such that the extracted ions cannot reach the detector, resulting in a pronounced minimum in the number of detected ions. Axial sidebands at the frequencies $\nu_+ \pm \nu_z$ (at 2.6 and 2.8 MHz) are clearly visible in Figure 5, which are due to anharmonic components of the electrostatic trapping potential.

Penning-trap mass spectrometers at ISOLDE/CERN [5], Stockholm [6], and Argonne [7] use a time-of-flight method [8] to measure eigen-frequencies of trapped ions. The ions are ejected out of the trap and their time-of-flight to a detector is recorded. Typically this requires the detection of a total number of 100–1000 ions. Assuming a production rate in the order of 10^{-3} ions/s, e.g., for Sg ($Z = 106$),

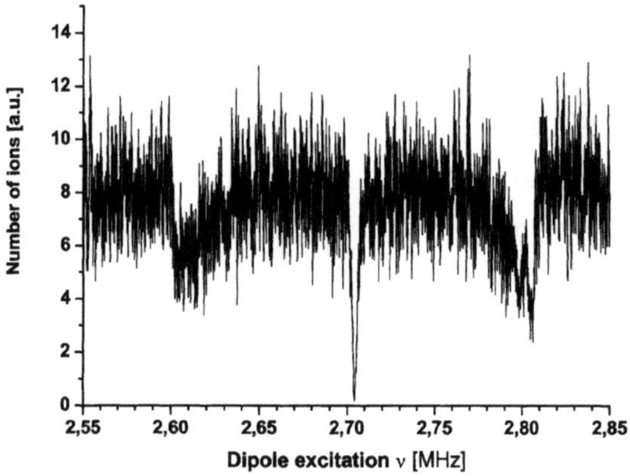

Figure 5. Dipole excitation of the cyclotron motion of stored Ar^{+} ions in the Penning trap. When the cyclotron motion (at about 2.7 MHz) is excited, the number of extracted ions is strongly reduced.

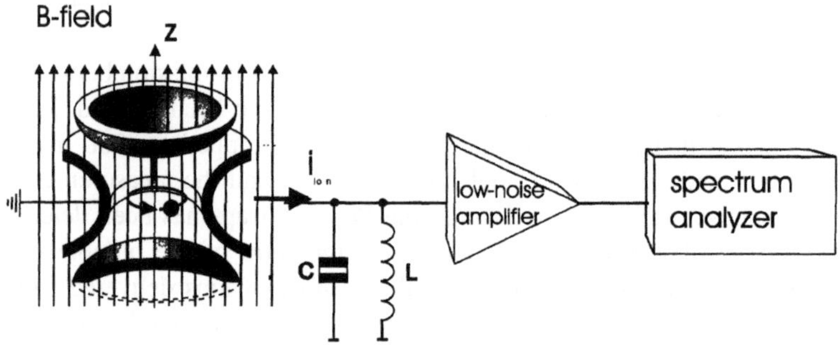

Figure 6. Principle of non-destructive FT-ICR technique. The image current of trapped ions is picked up on a split electrode, amplified and frequency-analyzed.

this corresponds to a required beam time of about 1–10 days. The measurement sensitivity can be enhanced by a non-destructive (Schottky) detection method via the image currents which are induced in the trap electrodes by the ion motion (Figure 6). This method is employed, e.g., at the University of Washington [9] and at MIT [10]. It allows for the non-destructive identification of one single ion. The required beam time for mass measurements on heavy elements would be strongly reduced. The image current of a single trapped ion is given by

$$I_{\text{ion}} = \frac{1}{\sqrt{2}} \frac{r_{\text{ion}}}{D} \cdot q \cdot \omega_{\text{ion}}. \tag{1}$$

This signal can be detected via a LC-resonance circuit of quality factor Q. With inclusion of the thermal noise of the circuit, the signal-to-noise ratio (S/N) is

$$\frac{S}{N} = \frac{\sqrt{\pi}}{2} \cdot \frac{r_{\text{ion}}}{D} \cdot q \cdot \sqrt{\frac{\nu}{\Delta\nu}} \sqrt{\frac{Q}{kT \cdot C}}, \tag{2}$$

where r_{ion} is the radius of the ion orbit in the trap, D characterizes the trap size, T is the temperature, and C is the stray capacitance of the system. The signal-to-noise ratio can be optimized by a large quality factor Q, a low temperature T, and by minimizing the capacitance C. The set-up, which is under development at SHIPTRAP, will consist of a cryogenic system with the trap environment at liquid nitrogen temperature and with an external superconducting resonance circuit at 4 K. With this set-up, single-particle detection sensitivity can be reached.

5. Summary

The SHIPTRAP facility at GSI Darmstadt is designed to slow down heavy-ion projectiles from the velocity filter SHIP to thermal energies, to accumulate and cool them in an ion-trap system and to deliver these ions as isobarically pure ion bunches with low emittance to physics experiments. After an intense simulation and construction phase all components are presently under test. SHIPTRAP will undergo a test beam time in winter 2001 and in summer 2002. The experimental programme which is envisaged by the SHIPTRAP user community promises to give new insights into the nuclear, atomic and chemical properties of elements beyond uranium.

Acknowledgement

We acknowledge financial support by the European Union (Network EXOTRAPS).

References

1. Dilling, J. *et al.*, *Hyp. Interact.* **132** (2001), 495.
2. Münzenberg, G. *et al.*, *Nucl. Instrum. Meth.* **161** (1979), 65.
3. Engels, O. *et al.*, *Hyp. Interact.* **132** (2001), 505.
4. Marx, G. *et al.*, *Hyp. Interact.* **132** (2001), 463.
5. Bollen, G. *et al.*, *Nucl. Instr. Meth. A* **368** (1996), 675.
6. Carlberg, C. *et al.*, *Phys. Rev. Lett.* **83** (1999), 4506.
7. Savard, G. *et al.*, *Nucl. Phys. A* **654** (1999), 961c.
8. Gräff, G. *et al.*, *Z. Phys. A* **297** (1980), 35.
9. Farnham, D. L. *et al.*, *Phys. Rev. Lett.* **75** (1995), 3598.
10. Bradley, M. P. *et al.*, *Phys. Rev. Lett.* **83** (1999), 4510.

Hyperfine Interactions **146/147:** 253–258, 2003.

Design Studies of an Electrostatic Storage Ring

CARSTEN P. WELSCH[1], ALWIN SCHEMPP[1] and
HORST SCHMIDT-BÖCKING[2]
[1]*Institut für Angewandte Physik, Johann Wolfgang Goethe Universität, Frankfurt am Main, Germany; e-mail: c.welsch@iap.uni-frankfurt.de*
[2]*Institut für Kernphysik, Johann Wolfgang Goethe Universität, Frankfurt am Main, Germany*

Abstract. The design of a small electrostatic storage ring for all different kinds of ions at energies up to 50 keV is presented in this paper. One quarter of such a ring is presently being build up at IAP to study injection, optimize the optical elements and integrate the diagnostic system.

1. Project description

1.1. GENERAL

In the last decades, many activities in the field of accelerator physics have focused on the development of machines with higher and higher final energy. The enormous amount of money needed to finance these projects set a limit to new ideas. As a result, more and more experiments nowadays point into more exact measurements in energy regions of some keV to MeV.

In order to extend possibilities to the fields of decay processes, beam interaction and storage of large biomolecules, the design of a small electrostatic storage ring for ions with energies up to 50 keV was proposed [1, 2].

Such a device may be seen as a cross between an electromagnetic trap and "classical" rings.

Low costs, small size, combined with energy independence of the necessary fields and good accessibility of the experimental sections are only some of the interesting properties. Possible experiments with such a machine cover a wide range: Especially the advantage of being able to store heavy biomolecules for an extended period of time and the absence of magnetic fields make experiments possible one cannot cover with magnetic rings.

In principle, the geometrical shape of an electrostatic machine is not limited by any rules. However, beam energy, aperture and required space for later experiments do reduce the number of possible layouts to a few.

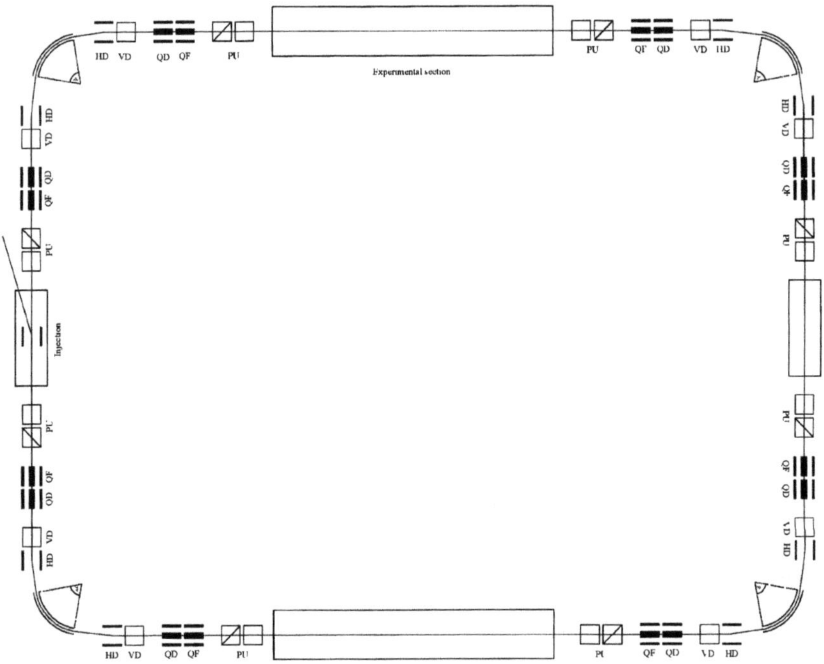

Figure 1. Layout of the complete ring.

1.2. BEAM PROPERTIES

The beam will leave an ECR ion source [3] with an energy of up to 50 keV, corresponding to a velocity of protons of $0.01c$. Thus relativistic effects can be neglected in all calculations.

The aperture defines the maximum size of the circulating beam and is at a minimum of 30 mm in the cylindrical deflectors.

One main difference to magnetic type rings is the mass-independence of the necessary fields. Theoretically, one can set up the fields for one charge/energy relation and switch from one ion type to another without changing any of the rings' parameters.

This advantage could already be verified at the two existing electrostatic rings at ISA [4] and at KEK [5].

2. Lattice

2.1. INJECTION

The beam will enter the ring at an angle of 10° and is bend towards the axis by an electric "fast kicker" inflector of two 100 mm $\times$ 70 mm copper plates placed at a distance of 50 mm.

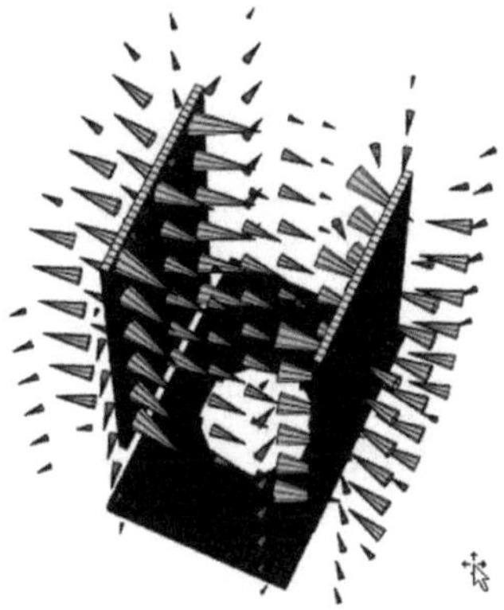

Figure 2. MAFIA plot of the calculated field in a 10° electrostatic deflector.

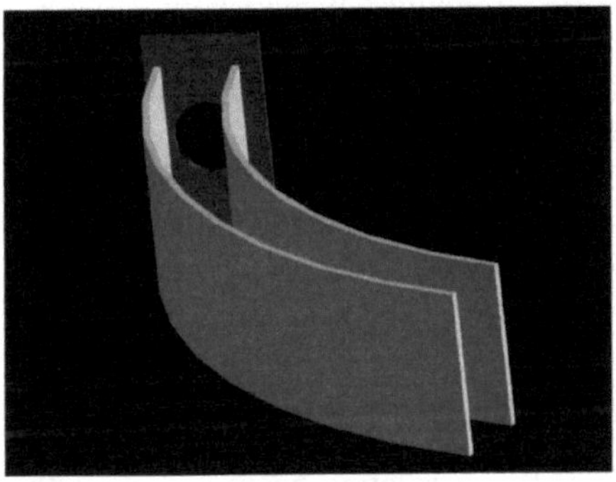

Figure 3. Rendered image of 70° cylindrical deflector with zero Volt shield in the background.

The field of 1.8 kV/cm is switched off after the beam has traveled half the circumference of the machine, i.e. 1.5 μsec (p).

2.2. BENDING SECTIONS

The bending is done in two steps: First, a 10° electro-static deflector bends the beam away from the straight section, afterwards a 70° cylindrical deflector and another 10° parallel plate deflector add up to a 90° bend.

This allows to detect neutral particles at the end of the straight sections. In addition, interaction processes with laser or electron beams can be studied.

The cylindrical deflector will be placed in a pillbox cavity, which guarantees good accessibility and space for diagnostic elements. Its two plates are 3 cm apart, requiring a field of 4 kV/cm to keep the beam on a mean radius of 25 cm.

Zero Volt shields at the entrance and exit of the deflector limit the maximum size of the beam to 30 mm and reduce fringe fields.

Figure 4. Quadrupole mounted on support.

It should be noted that, in contrast to magnetic bending sections, the longitudinal energy is not conserved in the electrostatic case. Particles entering the cylindrical deflector off-axis do see a longitudinal field component and therefore get accelerated or decelerated.

Finally, there is a stronger focusing in the horizontal plane as compared to magnetic bends, but no focusing in the vertical plane. Closed orbit correction is done by a small vertical deflector.

2.3. FOCUSING ELEMENTS

Transverse confinement of the circulating beam is done with pairs of quadrupole doublets. Electrodes of 10 cm length produce a field homogeneous enough to allow an aperture of about 5 cm.

A total number of at least 4 doublets will give the possibility to change the properties of the beam in a wide range. The required voltages are up to ± 1 kV, depending on the desired shape of the beam.

The ideal shape of hyperbolic electrode surfaces is approximated by cylindrical electrodes of radius $r = 1.1468 \cdot r_{\text{aperture}}$, which is no problem at these low field levels.

3. Vacuum system

The lifetime of the circulating beam is mainly limited by collisions with the residual gas and therefore a vacuum pressure in the order of 10^{-12} mbar is envisaged. Therefore, all vacuum chambers will be built from stainless steel. Oxygen-free, gold-plated copper was chosen for the electrodes and aluminum-oxide for insulation purposes.

A combination of turbomolecular pumps and NEG foils placed directly inside the vacuum chambers will be used.

Table I. List of design parameters

General parameters	
Maximum energy	50 keV
Circumference	17.91 m
Revolution time	3.5 μs (p)
10° deflectors	
Plate area	100 mm × 70 mm
Plate distance	50 mm
Voltage	±4.5 kV
70° deflectors	
Height	70 mm
Radii	235 mm and 265 mm
Voltage	±6 kV
Quadrupoles	
Length	100 mm
Outer radius of electrodes	29 mm
Voltage	±1 kV

The whole system, including diagnostics and control system is bakeable to 250°C.

4. Diagnostics

Diagnostics will include beam-position monitors, Faraday cups, scrapers, scintillators with CCD cameras and single particle detectors. Since only about 10^5 ions will circulate the ring, a good signal/noise ratio of the amplifiers is needed [6].

The whole machine will be controlled by a LabView/Group3 system.

5. Present status and outlook

The present parameters of the planned storage ring are given in Table I.

The construction of a small pillbox cavity has just been completed. Its size has been chosen to be big enough to house all the different electrodes and diagnostic elements.

The layout of the optical elements is finished and the electrodes have been manufactured.

The effects of electrode displacements, surface errors and fringe fields will be studied as well as questions arising when dealing with XHV. Furthermore, the control system will be tested.

The next step will be the construction of one quarter of the ring.

Figure 5. Drawing of pillbox-type test cavity.

Acknowledgements

The authors would like to thank Søren Pape Møller for his support throughout the design process. This work was supported by BMBF and Studienstiftung des deutschen Volkes.

References

1. Welsch, C. P., An electrostatic storage ring at IAP, Graduiertenkolleg, Annual Report, Frankfurt, Darmstadt, Germany, 2000.
2. Welsch, C. P., An electrostatic storage ring at IAP, PAC, Chicago, IL, USA, 2001.
3. Schmidt-Böcking, H. *et al*, *Rev. Sci. Instrum.* **63**(4) (1992).
4. Møller, S. P., *NIM A* **394** (1997), 281.
5. Tanabe, T., *An Electrostatic Storage Ring for Molecular Science*, EPAC, Vienna, Austria, 2000.
6. Søby, L. *et al.*, The closed orbit measurement system for the CERN antiproton decelerator, DIPAC99, Daresbury Lab, UK.

Hyperfine Interactions **146/147:** 259–263, 2003.

First Tests of a Linear Radiofrequency Quadrupole for the Cooling and Bunching of Radioactive Light Ions

G. BAN, G. DARIUS, D. DURAND, P. DELAHAYE, E. LIÉNARD, F. MAUGER, O. NAVILIAT-CUNCIC and J. SZERYPO*
Laboratoire de Physique Corpusculaire de Caen, CNRS-ISMRA, 6, bd du Maréchal Juin, 14050-Caen, France; e-mail: naviliat@caelav.in2p3.fr

Abstract. A linear radiofrequency quadrupole has been built at the LPC-Caen for the manipulation of radioactive ions. The system was designed for the cooling and bunching of a continuous beam using the buffer gas cooling technique. The technique is being extended to its ultimate limit to apply it for the cooling of light ions using H_2 as buffer gas. The bunches are to be injected into a transparent Paul trap for further decay studies. The cooling and bunching section of the system has been successfully tested off-line with a source of stable ions. We describe here the specifications of the device and present the results of the first tests.

Key words: ion manipulation, beam cooling.

1. Introduction

Linear radiofrequency quadrupole (RFQ) ion guides have extensively been used for applications in mass spectrometry. Gas-filled RFQ structures provide a fast cooling scheme which is particularly well suited for radioactive ion species [1]. For many applications the cold ions have to be further injected in the form of short pulses into other trapping devices and it is therefore necessary to transform the primary continuous radioactive beam into cold bunches. Several such devices which simultaneously cool and bunch ion beams are presently under operation [2, 3] or under construction [4, 5] near radioactive beam facilities. The reduction of the beam emittances achieved by the ion guide strongly improve the injection and the confinement efficiency into subsequent trapping devices.

We present here the specifications and the first tests of a RFQ cooler and buncher which has been designed and built at the LPC-Caen. The device will be adapted to the low energy radioactive beam line presently under construction near the SPIRAL facility at GANIL [6]. As a first application the RFQ is expected to deliver cold bunches to an experiment which aims to detect the β-decay products (electron and recoiling ion) from ^{6}He ions confined inside a transparent Paul trap [7]. In addition

* Present address: Department of Physics, University of Jyväkylä, 40351 Jyväkylä, Finland.

to the constraints imposed by the primary beam emittance, the ion guide should then also be able to extend the buffer gas cooling technique down to very light ions, what has not been proved so far experimentally.

2. Constraints and specifications

2.1. PRINCIPLE

The emittance reduction with the buffer gas cooling technique [1] is obtained by collisional energy losses of the initial ions with the atoms of a gas contained inside the RFQ guide. The quadrupole field confines the ions in the transverse directions while the presence of an axial electric field allows to drive the ions towards the end of the structure. The ions can then eventually be stored in a potential well for later extraction in the form of bunches. One of the conditions for an efficient cooling and transmission of the primary ions is that the mass of the buffer gas atoms be much smaller than the mass of the ions. When considering the cooling of ^{6}He ions the number of possible buffer gases is very much limited. ^{4}He is primarily excluded because of the strong resonant charge exchange mechanism in the collisions between two atoms of He isotopes, which would result in a dramatic loss of the beam transmission through the device. Monte-Carlo simulations have shown that, even with a large ratio of $1/3$ between the masses of H_2 molecules and ^{6}He ions, the typical expected transmissions are of the order of 10% with ions fully thermalized at the end of the structure.

2.2. SOURCE EMITTANCE

The device was designed to ultimately be incorporated to a low energy beam line [6] where the radioactive ions are produced by fragmentation of a stable ion beam followed by ionization in an ECR source and isotope separation. The energies of the ions extracted from the source are between 10 and 30 keV/q, where q is the charge state of the ion, and the beam emittances are up to 100π mm·mrad. Such a large emittance figure imposes an additional constraint on the transverse dimension of the RFQ rod structure to obtain a large acceptance of the incident radioactive beam.

2.3. MECHANICAL SPECIFICATIONS

The mechanical structure (Figure 1) consists of four rods with a total length of 460 mm. The total gap between the rods is $2r_0 = 30$ mm. The rods are segmented in order to distribute the voltage which produces the axial driving field. A first section of 400 mm is divided into segments of 20 mm in length while the 60 mm exit section, where the bunch is formed, is made of shorter segments of 10 mm in length. The gap between adjacent segments is 0.3 mm.

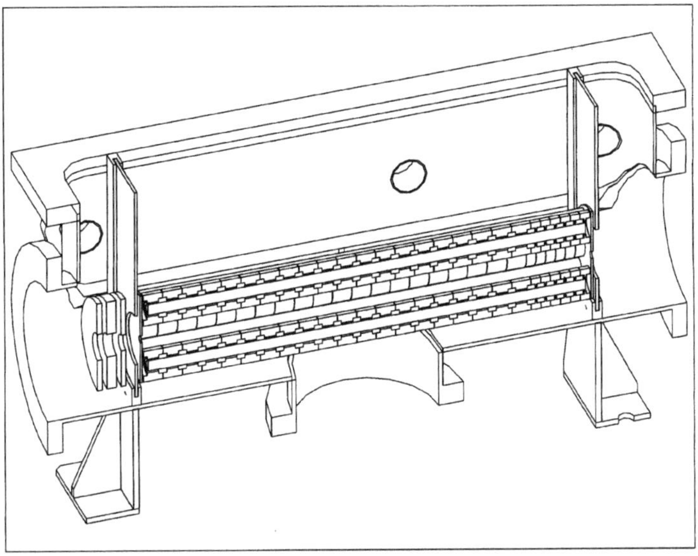

Figure 1. Longitudinal section of the cooler and buncher inside the vacuum chamber. The segmented structure of the rods is clearly visible.

The operating pressure inside the ion guide is adjusted between 0.1 and 10 Pa. The pumping of the whole volume is realized by three turbo molecular pumps, two of them, with a pumping power of 1000 l/s are located on each end and one pump, of 500 l/s, is located in the middle of the device. At the entrance and at the exit of the structure a ∅4 mm aperture separates the vacuum in the RFQ with those of the neighboring sections.

2.4. ELECTRICAL CONFIGURATION

For technical reasons not all the segments have a different voltage. Those in the first section have been grouped together producing 4 different voltage steps. The segments of the buncher section were also grouped together and the voltage applied to them produce the potential well in which the ions are stored. This voltage (of about 50 V) can then be switched off fast for the extraction of the bunches. The driving voltage varies between 0 and 100 V. The RF system has a wide dynamic range, with a maximum voltage amplitude of 250 V over a frequency band $\upsilon =$ 0.5–2.5 MHz.

3. Tests and results

The cooling performance of the ion guide has been assessed by measuring the time distribution of the bunches extracted from the RFQ using an off-line ion source.

3.1. THE SOURCE

Ions are produced by electron bombardment of the gas located in a ⌀50 mm cylindrical cell. The ions are extracted with a voltage applied between two plates through a ⌀5 mm extraction aperture. There is no ion heating inside such sources so that the ion distribution just after ionization is nearly thermal. However, the presence of stray fields, possible misalignments and the finite size of the source produce a transverse energy distribution larger than thermal. The energy spread is estimated to be of the order of few eV. For singly charged ions the extraction energy can be adjusted between 150 and 1000 eV. With beams up to 200 eV it is not necessary to set the RFQ system at high voltage as the DC applied to the rods is sufficient to decelerate the incident beam down to 100 eV.

3.2. MEASURING SEQUENCE

The ions from the source can be injected inside the RFQ in pulses of 1 to 5 ms width, with a repetition rate between 10 and 100 Hz. This is achieved by electrostatic steering plates located between the ion source and the RFQ. After an adjustable interval of up to 50 ms following the injection, the potential of the end plate can be switched off and the ions can then be extracted using a negative biased pulsed electrode. The ions are detected with a MCP and the time interval between the extraction pulse and the arrival of the bunch on the MCP is recorded.

3.3. RESULTS

The results presented here have been obtained by injecting N_2 gas in the source and using He as buffer gas in the ion guide. Figure 2 shows the time of flight distributions obtained with the measuring sequence described above. The left panel (a) shows the full distribution obtained with N_2 ions injected into the RFQ structure.

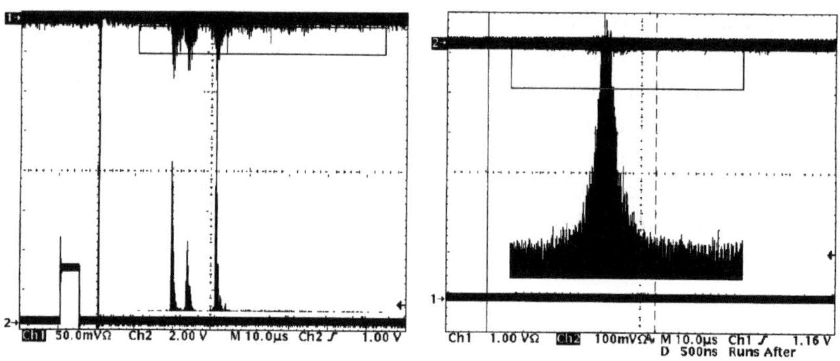

(a) (b)

Figure 2. Time structure of the bunches extracted from the RFQ cooler. See text for details.

The rectangular signal is the trigger for the time of flight measurement and corresponds to the duration of the voltage applied to the end plate. The three observed peaks are attributed to N ions and to H_2O and N_2 singly charged molecules. Their positions are in good agreement with the results obtained from simulations. It is seen that water is an important contaminant. The analysis of the residual gas with a vacuum mass spectrometer showed that water is in fact the main residual component. The right panel (b) is a zoom showing the details of the structure of the N_2 peak. The FWHM is about 200 ns. This is also in agreement with simulations and constitutes a convincing proof that the ions have been cooled down inside the RFQ ion guide.

4. Conclusions and outlook

A new linear RFQ cooler and buncher has been built to be installed at the new low energy beam line near the SPIRAL facility at GANIL. First tests of the device with a stable ion source have been successful and are in good agreement with the results from Monte-Carlo simulations. So far Ar and N_2 ions have been cooled and extracted as bunches using He as buffer gas.

Future tests include the cooling of ^{4}He ions with H_2 as buffer gas, the injection of cold ions inside a transparent Paul trap and the transmission tests of the device with ions from an ECR source.

Acknowledgements

We warmly thank the assistance of the technical staff at the LPC-Caen, and in particular the valuable contributions of Ph. Vallerand for the electronic equipment, Y. Merrer for the mechanical design and Ph. Desrues for the installations. The work described here was partly carried out within the EXOTRAP network of the European Union LSF-RTD program under contract number ERBFMGECT980099. Enlighting discussions with B. Bruyneel (IKS-Leuven) and D. Lunney and S. Henry (CSNSM-Orsay) are gratefully acknowledged.

References

1. Lunney, M. D. and Moore, R. B. *Int. Jour. Mass Spec.* **190/191** (1999), 153.
2. Nieminen, A. *et al.*, *Nucl. Instr. Meth. Nucl. Res. A* **469** (2001), 244.
3. Herfurth, F. *et al.*, *Nucl. Instr. Meth. Nucl. Res. A* **469** (2001), 254.
4. Proposal for SHIPTRAP, 1998, http://www-aix.gsi.de/shiptrap/Proposal.htm; see also Marx, G. *et al.*, this volume.
5. Argonne Nat. Lab., http://www.phys.anl.gov/atlas/cpt.html.
6. Auger, G. and Peghaire, A., *Nouvelles du GANIL* **68** (2001), 5.
7. Ban, G. *et al.*, Internal Report LPC-Caen, LPCC 99-16, 1999.

Hyperfine Interactions **146/147:** 265–268, 2003.

Molecular Physics in Storage Rings: From Laboratory to Space

D. ZAJFMAN[1,2], D. SCHWALM[2] and A. WOLF[2]
[1]*Department of Particle Physics, Weizmann Institute of Science, Rehovot, 76100, Israel*
[2]*Max-Planck-Institut für Kernphysik and Physikalisches Institut der Universität Heidelberg, D-69029 Heidelberg, Germany*

Abstract. The advantages of using the heavy-ion storage ring technique for measuring rate coefficient of reaction relevant to the evolution of the interstellar medium is presented, with the dissociative recombination of LiH^+ as a specific example.

Key words: molecular astrophysics, dissociative recombination, molecular ions.

1. Introduction

The physical processes that lead to the formation of most interstellar molecules are thought to occur within the interstellar clouds, since outside of these sources molecules would be destroyed in a relatively short time by the harsh ultraviolet radiation field that pervades interstellar space. The dominant gas phase reactions that occur at the low temperature (10–100 K) characteristic of interstellar clouds are ion–molecule and electron–molecular ion reactions [1]. A number of model calculations have shown that cosmic ray bombardment leads only to a small fractional ionization (10^{-7}–10^{-8}) but that the ions produced are quite efficient in leading to the synthesis of polyatomic molecules [2, 3].

One of the difficulties in reproducing in the laboratory some of the reactions where molecular ions are involved in the interstellar environment, is the low temperature. This temperature affects both the internal excitation of the molecular ions, as produced in standard (hot) ion sources, as well as the relative interaction energy in collisions. In the following, we show how the heavy ion storage ring technique has been very successful in solving these two issues simultaneously, specially for interaction between molecular ions and free electrons.

2. Dissociative recombination of cold molecular ions

The interaction between low-energy electrons and molecular ions is of fundamental importance in various natural environments, such as astrophysical and laboratory plasmas, planetary ionospheres and others [4–6]. Very often, this interaction causes the recombination and dissociation of the molecular ion, a process which is called dissociative recombination (DR), and leads to the production of neutral fragments with relatively high kinetic energy (eV's).

For many years, the main problem in laboratory study of this process has been the extreme sensitivity of DR to the initial vibrational state v of the molecular ion AB^+. It is well known, that it is often difficult to produce molecular ions in a well define initial vibrational state. This problem has led to experimental results (i.e. rate coefficients) which are very much dependent on ion source conditions, and thus difficult to compare with existing theoretical predictions.

During the last decade, the heavy-ion storage ring technique has been used to produce infrared active molecular ion beam in their ground vibrational state [7]. In this technique, a vibrationally excited molecular ion beam is generated by a standard (hot) ion source, and injected into the storage ring where it is stored for a time which is long enough to allow for complete vibrational relaxation through infrared transitions between the various vibrational states. Rotational cooling has also been demonstrated using this technique, the limit here being the blackbody radiation of the storage ring walls (300 K) [8]. After full relaxation is obtained, the beam is merged with an intense, cold electron beam, which is produced within an electron cooler device [9] at a velocity similar to that of the ion beam. Due to the kinematical transformation between the energies in the laboratory frame of reference to the center of mass frame of reference, a strong reduction is obtained in energy spread, and resolution down to few meV is usually obtained in the study of DR reactions. The heavy-ion storage ring technique has led to many breakthroughs in the field of DR, and has considerably enhanced the theoretical understanding of this process [7].

3. Example: DR of LiH^+

Among the molecules which were created in the early universe during the post-recombination era, LiH is of particular interest because of its contribution to the cooling of the primordial gas and because of its possible effects on the cosmic blackbody background radiation. However, because of its low ionization potential, lithium almost fully ionized at the time of hydrogen recombination. Thus, lithium chemistry was most likely initiated by radiative association process which produced LiH^+ molecules [10, 11]. On the other hand, the LiH^+ abundance is limited by photoionization, collisions and dissociative recombination (DR) [12]. This latter process, which for many molecular ions is usually fast, can be described as

$$\mathrm{LiH} + (v) + e^-(E) - \mathrm{Li}(nl) + \mathrm{H}(n'l') + E_{\mathrm{ker}}, \quad (1)$$

where v denotes the initial vibrational state of the molecular ion, n, n' and l, l' the principal and orbital quantum numbers of the atomic products, respectively, and E_{ker} is the kinetic energy release. The absolute rate coefficient for DR is unknown and a rather low estimated value of 2.6×10^{-8} cm^3 s^{-1} at $T = 300$ K has been used in early universe models [13].

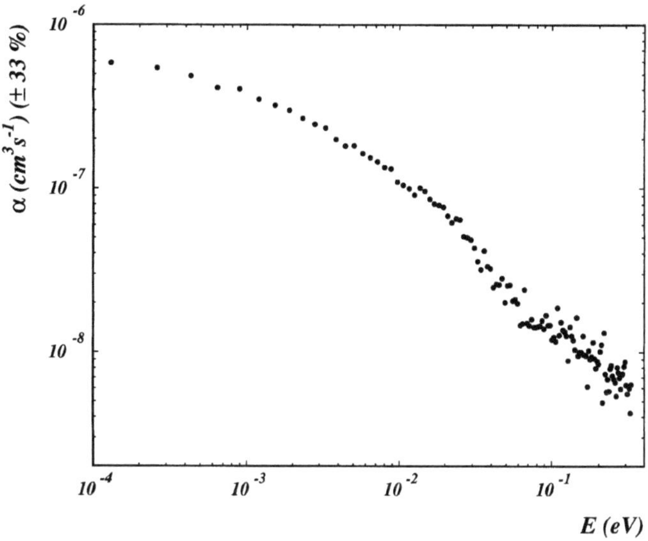

Figure 1. Measured rate coefficient for the DR of LiH^+ as a function of the electron energy.

Using the test storage ring (TSR) located at the Max-Planck-Institut für Kernphysik, Heidelberg, Germany, we have measured the DR rate coefficient for the DR of LiH^+. The LiH^+ ion beam was produced by a Tandem accelerator from gas stripping and dissociation of LiH^+ ions which were delivered by a cesium sputter source. After acceleration to an energy of $E_i = 6$ MeV and magnetic mass selection, the beam was injected into the TSR and stored for about 17 s. The beam was merged with an electron beam in the electron-cooler region over a length of 1.5 m. The electron beam was characterized by a density of $n_e = 8.5 \times 10^6$ cm^{-3} at $E = 0$ eV relative electron energy and by a longitudinal and transversal temperatures of $kT_{\|} = 0.1$ meV and $kT_{\perp} = 11.9$ meV, respectively. During the storage time, the LiH^+ ions, which were produced in high vibrational states by the stripping process, relaxed to a vibrational population distribution which is in equilibrium with the blackbody radiation of the storage ring walls ($T = 300$ K). At this temperature, about 78% of the LiH^+ molecules are in the ground vibrational state ($v = 0$) and about 14% are in $v = 1$. The remaining 8% are distributed over the three additional vibrational states.

Because the final states n of the Li fragments (see Equation (1)) were found to be very high Rydberg states ($n \geqslant 9$) [14], all of them were field ionized by the dipole magnet of the TSR, and very few pair of nascent fragments (Li,H), produced by the DR reaction, reached the detector located downstream of the electron cooler, after the dipole magnet. However, by measuring the change in the beam lifetime in the ring, with and without electrons, the rate coefficient could be extracted [14], and it's value is shown as a function of the electron energy in Figure 1. From this figure, and using a simple calculation, it is possible to extract the rate coefficient convoluted with a Maxwell–Boltzmann distribution at $T = 300$ K, which is found to be $\alpha(E = 0) = 3.9 \times 10^{-7}$ cm^3 s^{-1}.

The value of the rate coefficient obtained here is very large, and is more than one order of magnitude larger than the assumed rate in the early universe model referred to above [13]. However, it is important to remember that the measured value is valid for LiH^+ in a superposition of vibrational states as described above. On the other hand, the present rate coefficient suggest that this DR process is the fastest reaction among all those which are taken into account in the various different early universe models (see Table I in [13].) Also, it is an interesting question as to how such a large rate coefficient can be obtained, since no direct curve crossings between the ionic ground state and doubly excited neutral states are available for the process.

This work has been funded by the German Federal Minister of Education, Science, Research and Technology (BMBF) under Contract No. 06HD 8511, and within the framework of the German–Israeli Project Cooperation in Future-Oriented Topics (DIP).

References

1. Herbst, E., In: T. J. Millar and D. A. Williams (eds), *Rate Coefficients in Astrochemistry*, Kluwer, Dordrecht, 1988, p. 239.
2. Leung, C. M., Herbst, E. and Huebner, W. F., *Astrophys. J. Suppl. Ser.* **56** (1984), 231.
3. van Dishoek, E. F., In: T. W. Hartquis and D. A. Williams (eds), *The Molecular Astrophysics of Stars and Galaxies*, Oxford Science Publications, Clarendon Press, Oxford, 1998.
4. Bates, D. R. and Massey, H. S. W., *Proc. Roy. Soc. (Lond.)* **192** (1947), 1.
5. McElroy, M. B., Prather, M. J. and Rodriguez, J. M., *Science* **215** (1982), 1614.
6. Andersen, L. H. *et al.*, *Phys. Rev. Lett.* **77** (1996), 4891, and references therein.
7. Larsson, M., *Ann. Rev. Phys. Chem.* **48** (1997), 151, and references therein.
8. Hechtfischer, U. *et al.*, *Phys. Rev. Lett.* **80** (1998), 2809.
9. Habs, D. *et al.*, *Nucl. Instr. Meth. B* **43** (1989), 390.
10. Dalgarno, A. *et al.*, *ApJ* **458** (1996), 397.
11. Gianturco, F. A. and Giorgi, G., *ApJ* **479** (1997), 560.
12. Kirby, K., *Physica Scripta T* **59** (1995), 59.
13. Stancil, P. C. *et al.*, *ApJ* **458** (1996), 401.
14. Krohn, S. *et al.*, *Phys. Rev. Lett.* **86** (2001), 4005.

Hyperfine Interactions **146/147:** 269–273, 2003.

Astrophysically Motivated, Forbidden-Line Lifetime Measurements on Iron Ions (Fe^{9+}–Fe^{12+}) Using a Heavy-Ion Storage Ring

ELMAR TRÄBERT
I.P.N.A.S., Université de Liège, B-4000 Liège, Belgium, and Experimentalphysik III, Fakultät für Physik und Astronomie, Ruhr-Universität Bochum, D-44780 Bochum, Germany

Abstract. Atomic lifetimes in the millisecond range, of low-lying levels in iron ions of primary astrophysical interest, have been measured at a heavy-ion storage ring. The precision reached clearly exceeds that of the available predictions.

Key words: atomic data, spectroscopic techniques, heavy-ion storage ring, solar corona.

1. Introduction

Astrophysical plasmas like the solar corona combine high-temperature (leading to the collisional production of highly charged ions) and low-density (permitting radiative decays even via "slow" electric-dipole forbidden magnetic dipole (Ml), electric quadrupole (E2) and magnetic quadrupole (M2) transitions). Knowledge of the various transition rates is essential for astrophysical plasma diagnostics which make use of such forbidden lines in radiative-collisional models.

Since a few years, the SOHO spacecraft watches the sun. Transitions continually observed include electric dipole (E1) transitions in various Fe ions that give rise to spectral lines in the EUV. In the same spectral range, these ions also give rise to Ml and M2 transitions, and some of these feed the Ml transitions in the ground configurations of Fe X (Fe^{9+}) and Fe XIV (Fe^{13+}), the prominent "red" (637.46 nm) and "green" (530.286 nm) solar corona lines of iron.

The predicted rates of the forbidden lines lead to upper level lifetimes in the millisecond range. Such lifetimes are about a million times longer than the range accessible by classical beam-foil spectroscopy. In order to measure such long lifetimes in the laboratory, the ions have to be stored in traps [1].

2. Measurements

Our experiments employed a heavy-ion storage ring TSR at the Max Planck Institute for Nuclear Physics, at Heidelberg, Germany. The technique is an extension of beam-foil spectroscopy, using an ion beam tube that curves back on itself so

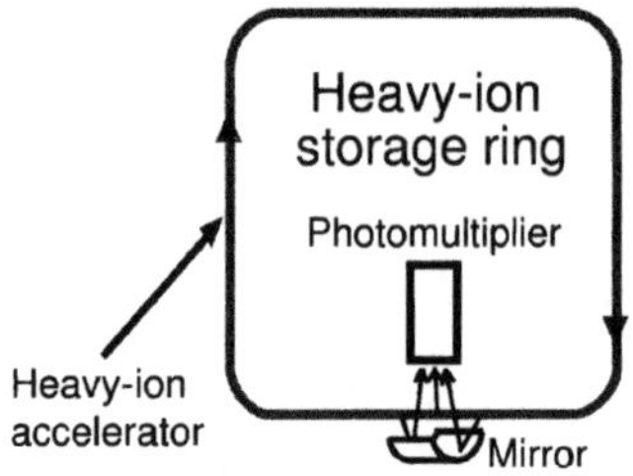

Figure 1. Simplified lay-out of the experiment.

that the same ions pass the same detector over and over again (Figure 1). The solid angle of detection in such a scheme, observing light from a 5 cm long section of a 55-m circumference storage ring, may seem small, but experiment shows that it is quite sufficient. The procedures employed have been described previously [2, 3]. The experiments of this series covered transitions within the ground complex of Fe^{10+} (S-like), Fe^{11+} (P-like) and Fe^{12+} (Si-like), as well as decays of long-lived 3d levels in Fe^{9+} (Cl-like) and Fe^{10+}.

All ion beams were produced as negative Fe ions from a sputter-type ion source, accelerated in the first half of a tandem accelerator, stripped to the desired charge state in a foil stripper, and accelerated further to final energies of order 30 to 150 MeV. Only a selected-charge state ion beam was transported to and injected into the storage ring. Multiturn injection and stacking of the ions over about 30 turns increased the number of stored ions, so that the ion currents in the ring reached up to 8 μA particles. The ions were left coasting for 200 ms to 1 s. Then the beam was dumped and the procedure repeated. Each detection cycle was started about 1 ms before injection, and events were sorted into 1000 bins of 0.2 ms to 1 ms width each.

The ion beam travels about 100 m from the injector to the ion storage ring, which at these ion energies takes about 6 μs, that is about twice the revolution time of the ions in the storage ring. A few-percent fraction of the ion beam was expected to be in excited levels from the stripping and excitation processes that take place inside the accelerator. The ion storage time constants (limited by collisional losses) depend on the background gas pressure (here a few times 10^{-11} mbar); they ranged from about 3 s (at relatively low ion beam energies) to 65 s for Fe^{12+} at the highest energies used. The ion beam lifetime is important as a systematic correction of the apparent optical decay data, but in the present case this correction remains well below 1%. The full injection and settling time is faster than the shortest of the expected radiative lifetimes of present interest, but very long compared to all cascade transitions from higher-lying levels except for those that involve some high-J $3d$ levels that only decay via Ml, E2, or M2 transitions.

We used optical observation in a side-on geometry, supported by a light collection system [2]. The light was detected by low dark-rate photomultiplier tubes (solar-blind PMT for UV light, bi-alkali PMT for visible and near-UV light). Interference filters were used only with the latter device. The width of the wavelength interval seen by the solar-blind detector led to the simultaneous observation of several decay components (all in the same ion species and charge state).

3. Data evaluation and results

The data sets (for a sample, see Figure 2) were collected and evaluated individually. Non-linear least-squares fits of one, two, or three exponential components (plus a constant background) were tried, as well as various schemes to search for systematic error in the data. The statistical uncertainty of some of the fit results is very small, owing to the high number of signal counts collected. A considerable uncertainty arises from the observed shape of some of the decay curves that reveals the presence of several decay components. Some of these are relatively easy to deal with, because they are sufficiently different in time constant, as is the case in Fe X, with three major components of predicted lifetimes near 5, 14 and 58 ms.

Table I lists a few examples of our lifetime results. (The full results, along with a comparison to the numerous predictions and the few other measurements, will be presented elsewhere [4].) Among them is the measurement on the $3s^2\,3p^2\,{}^1D_2$ level of the Si-like ion Fe^{12+}. This decay appears as a single-exponential decay (the cascade from the $3s^2\,3p^2\,{}^1S_0$ level has a very small branch fraction only) and is easily evaluated. Our decay curves yield a lifetime of 8.0 ms with a statistical uncertainty of about 1%. In contrast, Moehs *et al.*, working with a Kingdon ion trap, find a lifetime of just under 7 ms and assign a small uncertainty to it [5, 6]. A similarly large discrepancy of strage ring and Kingdon trap results is apparent for the rather similar case of the $3s^2\,3p^4\,{}^1D_2$ level of the S-like ion Fe^{10+}.

Comparing the intrinsic features of the various experiments (e.g., with [5, 6]), the signal levels obtained and the systematic error studies conducted, the heavy-ion storage ring technique appears to be superior for this kind of work. It employs the storage of a single charge state of a single element, and it monitors the ion storage

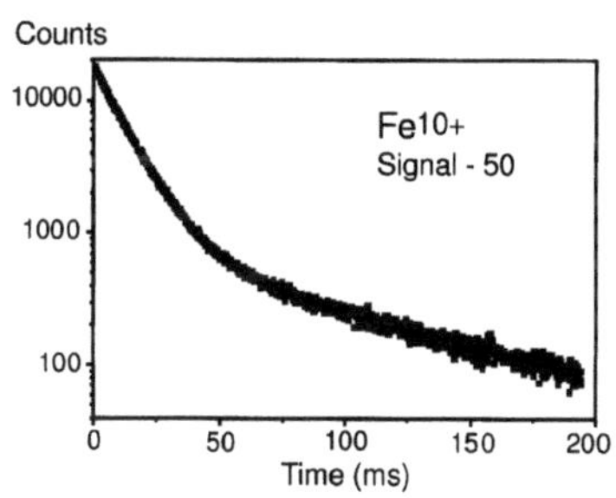

Figure 2. Photon signal (logarithmic scale) obtained with Fe^{10+} ions in the wavelength interval from 190 to 290 nm. The dominant contribution to the decay curve is from the $3s^2\,3p^4\,{}^1D_2$ level. The slower decay component represents a superposition of decays from the $3s^2\,3p^3\,({}^2D^\circ)\,3d\,{}^{1,3}G^\circ_4$ levels.

Table I. Measured lifetimes τ for levels in various ions of Fe

Upper level	Lifetime (ms) this work	Other experiments [5, 6]
Fe^{9+}		
$3s^2\,3p^4\,(^3P)\,3d\,^2F_{7/2}$	17.0 ± 1.7	
$3s^2\,3p^4\,(^1D)\,3d\,^2F_{7/2}$	4.9 ± 0.4	
$3s^2\,3p^4\,(^3P)\,3d\,^4F_{7/2}$	58 ± 10	93 ± 30
$3s^2\,3p^4\,(^3P)\,3d\,^2G_{7/2}$	not distinguished from above in the joint decay curve	
Fe^{10+}		
$3s^2\,3p^4\,^1D_2$	11.05 ± 0.1	$9.86 \pm 0.0.22$
$3s^2\,3p^3\,(^2D^\circ)\,3d\,^3G^\circ_4$	68 ± 4	
$3s^2\,3p^3\,(^2P^\circ)\,3d\,^1G^\circ_4$	not recognized separately from above in joint decay curve	
Fe^{11+}		
$3s^2\,3p^3\,^2D^\circ_{5/2}$	306 ± 10	
$3s^2\,3p^3\,^2D^\circ_{3/2}$	18.0 ± 0.1	20.35 ± 1.24
$3s^2\,3p^3\,^2P^\circ_{3/2}$	1.70 ± 0.02	1.85 ± 0.24
$3s^2\,3p^3\,^2P^\circ_{1/2}$	4.1 ± 0.12	4.38 ± 0.42
Fe^{12+}		
$3s^2\,3p^2\,^1D_2$	8.0 ± 0.1	6.93 ± 0.18

on-line, notwithstanding operation under notably better vacuum conditions than available in the other experiments.

The resulting lifetime values are in reasonable agreement with some of the predictions from theoretical calculations. In some cases theoretical predictions scatter by a factor of three; in other cases the scatter is small, but the typical predictions do not necessarily agree with the experimental findings. The experimental data, even with their scatter of some 10% from one experiment to the other, mark a significant step forward and permit a re-balancing of the error budgets in astrophysical analyses that involve forbidden transitions.

Acknowledgements

The Fe lifetime measurements have been done in collaboration with G. Gwinner and A. Wolf (Heidelberg), H.-P. Garnir and X. Tordoir (Liège), and E. J. Knystautas (Laval, Québec, Canada). Support by DFG (Germany), FNRS (Belgium), NSERC (Canada), and MPI-K Heidelberg (Germany) is gratefully acknowledged.

References

1. Träbert, E., *Phys. Scr.* **61** (2000), 257.
2. Doerfert, J., Träbert, E., Wolf, A., Schwalm, D. and Uwira, O., *Phys. Rev. Lett.* **78** (1997), 4355.
3. Träbert, E., Wolf, A., Linkemann, J. and Tordoir, X., *J. Phys. B* **32** (1999), 537.
4. Trabert, E., Gwinner, G., Wolf, A., Knystautas, E. J., Garnir, H.-P. and Tordoir, X., *J. Phys. B* **35** (2002), 671.
5. Moehs, D. P., Church, D. A., Bhatti, M. I. and Perger, W. F., *Phys. Rev. Lett.* **85** (2000), 38.
6. Moehs, D. P., Bhatti, M. I. and Church, D. A., *Phys. Rev. A* **63** (2001), 032515.

Hyperfine Interactions **146/147:** 275–281, 2003.

Laser Investigations of Stored Metal Cluster Ions

L. SCHWEIKHARD[1,*], K. HANSEN[2], A. HERLERT[3],
M. D. HERRÁIZ LABLANCA[3], G. MARX[1] and M. VOGEL[3]
[1]*Institut für Physik, Universität Greifswald, D-17487 Greifswald, Germany*
[2]*Department of Experimental Physics, Gothenburg University and Chalmers University of Technology, SE-41296 Gothenburg, Sweden*
[3]*Institut für Physik, Universität Mainz, D-55099 Mainz, Germany*

Abstract. The combination of ion storage in a Penning trap and photoexcitation by pulsed lasers has proven to be a versatile instrument in metal cluster research. Recent experiments which make use of both components allow a detailed study of the clusters' properties. In particular, a new method to measure dissociation energies is reviewed and preliminary results on the competition between electron emission and neutral monomer evaporation from dianionic metal clusters are presented.

Key words: metal clusters, Penning trap, laser excitation.

1. Introduction

Ion storage devices have proven to be an important tool in the study of clusters for several reasons. In particular, they allow to size-select the clusters of interest and to vary the duration of experiments over many orders of magnitude. Furthermore, they are versatile with respect to the clusters' reaction partners (neutral atoms or molecules, electrons, photons, etc.). In addition, it is possible to generate ensembles of thermalized cluster ions which, in combination with photoexcitation, allows detailed studies of unimolecular reactions under well controlled conditions.

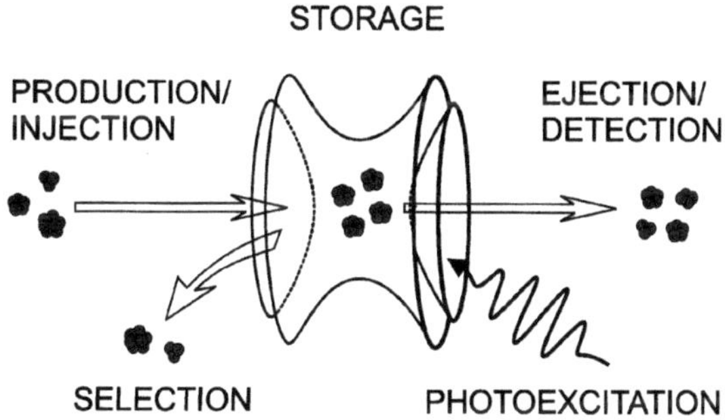

Figure 1. Schematical drawing of the injection, storage (including selection and laser excitation) and ejection of cluster ions in a Penning trap.

* Corresponding author.

In the following, experiments on metal clusters in a Penning trap are described. This device combines static electric and magnetic fields for the storage of charged particles [1]. Reaction pathways can be followed for a long range of mass-over-charge (m/z) values. Alternatively, particular species of interest can be isolated by application of rf fields.

Injection, storage and ejection of the clusters are shown schematically in Figure 1. The ions are produced in a Smalley-type source [2, 3] and transferred to the trap where they are captured in flight [4]. To purge the ion ensemble the cyclotron motion of the unwanted species is excited and they are radially ejected from the trap. For photoexcitation a pulsed (Nd:YAG or Nd:YAG-pumped dye) laser beam is directed axially through the setup. The charged reaction products remain stored and are finally analyzed by time-of-flight (TOF) mass spectrometry (MS). (For further details, see [5–7].)

2. Photodissociation at high laser pulse energies

As a first example, Figure 2 shows a TOF spectrum of size-selected Ag_{65}^+ clusters after photoexcitation by an intense laser pulse. The excitation leads to multiple photoabsorption and multiple decay. The abundances of the product clusters can be related to their stability and the spectrum therefore contains information about the structure of the clusters [8]. A number of local abundance maxima are observed at electron numbers consistent with a jellium/mean field model of the electronic structure of metal clusters [9, 10]: $n_e = 2, 8, 20, 34, 40, 58$, where n_e equals the cluster size n minus one, since silver is a monovalent element and the clusters studied are singly charged cations.

In addition to these maxima expected from mean field theory, two peaks are observed at $n = 49$ and 55. This pattern reproduces earlier results [11], where silver clusters were produced by sputtering and detected by Fourier Transform Ion Cyclotron Resonance (FT-ICR) MS (a widespread alternative detection method

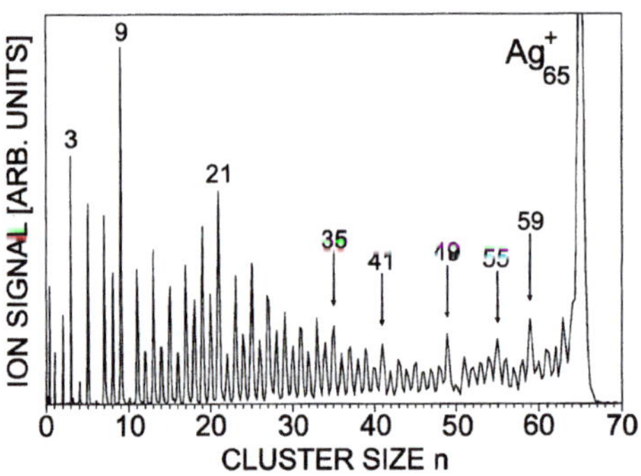

Figure 2. TOF spectrum after photoexcitation of size-selected Ag_{65}^+ clusters (laser pulse energy 1 mJ, wavelength $\lambda = 355$ nm, delay between laser pulse and TOF analysis $\Delta t = 100$ ms).

for ions stored in Penning traps, see e.g., [12–14] and references therein). Neither $n = 49$ nor $n = 55$ are consistent with the simple shell structure prediction. However, a comparison with earlier electron impact ionization/dissociation experiments on selected Ag_n^+ clusters allows an identification of the origin of these peaks. These experiments showed increased abundances for Ag_{50}^{2+}, Ag_{51}^{3+}, Ag_{55}^{2+}, and Ag_{55}^{3+} [15]. Its charge independence indicates that the extra stability of $n = 55$ is due to an atomic shell structure, i.e., the packing of the atoms in a particular geometrical (e.g., icosahedral) structure, which is well known from noble gas clusters [16] and large, cold clusters of alkali and earth alkali metal clusters [17]. Similarly, we can ascribe the enhanced abundances of clusters with electron number 48 to the electronic structure. It is difficult to reconcile the valence electron number 48 with the simple shell model since it requires a major rearrangement of levels around the Fermi energy. The origin of this shell closing is still an open question.

3. Model-free measurements of dissociation energies

In the example of the previous section a laser pulse of relatively high energy has been applied to induce the dissociation of a cluster of Ag_{65}^+ down to very small cluster sizes. When the laser pulse energy is reduced, only one or a few evapora-

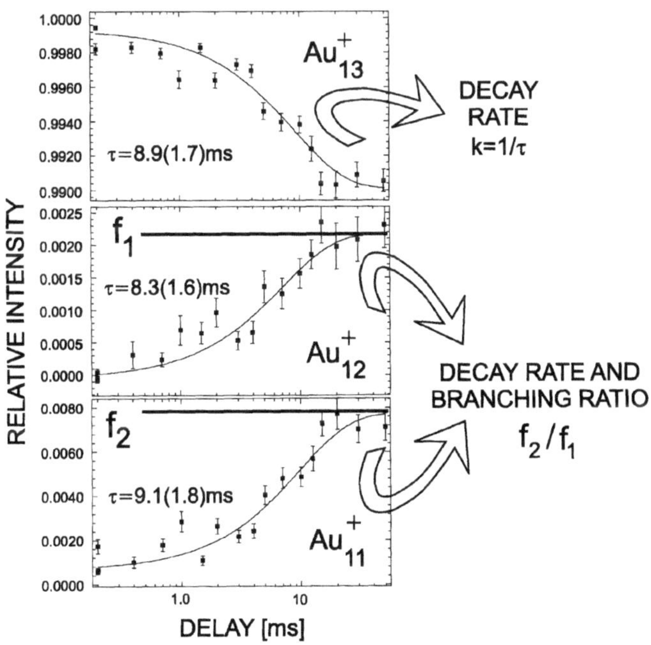

Figure 3. Cluster ion intensity of Au_{13}^+ (*top*), Au_{12}^+ (*center*), and Au_{11}^+ (*bottom*) as a function of delay between laser excitation and TOF mass analysis for size-selected Au_{13}^+ irradiated with a 75 μJ pulse at $\lambda = 264$ nm.

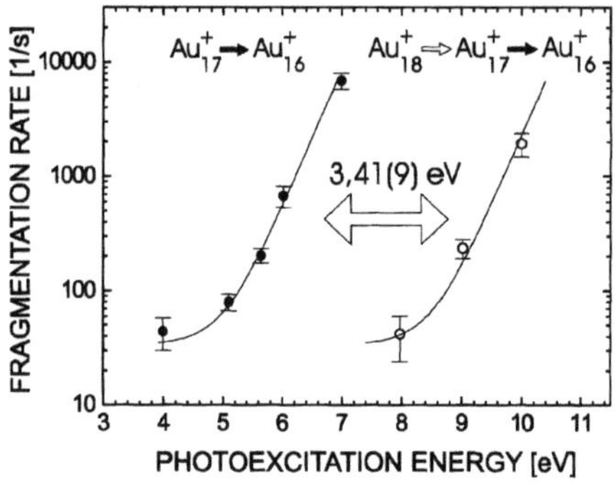

Figure 4. Fragmentation rates for the decay $Au_{17}^+ \rightarrow Au_{16}^+$ as a function of photoexcitation energy after excitation of Au_{17}^+ (*left*) and Au_{18}^+ (*right*).

tion steps are observed. The last evaporation step may be monitored time resolved. Thus, rates of evaporation can be determined by repeated measurements at different delay periods between the laser pulse and the TOF analysis.

Figure 3 shows an example where the decay of the gold cluster Au_{13}^+ and the appearance of the fragments Au_{12}^+ and Au_{11}^+ have been monitored as a function of the delay between the laser pulse and the TOF analysis. It is possible to distinguish between sequential monomer evaporation and the present reaction, where monomer and dimer evaporation are competing decay channels [18]. As reported recently [19], the rate information of such simple decays can be combined with sequential decays $Au_{n+1}^+ \rightarrow Au_n^+ \rightarrow Au_{n-1}^+$ to determine the dissociation energy of Au_{n+1}^+ without further assumptions on the details of the decay. In short, the excitation energy E_{n+1} of the sequential decay is compared with the excitation energy E_n of the simple decay of its intermediate product Au_n^+. The energies E_{n+1} and E_n may be adjusted to result in the same decay rates. Thus, the simple decay serves as an uncalibrated thermometer of the intermediate product. Apart from the small kinetic energy of the evaporated atom in the first step of the sequential decay, the dissociation energy of Au_{n+1}^+ is then given by the difference $E_{n+1} - E_n$. In practice, the energy difference may not only be taken at one particular decay rate but the energies can be compared at several decay rates, i.e., the rates are measured as a function of excitation energy, see Figure 4. Such measurements have been performed for a series of gold clusters Au_n^+, $n = 14$–24 [19].

In addition, the method has been supplemented with another thermometer: The energy-dependent branching ratio [20] between monomer and dimer evaporation has been used, and the earlier results from the decay rates were confirmed [21].

One interesting aspect of these results is the small magnitude of the odd–even effect in the dissociation energies compared with the abundance spectra (Figure 2). The different magnitude confirms the theoretical expectation from the analysis of evaporative equilibrium that the variations in the abundance spectra in general are larger than in the dissociation energies [8]. In evaporative equilibrium variations in dissociation energies cause variations in the abundances due to the dependence of the rate constants on the dissociation energies. Since abundances reflect the amount of energy one can store in a cluster and rate constants are determined by the microcanonical temperature, abundance variations are magnified with the heat capacity (in units of k_B) when compared with variations in the dissociation energy. As shown in [8] this magnification actually involves the differences between two consecutive dissociation energies and must be multiplied by the maximum sustainable temperature and divided by the dissociation energy, the ratio of which is approximately $1/30$. The resulting abundance variations exceed the variations in the dissociation energies when the cluster size is larger than about $n = 10$. A quantitative data analysis will be the subject of future work.

4. Photoexcitation of dianionic metal clusters

Another application of trapping is the production of dianionic metal clusters. These had not been observed until recently, when singly-charged anions were bathed in a sea of simultaneously stored electrons [22]. Already the efficiency of electron attachment gives information about the clusters' properties [23, 24].

A new series of measurements has now been started, where the dianionic clusters are probed by laser excitation. The preliminary results show at least two decay channels: (i) detachment of an electron: $Au_n^{2-} \rightarrow Au_n^{1-}$ and (ii) neutral monomer evaporation: $Au_n^{2-} \rightarrow Au_{n-1}^{2-}$. Small clusters preferentially emit electrons while the larger ones evaporate neutral monomers, as is indeed expected from simple liquid-drop-model estimations of the dissociation energy, the electron affinity and the Coulomb barrier for the emission of the second excess electron [23].

As an example, Figure 5 shows TOF spectra where gold cluster dianions Au_{35}^{2-} and Au_{50}^{2-} (both size and charge-state selected) have been photoexcited. The laser-pulse energy was high enough to induce a number of sequential decays. In the case

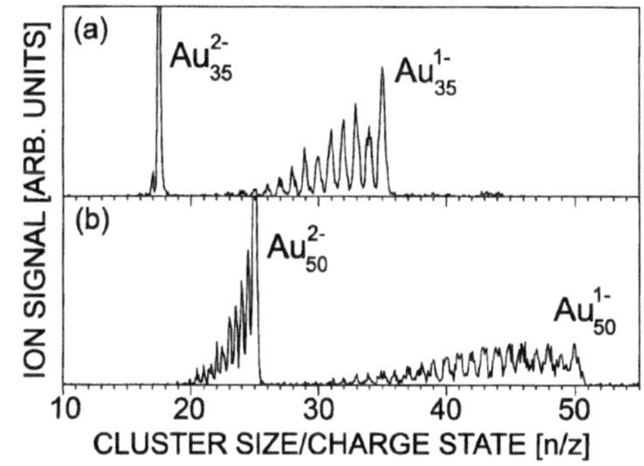

Figure 5. TOF spectra of size and charge-state selected clusters (a) Au_{35}^{2-} and (b) Au_{50}^{2-} after photoexcitation ($\lambda = 355$ nm, laser-pulse energy about 10 mJ, delay between laser pulse and TOF analysis $\Delta t = 100$ ms).

of Au_{50}^{2-} chains of fragmentation products are observed for both di- and monoanions. In contrast, only a small fraction of Au_{35}^{2-} decays by monomer evaporation to Au_{34}^{2-}. The main decay pathway is electron emission, thus Au_{35}^{1-} and its singly charged fragmentation products appear in the spectrum.

5. Conclusion

Ion trapping techniques and in particular the Penning trap allow to prepare ensembles of specific cluster size and charge state and to expose them to well-defined interactions. The products may be analyzed by mass spectrometry in a time-resolved manner. New methods, such as the direct determination of dissociation energies, have been developed and new species, such as dianionic metal clusters, have been produced. The combination of the wide range of time scales, the possibility of storing several species simultaneously and the application of laser excitation make the trap a leading tool in the quantitative study of the properties of small unsupported particles.

Note added in proof

Further developments and applications of the method of comparison between different decay chains have been summarized recently [25]. The studies on polyanionic clusters have also been extended [26, 27] and the Cluster Trap as a whole has been described in a new review [28].

Acknowledgements

This work was funded by the Deutsche Forschungsgemeinschaft, the EU network "CLUSTER COOLING" and the Fonds der Chemischen Industrie.

References

1. Brown, L. S. and Gabrielse, G., *Rev. Mod. Phys.* **58** (1986), 233.
2. Powers, D. E., Hansen, S. H., Geusig, M. E., Puiu, A. C., Hopkins, J. B., Dietz, T. G., Duncan, M. A., Langridge-Smith, P. R. R. and Smalley, R. E., *J. Phys. Chem.* **86** (1982), 2556.
3. Weidele, H., Frenzel, U., Leisner, T. and Kreisle, D., *Z. Phys. D* **20** (1991), 411.
4. Schnatz, H., Bollen, G., Dabkievicz, P., Egelhof, P., Kern, F., Kalinowsky, H., Schweikhard, L., Stolzenberg, H. and Kluge, H.-J., The ISOLDE Collaboration, *Nucl. Instrum. Methods A* **251** (1986), 17.
5. Schweikhard, L., Becker, St., Dasgupta, K., Dietrich, G., Kluge, H.-J., Kreisle, D., Krückeberg, S., Kuznetsov, S., Lindinger, M., Lützenkirchen, K., Obst, B., Walther, C., Weidele, H. and Ziegler, J., *Physica Scripta* **T59** (1995), 236.
6. Schweikhard, L., Krückeberg, S., Lützenkirchen, K. and Walther, C., *Eur. Phys. J. D* **9** (1999), 15.

7. Schweikhard, L., Herlert, A. and Vogel, M., In: E. E. B. Campbell and M. Larsson (eds), *The Physics and Chemistry of Clusters (Proceedings of the Nobel Symposium 117)*, World Scientific, Singapore, 2001, p. 267.
8. Hansen, K. and Näher, U., *Phys. Rev. A* **60** (1999), 1240.
9. Brack, M., *Rev. Mod. Phys.* **65** (1993), 677.
10. de Heer, W. A., *Rev. Mod. Phys.* **65** (1993), 611.
11. Selinger, A., Schnabel P., Wiese, W. and Irion, M. P., *Ber. Bunsenges. Phys. Chem.* **94** (1990), 1278.
12. Amster, I. J., *J. Mass. Spectrom.* **31** (1996), 1325.
13. Marshall, A. G., Hendrickson, C. L. and Jackson, G. S., *Mass. Spectrom. Rev.* **17** (1998), 1.
14. Marshall, A. G., *Int. J. Mass. Spectrom.* **200** (2000), 331.
15. Krückeberg, S., Dietrich, G., Lützenkirchen, K., Schweikhard, L., Walther, C. and Ziegler, J., *Eur. Phys. J. D* **9** (1999), 169.
16. Echt, O., Sattler, K. and Recknagel, E., *Phys. Rev. Lett.* **47** (1981), 1121.
17. Martin, T. P., Bergmann T., Göhlich H. and Lange, T., *Chem. Phys. Lett.* **172** (1990), 209.
18. Vogel, M., Hansen, K., Herlert, A. and Schweikhard, L., *Eur. Phys. J. D* **16** (2001), 73.
19. Vogel, M., Hansen, K., Herlert, A. and Schweikhard, L., *Phys. Rev. Lett.* **87** (2001), 013401.
20. Vogel, M., Hansen, K., Herlert, A. and Schweikhard, L., *J. Appl. Phys. B* **73** (2001), 411.
21. Vogel, M., Hansen, K., Herlert, A. and Schweikhard, L., *Chem. Phys. Lett.* **346** (2001), 117.
22. Herlert, A., Krückeberg, S., Schweikhard, L., Vogel, M. and Walther, C., *Physica Scripta* **T80** (1999), 200.
23. Schweikhard, L., Herlert, A., Krückeberg, S., Vogel, M. and Walther, C., *Philos. Mag. B* **79** (1999), 1343.
24. Yannouleas, C., Landman, U., Herlert, A. and Schweikhard, L., *Phys. Rev. Lett.* **86** (2001), 2996.
25. Vogel, M., Hansen, K., Herlert, A. and Schweikhard, L., *J. Phys. B* **36** (2003), 1073.
26. Herlert, A., Jertz, R., Alonso Otamendi, J., Gonzales Martinez, A. J. and Schweikhard, L., *Int. J. Mass Spectrom.* **218** (2002), 217.
27. Herlert, A. and Schweikhard, L., *Int. J. Mass Spectrom.* (2003), in print.
28. Schweikhard, L., Hansen, K., Herlert, A., Marx, G. and Vogel, M., *Eur. Phys. J. D* (2003), in print.

Hyperfine Interactions **146/147:** 283–291, 2003.

Studies of Clusters and Biomolecules in ELISA

J. U. ANDERSEN[1], L. H. ANDERSEN[1], P. HVELPLUND[1], A. LAPIERRE[1], S. P. MØLLER[2], S. B. NIELSEN[1], U. V. PEDERSEN[2] and S. TOMITA[1]
[1]*Department of Physics and Astronomy, University of Aarhus, DK 8000 Aarhus C, Denmark*
[2]*Institute for Storage Ring Facilities (ISA), University of Aarhus, DK 8000 Aarhus C, Denmark*

Abstract. Application of the electrostatic ion storage ring ELISA to studies of clusters and biomolecules is discussed. Ions injected from a plasma source or a sputter source are hot, and at short times the yield of neutrals is usually dominated by decay of metastable ions. We have demonstrated that the decay function is close to a $1/t$ dependence when the internal energy of the ions is conserved, i.e., when photon emission can be ignored. Deviations from a $1/t$ distribution therefore gives information about the radiative lifetime or, for larger systems, about the intensity of the emitted radiation. Systematic measurements have been carried out for fullerene anions C_N^-, for even values of N from 36 to 96, to test a classical dielectric model. Recently we have installed an electrospray ion source with a Paul trap for bunching, which can be used to inject biomolecular ions from solution, and the first experiments on laser spectroscopy of biomolecules have been carried out. Also lifetimes of excited states have been measured for stored biomolecular ions excited by a laser pulse.

Key words: storage ring, cluster, biomolecule, green fluorescent protein, nucleotide, dissociation, spectroscopy, lifetime.

1. Introduction

The ion storage rings in Aarhus, ASTRID and ELISA, have turned out to be very useful facilities for the study not only of atoms and small molecules but also of clusters and large biomolecules [1, 2]. In particular, the purely electrostatic ring ELISA (Electrostatic Ion Storage Ring, Aarhus) [3] is well suited for such studies because the ion trajectories depend only on the ion charge and energy and not on the ion mass. This implies that the parameters of the ring can be optimized with intense beams of ions which are easy to produce before storage of weak beams for which the diagnostics is poor. Until recently we have only used 'hot' ion sources at ELISA. The Nielsen source [4] is a versatile plasma source for atomic and molecular ions, which also gives very good beams of fullerene ions, both positive and negative, and a Cs sputter source [5] has been used to produce negative metal cluster ions. Recently we have implemented an injection system based on an electrospray source [6] combined with a Paul trap for beam bunching, and this system has been used to study biomolecules in ELISA [7]. In the following we give examples of results obtained with these different sources.

The electrostatic storage ring ELISA is illustrated in Figure 1. After acceleration by a 22 kV potential and mass selection by a magnet, a bunch of ions is injected

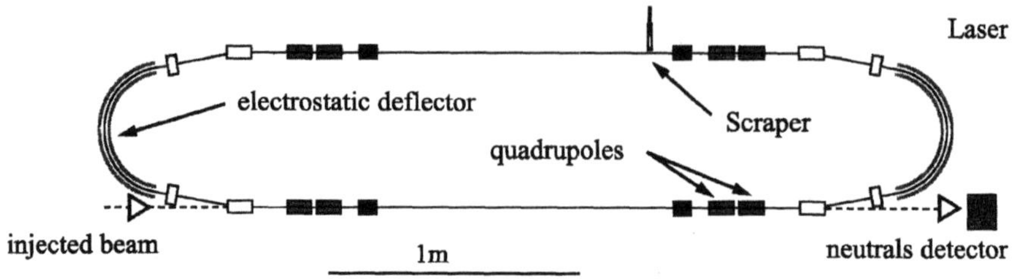

Figure 1. Schematic illustration of the electrostatic ion storage ring ELISA.

into ELISA, as indicated by the arrow in the lower left side of the figure, typically with a 10 Hz repetition rate. The ions are stored with a circulation period varying from a few to hundreds of microseconds, depending on the ion mass. In all the measurements to be presented, the signal detected was the yield of neutral particles in the MCP detector positioned at the lower right hand side in the figure. For stable ions the signal is due to collisions with rest gas atoms, and with a pressure of a few times 10^{-11} mbar the lifetime is of order 10 s for negative ions. At short times, the signal can be dominated by fragmentation or electron loss from metastable ions. Such a signal can also be used to monitor absorption of photons from a laser beam overlapping with the ions in the other side of ELISA, as illustrated in Figure 1. The MCP detector is large enough to collect essentially all neutral particles from the straight ring section in front of it but the efficiency of the detector is not constant over the surface area. Hence there are small variations with time of the count rate which are associated with betatron oscillations of the beam. These variations often dominate over the uncertainty due to counting statistics, and we have therefore not included statistical error bars in the figures.

2. Injection from 'hot' sources and $1/t$ decay law

A difficulty in the interpretation of experiments with ions from a plasma source or a sputter source is the large variation in internal excitation of the ions. This is of particular importance for clusters and large molecular ions, which have a large vibrational heat capacity. However, for ensembles of ions with a very broad distribution in internal energy, the exponential decay law, valid for systems in a well defined excited state, is replaced by a new decay law [8]: Denote by $g(E)$ the initial energy distribution in the ensemble and assume that the internal energy E of individual ions is conserved and that the decay rate is a rapidly increasing function, $k(E)$. This is usually a realistic picture because the internal state has reached a statistical equilibrium on the time scale of microseconds and the decay rate is of Arrhenius type. The decay yield at time t after the exit of the ions from the source can then be expressed as

$$I(t) = \int k(E) \exp\bigl(-k(E)t\bigr) g(E) \mathrm{d}E. \tag{1}$$

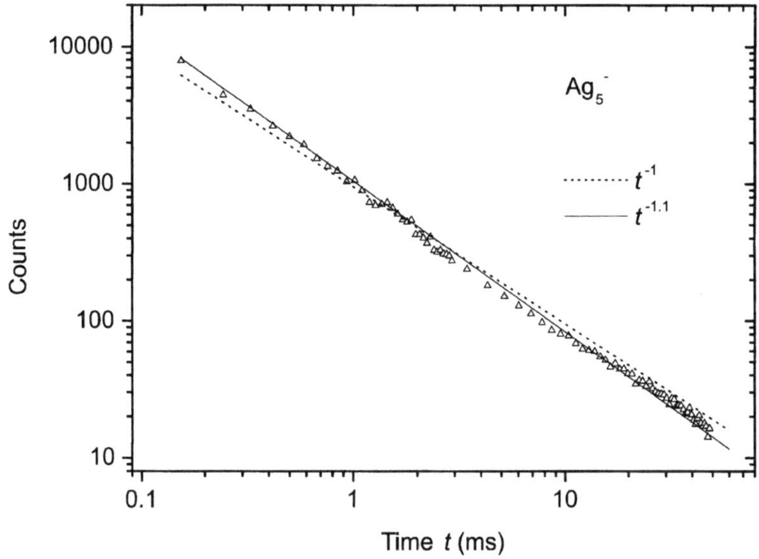

Figure 2. Yield of neutrals detected turn by turn after injection of Ag_5^- clusters into ELISA from a sputter source [8]. For $t > 3$ ms the counts have been averaged over 10 revolutions. The lines are power-law fits.

The weight function $k(E)\exp(-k(E)t)$ peaks strongly at a value of E corresponding to $k(E) = 1/t$ and the maximum value of the function scales with $1/t$. Since the width of the peak varies slowly with t the integral then varies approximately as $I(t) \propto 1/t$ if $g(E)$ is broad.

A storage ring is very well suited for investigation of this new decay law because the signal from the metastable ions can be followed over several decades in time. We have observed a decay close to a $1/t$ dependence for many molecules and clusters, and an example for negative silver cluster ions created in a sputter source is shown in Figure 2. These ions are known to decay mainly by emission of electrons or monomers and dimers [9], resulting in neutral particles which can be recorded with the MCP detector. The time dependence of the yield is seen to be close to a straight line on a log–log plot, with a slope corresponding to the power law $I(t) \propto t^{-1.1}$. The small correction to the $1/t$ law can be explained by the variation with time of the width of the peak in the weight function in the integral in Equation (1) [8].

3. Radiative cooling

Photon emission can be an important competing channel for deexcitation of hot molecules or clusters [10–14]. For large systems such as fullerene ions, the radiation may be included in the description as a continuous cooling process [14].

However, for small systems, emission of a single photon is typically sufficient to quench the electron emission or fragmentation, and radiative decay can instead be included as an additional term $-t/\tau$ in the exponent in Equation (1), where τ is the lifetime for radiative decay. Normally τ varies much more slowly with E than $k(E)$ and we may for a limited time interval assume τ to be a constant. The integral in Equation (1) then gives the approximate time dependence

$$I(t) \propto e^{-t/\tau}/t. \tag{2}$$

The measurement illustrated in Figure 2 therefore shows that the radiative lifetime for the silver clusters is longer than ~100 ms.

A measurement for a metal cluster ion with a shorter radiative lifetime is illustrated in Figure 3 [8]. Here the reciprocal of the counts of neutrals is plotted against time, and the $1/t$ law then corresponds to a straight line while the radiative decay introduces a positive curvature. The data points are fitted well by Equation (2) with a lifetime $\tau = 2.3$ ms. There are two sets of data points, one from the decay of initially hot cluster ions and the other for decay after excitation by a laser pulse, and there is seen to be good agreement after a relative normalization. This demonstrates that the decay is independent of the way in which the clusters have been excited and thus supports the assumption of an internal statistical equilibrium of the decaying ions.

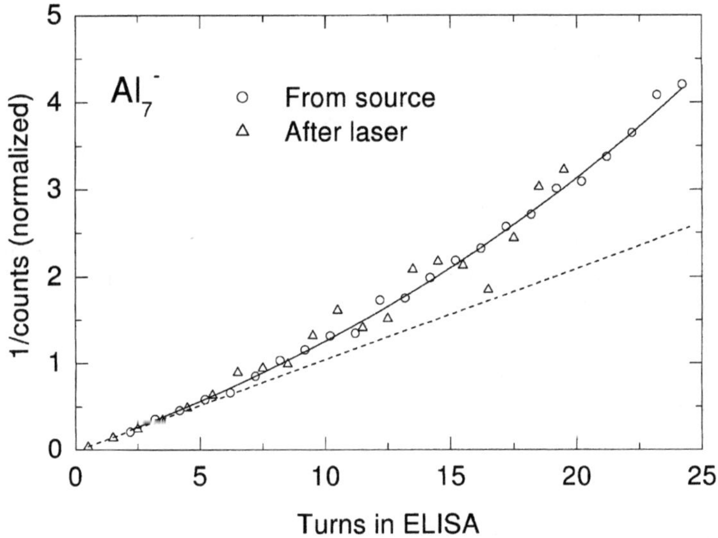

Figure 3. The reciprocal of the counts of neutrals recorded turn by turn after injection of Al_7^- clusters into ELISA [8]. The ions exit from a sputter source at $t = 0$. The triangles indicate the signal after excitation of the stored clusters with a laser pulse, and for these measurements the photon absorption defines $t = 0$. One turn corresponds to about 50 μs.

The most detailed and comprehensive studies of radiative cooling have been carried out for fullerene anions, C_N^- with even N values from 36 to 96, and we have determined the radiation intensity from measurements of the type illustrated in Figure 3 [14]. There is overall agreement within a factor of two with predictions from a classical dielectric model of the interaction of fullerenes with an electromagnetic field [15]. The model predicts a scaling of the radiation intensity with the number of atoms in the molecule and with the sixth power of the temperature, and the measurements also confirm this prediction.

Based on these studies of the radiative cooling of fullerene anions, we have carried out an investigation of the dissociation of fullerene cations at ELISA [16]. The main decay channel is emission of a C_2 fragment, and there has been a longstanding controversy over the magnitude of the dissociation energy [17]. For C_{60}^+ it is predicted to be near 10 eV but the experimental values have mostly been much lower. Also concerning the dependence of the dissociation energy on the number N of carbon atoms in the molecule there has been disagreement between theory and experiments and inconsistences between experiments. We have applied a novel experimental method based on the quenching of the dissociation by radiative cooling: For measurement on a fixed time scale, the Arrhenius decay law implies that the decay temperature scales with the magnitude of the dissociation energy. From theoretical arguments [15], it is expected that the dielectric model should describe the radiation from fullerene cations fairly accurately, and this model gives a connection between the temperature and the radiation intensity. This in turn gives a link between the dissociation energy and the characteristic time for quenching by cooling, and this time can be measured in ELISA. Such measurements have been carried out for $N = 48$–70 and 76, and both the magnitude of the dissociation energy obtained for C_{60}^+ and the variation with N of this energy are in good agreement with theory.

4. Spectroscopy on biomolecules

The combination of ELISA with an electrospray source allows spectroscopy and lifetime experiments on biomolecules in the gas phase [7]. Since the current from an electrospray source is low ($\sim 10^5$–10^6 ions per second), the ions are accumulated in a trap for about 100 ms prior to injection as a short pulse into the ring. As for clusters, the decay of the ions can be followed in time by detection of neutrals, and irradiation with laser light may result in an increased yield depending on the absorption cross section. Hence, spectroscopic information on the cross section may be obtained from the yield of neutrals, recorded as a function of the laser wavelength.

The local environment of a reaction centre inside a protein is a subject of high interest. For example, the development of fluorescent marker proteins relies on a detailed knowledge of the properties of the particular chromophore in play and how they can be changed in a controlled manner. We have used ELISA to study the isolated chromophore of the Green Fluorescent Protein (GFP) [18] and find that

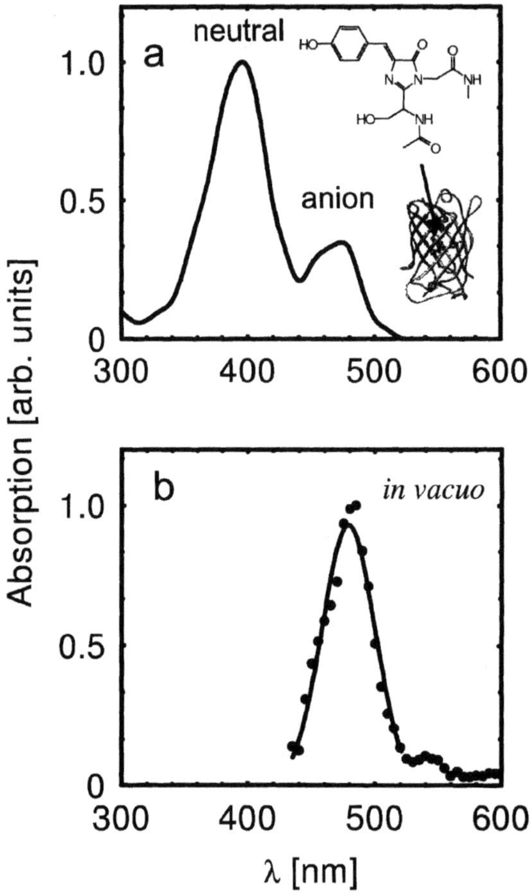

Figure 4. (a) Absorption spectrum of the wild-type *Aequorea victoria* GFP. The two peaks are ascribed to a neutral and a negative charge state of the chromophore in the protein. (b) Absorption spectrum of the model GFP chromophore shown in the inset, measured in ELISA [18].

the anion *in vacuo* has an absorption maximum at 479 nm (Figure 4). The yield of neutral fragments scales with the laser power squared, suggesting that the decay is a result of two-photon absorption. The absorption band essentially coincides with that of the two absorption peaks of GFP (at 477 nm) which is attributed to the anion form of the chromophore. The other absorption peak at 395 nm is due to the neutral form of the chromophore [19]. Hydration of the chromophore anion in alkaline solution causes a significant blue-shift of the band maximum from 479 to 426 nm, which clearly demonstrates that the electronic structure of the chromophore depends on its microscopic environment and that it can be seriously affected by hydrogen bonding. It is therefore very interesting that the absorption band is nearly the same for the anion in vacuum and in the local environment of the protein. This tells us that the special β-can structure of GFP (i.e., a hollow cylinder composed of eleven β strands as shown in the insert in Figure 4) [19] shields the chromophore from the surroundings without changing its electronic structure significantly. In other words, interactions with protein amino acid side-chains have only a small effect on the electron delocalization in the π-electron cloud of the chromophore anion.

ELISA provides an easy way of measuring the lifetime of excited ions, and recently we have studied the decay of mononucleotide anions and cations. Mononucleotides are the building blocks of RNA and DNA, "the molecules of life", and consist of a nucleobase (adenine, guanine, cytosine, uracil, or thymine), a furanose sugar, and a phosphate group. The nucleobase chromophores absorb ultraviolet (UV) radiation strongly and account for the absorption maximum of RNA and DNA around 260 nm [20]. The photophysics of nucleotides has been studied extensively in solution [21, 22] whereas only the smaller nucleobases have been studied *in vacuo* [23–25]. Gas-phase investigations may prove useful for our understanding of the important energy transfer processes in DNA and RNA. We have studied absorption of fourth-harmonic radiation from a Nd:YAG laser (266 nm) by both protonated and deprotonated adenosine 5′-monophosphate (AMP) [26].

From the experiments at ELISA, the lifetime of deprotonated AMP after absorption of a single 266-nm photon was found to be about 14 μs (Figure 5). As discussed in detail in [26], this lifetime is in accordance with a statistical frag-

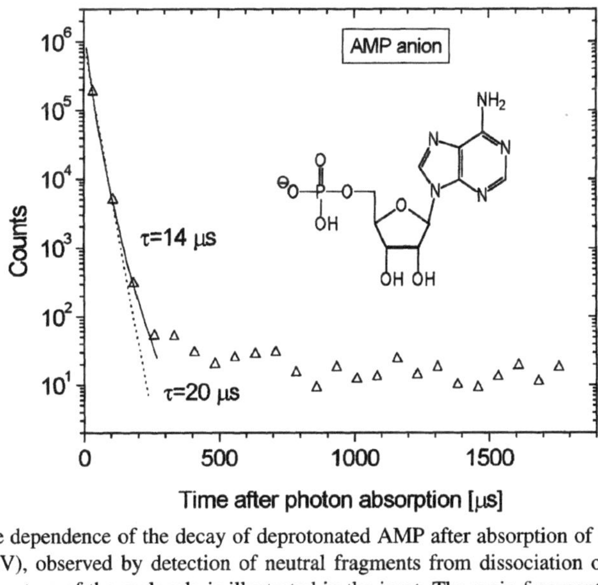

Figure 5. Time dependence of the decay of deprotonated AMP after absorption of a single 266 nm photon (4.66 eV), observed by detection of neutral fragments from dissociation of ions stored in ELISA. The structure of the molecule is illustrated in the inset. The main fragmentation channel is dissociation of a PO_3^- group. The full line is a fit by a transition-state calculation, while the dotted curve is an exponential fit.

mentation process (ergodic process) in which the energy is statistically distributed among the vibrational modes of the nucleotide before fragmentation occurs. The main decay channel is dissociation of a PO_3^- group with a dissociation energy of 1.26 eV [27]. The solid line in Figure 5 is from a simple transition-state calculation with a reduction factor ($\sim 10^{-2}$) which was adjusted to fit the data. Such a reduction may be expected since a concerted rearrangement of atoms is required for this dissociation process. The curvature comes from the initial spread in internal excitation energy at room temperature, as given by the canonical distribution, and the good fit to the data supports the interpretation as a process in an internal statistical equilibrium. In solution the fragmentation may be avoided due to interaction with the solvent, and the lifetime for fragmentation determines the time available for such vibrational relaxation [28]. For AMP the relaxation time has been measured to be less than 2 ps [22], and hence statistical fragmentation with microsecond lifetime is completely suppressed in solution.

Acknowledgement

This work was supported by the research center ACAP, funded by the Danish National Research Foundation.

References

1. Andersen, L. H., Andersen, T. and Hvelplund, P., *Adv. At. Mol. Opt. Phys.* **38** (1997), 155.
2. Hvelplund, P., Andersen, J. U. and Hansen, K., In: D. H. E. Dubin and D. Schneider (eds), *Trapped Charged Particles and Fundamental Physics, AIP Conference Proc. 457*, Woodbury, N.Y., 1999, p. 220.
3. Møller, S. P., *Nucl. Instr. Meth. A* **394** (1997), 281.
4. Almén, O. and Nielsen, K. O., *Nucl. Instrum. Methods* **1** (1957), 302.
5. Andersen, H. H. and Tykesson, P., *IEEE Nucl. Sci.* **22** (1975) 1632.
6. Gaskell, S. J., *J. Mass Spectrom.* **32** (1997), 677.
7. Andersen, J. U., Hvelplund, P., Nielsen, S. B., Tomita, S., Wahlgreen, H., Møller, S. P., Pedersen, U. V., Forster, J. S. and Jørgensen, T. J. D., *Rev. Sci. Instrum.* **73** (2002), 1284.
8. Hansen, K., Andersen, J. U., Hvelplund, P., Møller, S. P., Pedersen, U. V. and Petrunin, V. V., *Phys. Rev. Lett.* **87** (2001), 123401.
9. Shi, Y., Spasov, V. A. and Ervin, K. M., *J. Chem. Phys.* **111** (1999), 938.
10. Dunbar, R. C., *Mass Spectrom. Rev.* **11** (1992), 309.
11. Boissel, P., de Parseval, P., Marty, P. and Lefèvre, G., *J. Chem. Phys.* **106** (1997), 4973.
12. Andersen, J. U., Brink, C., Hvelplund, P., Larsson, M. O., Bech Nielsen, B. and Shen, H., *Phys. Rev. Lett.* **77** (1996), 3991.
13. Walther, C., Dietrich, G., Dostal, W., Hansen, K., Krückeberg, S., Lützenkirchen, K. and Sweikhard, L., *Phys. Rev. Lett.* **83** (1999), 3816.
14. Andersen, J. U., Gottrup, C., Hansen, K., Hvelplund, P. and Larsson, M. O., *Eur. Phys. J. D* **17** (2001), 189.
15. Andersen, J. U. and Bonderup, E., *Eur. Phys. J. D* **11** (2000), 413.
16. Tomita, S., Andersen, J. U., Gottrup, C., Hvelplund, P. and Pedersen, U. V., *Phys. Rev. Lett.* **87** (2001), 073401.
17. Lifshitz, C., *Int. J. Mass Spectrom.* **198** (2000), 1.

18. Nielsen, S. B., Lapierre, A., Andersen, J. U., Pedersen, U. V., Tomita, S. and Andersen, L. H., *Phys. Rev. Lett.* **87** (2001), 228102.
19. Tsien, R. Y., *Annu. Rev. Biochem.* **67** (1998), 509.
20. Stryer, L., *Biochemistry*, 3rd edn, W. H. Freeman and Co., New York, 1988.
21. Nikogosyan, D. N., Angelov, D., Soep, B. and Lindquist, L., *Chem. Phys. Lett.* **252** (1996), 322.
22. Pecourt, J.-M. L., Peon, J. and Kohler, B., *J. Am. Chem. Soc.* **123** (2001), 10370.
23. Kim, N. J., Jeong, G., Kim, Y. S., Sung, J., Kim, S. K. and Park, Y. D., *J. Chem. Phys.* **113** (2000), 10051.
24. Nir, E., Kleinermanns, K., Grace, L. and de Vries, M. S., *J. Phys. Chem. A* **105** (2001), 5106.
25. Lührs, D. C., Viallon, J. and Fischer, I., *Phys. Chem. Chem. Phys.* **3** (2001), 1827.
26. Nielsen, S. B., Andersen, J. U., Hvelplund, P., Pedersen, U. V. and Tomita, S., *Phys. Rev. Lett.* **91** (2003), 048302.
27. Ho, Y. and Kebarle, P., *Int. J. Mass Spectrom. Ion Processes* **165/166** (1997), 433.
28. Terpstra, P. A., Otto C. and Greve, J., *Biopolymers* **41** (1997), 751.

Hyperfine Interactions **146/147:** 293–306, 2003.

Molecular Ions and Nanoparticles in RF and AC Traps

D. GERLICH
Faculty of Natural Science, Technical University, 09107 Chemnitz, Germany

Abstract. This report summarizes recent advances in using innovative storage devices based on inhomogeneous, time-dependent electric fields. Electrode arrangements include quadrupoles, linear octopoles, higher-order multipoles, or a series of ring electrodes. Applying suitable oscillating voltages to such structures, with frequencies ranging from GHz via MHz to a few Hz, a variety of charged particles can be confined ranging from electrons via molecular ions to nanoparticles with masses of more than 10^{10} u. This contribution give a short summary of the theory, especially the effective potential approximation. The features of the technique are illustrated by several examples such as trapped ion beams, buffer gas cooling of ions in a 22-pole trap, and nanoparticle mass spectrometry in a quadrupole. Recent experimental results include growth of structures, isotope enrichment, and submonolayer gas ad- and desorption on a stored 500 nm SiO_2 particle. In the summary a few hints concerning ongoing experiments are given such as laser induced reactions, combination of traps with atomic beams, state-selected perturbation of a trapped low temperature ensemble with an infrared laser, and future trends in nanoparticle research.

Key words: multipole traps, inhomogeneous rf fields, effective potential approximation, mass spectrometry, ions, nanoparticles, reactions, low temperatures, growth of structures, isotope enrichment.

1. Introduction

Half a century ago researchers in various laboratories realized that charged particles can be confined in inhomogeneous, oscillatory electric fields. During those early times many ideas have been stimulated by methods developed in accelerator physics, e.g., the "strong focusing" principle which makes use of a series of electrostatic quadrupoles alternating in space. It was realized that the equation of motion describing fast charged particles traveling through such a structure, is the same for ions moving very slowly in a quadrupole field which alternates in time with an adequate repetition period. As demonstrated by the contributions to this conference *Atomic Physics at Accelerators*, also today, storage and accelerator techniques are still influencing each other. Trapping and guiding techniques have been and still are the basis of many new experiments in physics and chemistry. All these experiments make use of the unique advantages of these techniques such as long trapping times, the possibility to accumulate weak beams, phase space compression with lasers or buffer gas and many others.

The best-known examples are certainly Paul and Penning traps which are used in many laboratories. Their features and many applications have been summarized in several reviews [1, 2]. Less common are higher-order multipoles or other more complicated trap geometries, although also such arrangements have been covered thoroughly in a review [3]. The first rf octopole used in practice was the central element of the so-called Guided Ion Beam apparatus [4]. Since then, guiding, confinement and imaging properties of multipole arrangements have been improved significantly. The technique is now used routinely for measuring integral and differential cross sections at low kinetic energies [5, 6]. In this contribution a less known application is briefly described which is similar to the electrostatic mirror trap [7], the difference being that the oscillating beam, trapped in an octopole, has a kinetic energy of a few meV instead of keV. The development of other special temperature variable rf traps with wide field free regions was motivated by the aim (i) to study collision processes at very low energies, (ii) to produce internally cold ions, and (iii) to provide an ion source with *in situ* phase space compression via inelastic buffer gas collisions. The best tested arrangements which are now used in several laboratories, are 22-pole and ring electrode traps, but also ion labyrinths have been developed, which use more complicated geometries.

Today applications of traps extend into fields such as biology and modern material science. Due to their high sensitivity, storage devices become unique tools in analytical (bio-)chemistry, in aerosol research, or in studying whole cells and interstellar grain equivalents. One of the outstanding features is the high mass resolution in a mass range not accessible to any other existing techniques. Another one is the possibility to cool large molecules with many degrees of freedom to very low temperatures. This improves significantly the resolution in spectroscopy and of laser based diagnostic tools. Trapping allows one to synthesize new nanomaterials by adding atoms step by step (scaling up) or by long time annealing or evaporative reduction of single stored micro-particles (scaling down). This contribution describes several related activities going on in this laboratory ranging from low temperature collision dynamics to optical characterization and modification of a single trapped nanoparticle which may be probed for weeks.

2. Inhomogeneous rf fields

The method of ion guiding and trapping used in the present work is based on an inhomogeneous electric field $E_0(r, t)$ which usually has the following dependence on time

$$E_0(r, t) = E_0(r) \cos(\Omega t). \tag{1}$$

In most cases the spatial dependence, $E_0(r)$, is defined by the boundary conditions imposed by suitable electrodes; however, also nodal points or lines of standing or traveling electromagnetic waves can be of interest for traps. Basic parameters for characterizing operating conditions are the charge q, the mass m and the frequency Ω. Since many details concerning the mathematical treatment of oscillatory electric field can be found in [3], only a brief summary is given here.

For proper use of inhomogeneous time dependent fields a fundamental understanding of the adiabatic approximation is required and, in each practical case, one should calculate the effective potential V^* and the adiabaticity parameter η. Both are in general functions of the coordinate r and defined by

$$V^* = \frac{q^2 E_0^2}{4m\Omega^2}, \tag{2}$$

$$\eta = \frac{2q|\nabla E_0|}{m\Omega^2}. \tag{3}$$

In order to stay within the range of validity of the adiabatic approximation, one has to operate under conditions such that $\eta < 0.3$, i.e., the angular frequency $\Omega = 2\pi\nu$ has to be high enough for a given mass and charge combination. The method is quite general and can be used for confining electrons [8], ions, clusters and also charged microscopic particles [9] using frequencies ranging from GHz via MHz to a few Hz, respectively. As discussed in detail in [3] it is not difficult, at least in principle, to find operating conditions for simultaneously storing particles with quite different masses in the same trap. In the case of very large masses, where the gravitational force must be accounted for, superimposed electrostatic fields are commonly used.

Parameters for characterizing specific geometries are usually a scale length r_0 which is typically in the mm range, and the amplitude of the applied time dependent voltage, V_0, which can be below 100 V for many applications. The best characterized examples which are used in many applications are linear multipoles. To evaluate Equations (2) and (3) for multipoles, it is useful to introduce the parameter

$$\varepsilon = 1.036\, m\Omega^2 r_0^2/2n^2. \tag{4}$$

The practical units which are used in this formula are $[\varepsilon] = \text{eV}$, $[m] = \text{u}$, $[r_0] = \text{cm}$, and $[\Omega] = \text{MHz}$. $2n$ is the number of poles and r_0 the inscribed radius. With this the effective potential and the stability parameter can be written as

$$V^* = \frac{1}{8}\frac{(qV_0)^2}{\varepsilon}\left(\frac{r}{r_0}\right)^{2n-2}, \tag{5}$$

$$\eta = \frac{n-1}{n}\frac{qV_0}{\varepsilon}\left(\frac{r}{r_0}\right)^{n-2}. \tag{6}$$

The hyperbolic boundary conditions of multipoles are usually approximated by circular rods, the diameter d of which can be calculated in good approximation from

$$r_0 = (n-1)d/2. \tag{7}$$

Typical rod diameters for octopoles ($n = 4$) are $d = 2$ mm which must be equally spaced on a inscribed circle with $r_0 = 3$ mm. For constructing a multipole with

$d = 1$ mm rods and an inscribed circle of $r_0 = 5$ mm, one obtains from Equation (7) $n = 11$, i.e., the well-known 22-pole. For quadrupoles ($n = 2$) usually the slightly better approximation $r_0 = (1/1.1468)\ d/2$ is used [3]. Inspection of Equations (5) and (6) reveals that for $n = 2$, the effective potential V^* is harmonic and η has everywhere the same value, i.e. η does not depend on r. This leads to special properties, e.g., phase space conserving transmission [10]. Also for Paul type traps one gets harmonic eigenfrequencies. These are used to determine q/m of trapped nanoparticles [11].

3. Ions trapped in linear multipoles

3.1. TRAPPED ION BEAM

One interesting example illustrating several features of linear multipoles is the oscillating ion beam, shown in Figure 1. In this case radial confinement of ions is achieved using the rf field of an octopole. In axial direction the beam is influenced by two adjustable electrostatic barriers which are 3.2 cm apart. They are created by two cylindrical ring electrodes surrounding the octopole. A similar geometry is shown in Figure 30 of [3]. One way to monitor the trapped ion beam is to set the voltage of the entrance ring electrode at such a value that all ions are safely reflected while the exit barrier is so low that it is semipermeable for the trapped ion beam. Therefore, at each reflection, a small fraction of the trapped bunch of He^+ ions leaks out towards the detector as can be seen from the sequence of peaks in Figure 1. Due to this loss, the oscillating beam becomes weaker but the peak shape almost does not change. This is an experimental hint that the coupling between the axial and radial motion is rather weak, also in the region of the electrostatic mirrors.

The second measurement, also included in Figure 1, is the exponentially decreasing O^+ intensity which indirectly shows the slow loss of the trapped He^+ ion beam by a chemical reaction. For such a measurement the exit barrier is increased after the first round trip such that all He^+ ions are safely reflected on both mirror potentials. Adding oxygen into the trap leads to collisions between ions and these target molecules. The loss of He^+ is mainly via the dissociative charge transfer

$$He^+ + O_2 \rightarrow O^+ + O + He. \tag{8}$$

In this exothermic reaction, most O^+ ions formed have enough kinetic energy that they surmount immediately the exit barrier. From the observed time constant, $\tau = 2.7$ ms, and the measured number density of the target gas, $[O_2] = 6 \times 10^{11}$ cm^{-3}, the rate coefficient for reaction (8) can be derived,

$$k = \left(\tau[O_2]\right)^{-1}. \tag{9}$$

In the present example one obtains $k = 6 \times 10^{-10}$ $cm^3\ s^{-1}$. It must be noted, that in this experiment, k is not a thermal rate coefficient but a so-called effective rate

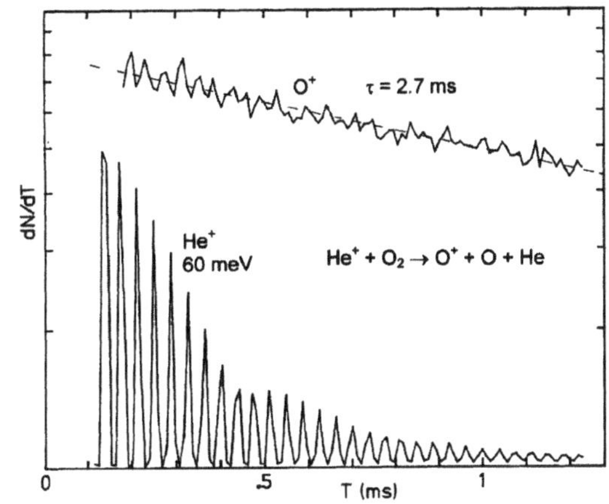

Figure 1. An He^+ ion beam, having a kinetic energy of $E_{lab} = 60$ meV, is trapped in an octopole between two barriers separated by 3.2 cm. For recording the lower signal, the exit barrier was tuned such that, after each round trip, a few He^+ ions leak out towards the detector. The upper curve shows O^+ ions produced by collisions of trapped He^+ ions with O_2 gas introduced at low number density into the chamber.

coefficient, since the collision energy distribution is determined by the ion beam energy (here $E_{lab} = 60$ meV in the laboratory frame) and the temperature of the oxygen (300 K). In the limit of a very slow ion beam, $E_{lab} \rightarrow 0$, one obtains indeed a thermal rate coefficient; however, at a temperature determined by temperature of the target gas and mass ratio of the reactants. In the example, one can reach a collision temperature of 33 K (4/36 $\times$ 300 K) with 300 K oxygen.

3.2. MINIMUM TRAPPING POTENTIAL

The method to inject, select and trap slow ion beams between two electrostatic barriers in a linear multipole, is an important diagnostic tool for characterizing axial potential distortions. Figure 2 shows a typical set of data measured with a linear 22-pole trap. In this case the electrostatic barriers have been created by thin electrodes surrounding the trap as shown in Figure 1 of [12]. Shown is the kinetic energy of the ion beam trapped between the entrance of the trap and a correction electrode close to the end, the voltage of which is varied. The linear fit indicates that the barrier height is proportional to the applied voltage; 1 V increase rises the barrier by 0.48 meV. The insert shows that it is possible to trap a beam with a mean axial energy of 2 meV.

Another important experimental parameter, characterizing the quality of rf based traps, is the minimum amplitude required for trapping. This value depends critically on surface potential distortions which are common for all types of materials. The best results obtained so far were with stainless steel rods whereas gold plated electrodes or coating with graphite was worse. The data shown in Figure 3 have

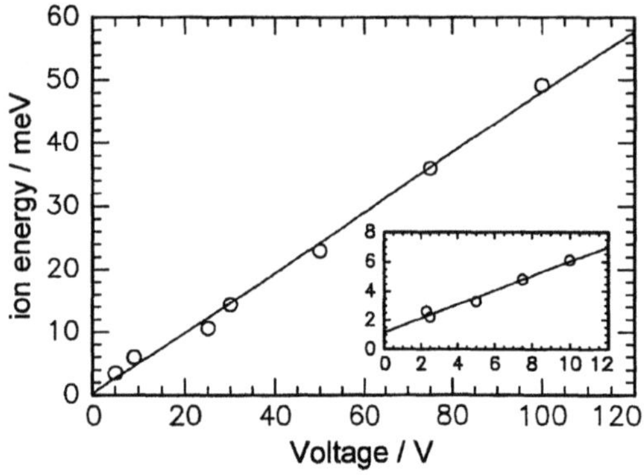

Figure 2. The homogeneity of the axial potential in a 22-pole trap is tested with the trapped ion beam method. The voltage on the ring electrode has been varied between 5 and 100 V. As can be seen from the linear fit, the resulting ion energy deduced from the flight time for one round trip, is proportional to the applied voltage, the factor being 0.48 meV per 1 V. With a very small offset of 1 mV this holds down to kinetic energies of 2 meV (see insert).

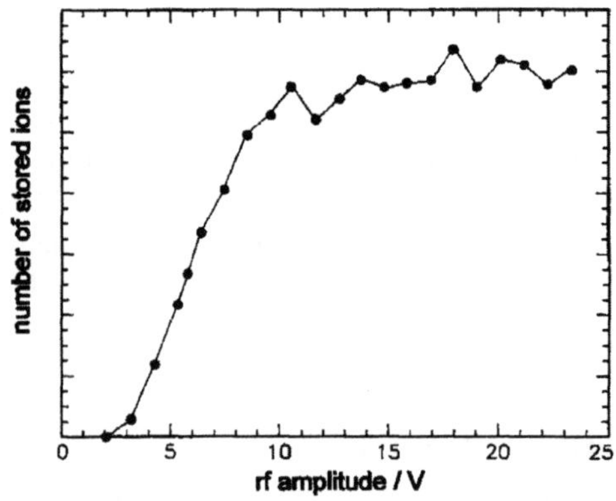

Figure 3. The number of stored ions depends on the rf amplitude applied to the electrodes. In this example, $C_2H_2^+$ ions ($m = 26$) are confined for some ms in a 22-pole ($n = 11$, $r_0 = 0.5$ cm) at $\Omega/2\pi = 16$ MHz. At the onset, $V_0 = 3$ V, the effective potential close to the rods is as low as $V^* = 4$ meV. For safe storage larger amplitudes must be used in order to avoid that the ions come to close to the rods. For example, for $V_0 = 20$ V, one obtains $V^*(r/r_0 = 0.9) = 5.4$ meV.

been measured for $C_2H_2^+$ ions ($m = 26$) in the 22-pole ($n = 11$, $r_0 = 0.5$ cm) operated with $\Omega/2\pi = 16$ MHz. With these parameters one can calculate from Equation (4) the characteristic parameter $\varepsilon = 281.23$ eV. With this, one gets easily the effective potential and the stability parameter at a given radius. For example for $V_0 = 3$ V one gets $V^*(r_0) = 4$ meV. From the data points one can conclude that safe trapping requires at least $V_0 = 10$ V. Since, with this amplitude, the stability parameter is still very small, $\eta < \eta(r_0) = 0.03$, there is no problem to further increase the rf amplitude. Stronger guiding fields have the advantage, that

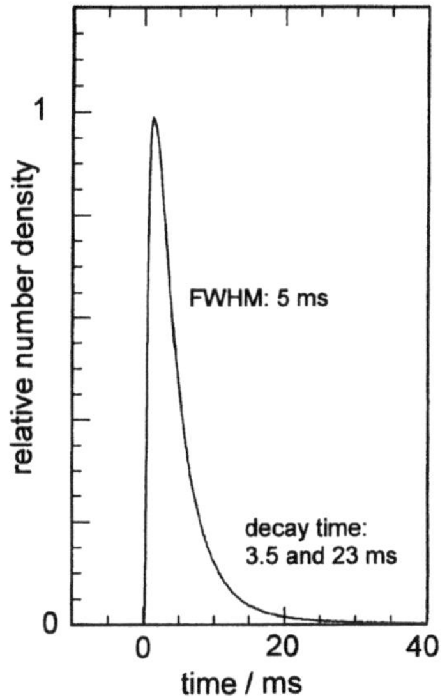

Figure 4. During their injection into the trap, ions are cooled with an intense pulse of buffer gas, in most cases He. The distribution (here H_2) has been measured *in situ* by passing a continuous fast Ar^+ ion beam ($E_{lab} = 8$ eV) through the trap and monitoring ArH^+ products. Since the piezo valve used has a response time in the μs range, the width of the pulse (FWHM = 5 ms) and the decay time is determined by the geometry of the box surrounding the trap and by the pumping speed.

the ions are kept further away from the surfaces with its potential distortions, see the numerical example given in the caption of Figure 3.

3.3. BUFFER GAS COOLING

In contrast to storage rings, fast beams trapped between electrostatic mirrors or Penning traps, one of the most important features of rf traps is the possibility to relax ions by collisions with buffer gas. In the 22-pole trap ions are cooled to the temperature of the cooled walls by using an intense and short pulse of He or H_2 buffer gas, injected into the trap via a fast piezo valve. As can be seen from Figure 4 the box containing the 22-pole (see Figure 1 of [12]) can be filled with a time constant of less than 1 ms. The width of 5 ms and also the decay is determined by two different time constants, a fast component of 3.5 ms, which is the time needed to transfer gas from the trap volume ($\sim$15 cm^3, conductance for $H_2 \sim 4$ l/s) into the main vacuum chamber and a slower component of 23 ms, which is the time for evacuating the chamber (volume $\sim$ 19 l) with the turbomolecular pump (800 l/s). Peak number densities up to 10^{15} cm^{-3} can be reached leading to thousands of collisions in 1 ms and, therefore, to a very efficient relaxation of all degrees of freedom. Important for many applications is that, after 100 ms, the density of the buffer gas is already so low that processes such as ternary association can be neglected.

3.4. CLUSTERING OF IONS

One application of long time trapping of ions at low temperatures and at high gas densities is to study formation of clusters or to synthesize special molecules such as van der Waals ions, isomers or isotopomers. A recently published example is the growth and fragmentation of $(CO)_n^+$, which was studied systematically as a function of temperature [11]. The results gave direct hints that formation of metastable isomers can be controlled by experimental parameters such as temperature or density of non-reactive buffer gas for cooling intermediate products. Another recent experimental development [13] is the adaptation of a trap to a linear time-of-flight mass spectrometer (TOF-MS) having the advantage that the masses of all ions contained in the trap are detected simultaneously. This significantly reduces statistical errors, especially in cases where many different masses are produced.

Figure 5 shows a complex mass spectrum, in this case measured by scanning a quadrupole mass spectrometer. Such a mass distribution is obtained, when H_3^+ ions are stored for 10 ms at a temperature of 10 K in the 22-pole trap filled with para-hydrogen having a number density of $[H_2] = 5 \times 10^{14}$ cm^{-3}. The details of forming hydrogen clusters and the competition between growth, fragmentation and deuteration if D_2 is added, was described in detail by Paul *et al.* [14, 15]. Such mass spectra recorded at different times contain a lot of information on structure, thermochemistry, and reaction dynamics. For example it is obvious that the clusters favor the structure $H_3^+ (H_2)_n$ since odd masses prevail. The fact that in Figure 5, H_9^+ and H_{11}^+ are the most abundant clusters is due to the transient nature of the situation.

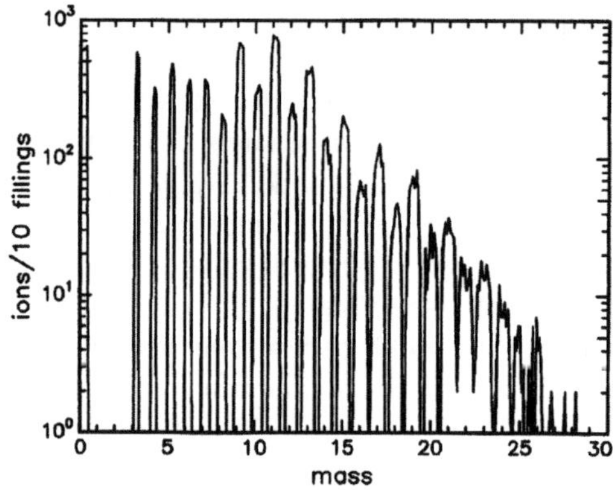

Figure 5. Mass spectrum of the trap content after storing H_3^+ for 10 ms in para-hydrogen having a number density of 5×10^{14} cm^{-3}. Under these conditions, H_9^+ and H_{11}^+ ions are dominant and odd masses have a larger weight than the even ones. The ions with even masses are formed by isotope fractionation, i.e., in collisions with traces of HD.

Extending the storage time, these clusters continue to grow and, finally, thermodynamic equilibrium prevails. At a temperature of 10 K, the mass distribution peaks at $n = 6$ or 7 [14].

Most, if not all clusters with even mass numbers are ions in which one H atom has been replaced by one D atom. Apparently isotopic fractionation, the enrichment of heavier isotopes in molecules due to a gain in zero-point energy, significantly influences the mass distribution at the low temperatures of the experiment. This process, the astrophysical importance of which has been discussed in recent publications [16, 17], is especially effective for D–H exchange. Although in the experiment HD is present only with its natural abundance, $[HD]/[H_2] = 3 \times 10^{-4}$ the ratio of deuterated ions to the non-deuterated ones is orders of magnitude larger, for example $[H_2D^+]/[H_3^+] > 0.5$. Comparison of other clusters reveals that the ratio depends on the cluster size. For a detailed analysis, also doubly deuterated ions have to be taken into account.

Structures, which are of fundamental interest but also of importance for characterizing the low temperature trap are rare gas cluster ions. For example the formation of He_2^+ dimers via the ternary association reaction

$$He^+ + 2He \rightarrow He_2^+ + He \tag{10}$$

has been studied over a wide temperature range in a SIDT apparatus (selected ion drift tube) by Böhringer *et al.* [18]. Their measurements, plotted in Figure 6 as open squares, have been extended towards lower temperatures operating the 22-pole trap at [He] number densities in the range of 10^{14} cm^{-3}. The results have been fitted with the temperature dependent function

$$k_3 = 1.4 \times 10^{-31} (300\ \mathrm{K}/T)^{0.6}\ \mathrm{cm}^6/\mathrm{s}. \tag{11}$$

Presently there are activities going on to use such a function for calibrating the He-buffer temperature at very low temperatures; however, more precise experiments are needed in order to separate the influence of ion temperature, density of

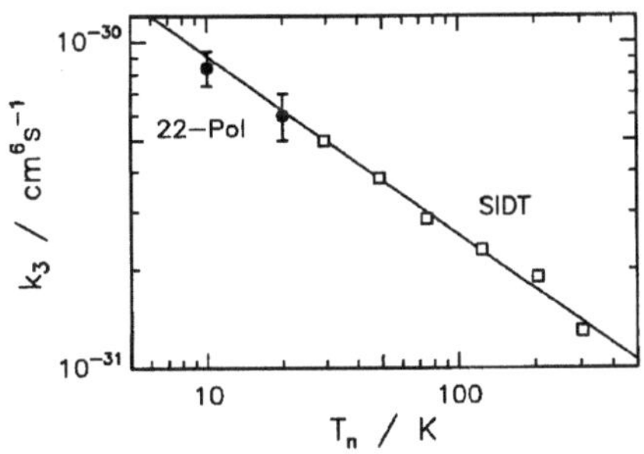

Figure 6. Temperature dependence of the ternary rate coefficients for the reaction $He^+ + 2He \rightarrow He_2^+ + He$. The filled circles are data obtained with the 22-pole trap, the open squares are from [18]. The line is a fit to the data using Equation (11).

the neutrals and the temperature dependence of the ternary association reaction approximated by Equation (11).

3.5. RADIATIVE ASSOCIATION

Another example demonstrating the unique sensitivity of the multi-electrode rf traps and the wide range of densities they can be operated under, are the systematic studies performed on radiative and ternary association. The astrochemical importance and many details of these types of reactions have been summarized in detail in [19]. Since all other reactions between the carbon ion C^+ and hydrogen are endothermic, there is only one reaction with which the formation of hydrocarbons in interstellar clouds can start: the radiative association process

$$C^+ + H_2 \rightarrow CH_2 + h\nu. \qquad (12)$$

In such a reaction the collision complex is stabilized by emission of a photon. This process has been studied with normal-hydrogen and para-hydrogen and the results have been reported in [19–21]. In order to learn more about the process, i.e., the competition between complex life time and radiative decay, reaction (12) has been studied with D_2. This isotopic substitution changes both the density of states determining the complex lifetime and the emission spectrum determining the radiative lifetime. Figure 7 shows for formation of CD_2^+ the measured effective rate coefficient, k^*, measured for target number densities ranging from 10^{11} to 10^{14} cm^{-3}. Fitting the data to the function

$$k^* = k_r + k_3[D_2] \qquad (13)$$

discussed in [19] leads to $k_r = 1.5 \times 10^{-15}$ cm^3 s^{-1} and $k_3 = 4.5 \times 10^{-28}$ cm^6 s^{-1}. Comparison with the corresponding results, measured for reaction (12) with para-hydrogen, $k_r = 1.7 \times 10^{-15}$ cm^3 s^{-1} and $k_3 = 1.8 \times 10^{-28}$ cm^6 s^{-1}, reveals that

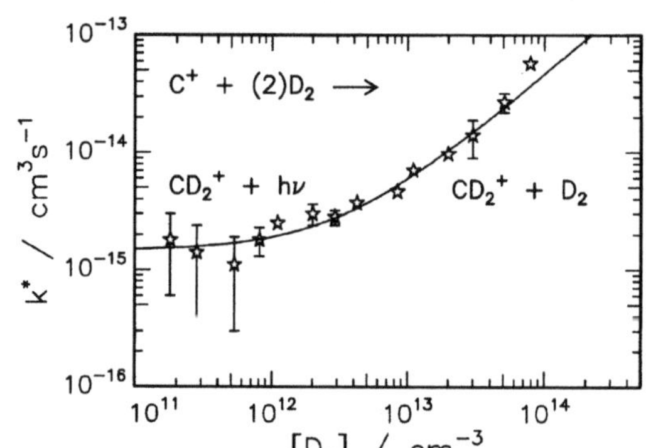

Figure 7. Effective rate coefficients for forming CD_2^+ ion as a function of the number density of $[D_2]$, measured in the 22-pole at 10 K. For $[D_2] < 10^{12}$ cm^{-3}, radiative association prevails, at large densities the dominant process is ternary association. The line is a fit to the data using Equation (13).

the probability for radiative stabilization is only slightly larger for D_2 while ternary stabilization is significantly more efficient with the heavier isotope.

4. Nanoparticles

It has been nicely demonstrated by Wuerker *et al.* [9] that Paul traps can confine micrometer sized particles with $m/q = 2 \times 10^5$ g/C if one uses a low-frequency (100 Hz) alternating field. By photographic detection, details of the trajectories of the trapped particles have been recorded. Also the formation of coulomb structures could be observed. Early experiments with charged droplets [22] have been summarized in [1]. In recent years there was a renaissance for such experiments [23] and especially for investigating stored particles due the growing interest in nanomaterials [24, 25]. Another motivation to study nm sized objects comes from the need to understand physical and chemical properties of interstellar grains [26, 27].

The trap used in this work and the features of the method such as reproducibility, long term stability, accuracy and linearity have been discussed in detail by Schlemmer *et al.* [25]. In the special design, described in that publication, compromises have been made in order to fulfill several experimental boundary conditions imposed by particle beams, lasers and imaging devices. Most experiments performed so far have used 500 nm diameter SiO_2 spheres which can be easily detected via light scattering. Fourier analysis of the recorded signal which is modulated by the particle's secular motion provides information on the q/m ratio. The charge of the particle can be changed by bombardment with an electron beam the current density of which is usually chosen so low that events occur with a rate less than 1 min^{-1}. By evaluating the statistics of a series of individual charge jumps, which can occur in both directions, quantitative values for the probability of sticking can be derived as well as the probabilities for emitting one or several secondary electrons. In such measurements one also obtains the absolute charge and thus the absolute mass. Presently a mass resolution $\Delta m/m$ of about 10^{-4} is obtained in a ten second measurement. Integration over longer periods of time can improve the resolution to the ppm regime.

Since results from the nanoparticle trap have been reported recently [25, 28] here only one example is presented which shows both the sensitivity and the limits of the presently used apparatus. In this experiment it has been tried to determine the net mass of a clean SiO_2 sphere by thermally removing impurities and adsorbed gas. For this purpose a CO_2 laser beam with an intensity of about 2 W/cm^2 irradiated the particle for half an hour leading to a temperature increase of about 100 K as derived from a simple estimate. After that, the laser was switched off for several hours. The process of heating and waiting has been repeated ten times over several days. Figure 8 shows the time dependence after the forth and fifth cleaning process. It can be seen that the mass drops very steeply by about 0.2 fg, when the laser is switched on (a). As long as the laser remains switched on (b) there is a further slow loss of mass with a rate of about −0.015 fg/h. This can

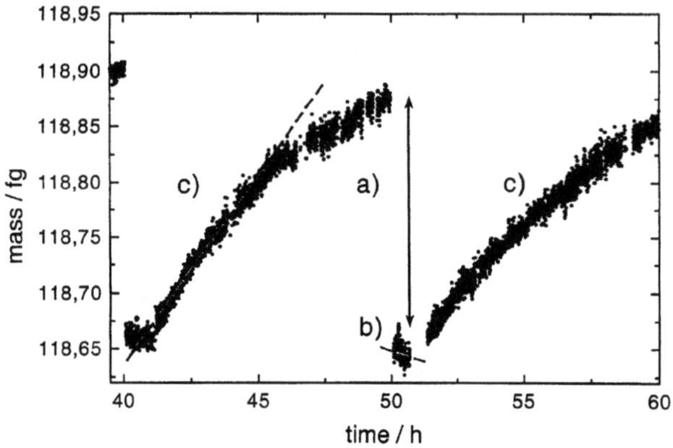

Figure 8. The mass of a 500 nm diameter SiO_2 sphere is periodically changed by gas ad- and desorption. During phase (a) and (b) a CO_2 laser heats with 2 W/cm^2, during (c) the laser is switched off for several hours. The process of heating and waiting has been repeated ten times over several days. The derived rates (dashed lines) are discussed in the text.

be explained by evaporation of some stronger bound molecules. When the laser is switched off (c), the particle regains mass with a rate of 0.033 fg/h. Quantitative tests have been performed to exclude H_2O or other rest gas molecules being usually present in ultrahigh vacuum systems, evacuated by turbomolecular pumps. The most probable candidates remaining are large organic molecules, e.g., pump oil. Assuming a molecular mass of 300 u, the rate corresponds to the adsorption of 17 molecules/s. With unit sticking efficiency this would correspond to a partial pressure of 2×10^{-11} mbar. It is obvious that the vacuum system has to be significantly improved.

Another experimental observation which is worthwhile to mention is that the 500 nm diameter SiO_2 sphere looses positive charges very quickly, if the CO_2 laser intensity is increased to 7 W/cm^2, corresponding to a temperature of about 500 K. Typical loss rates are a few e/s. This charge loss could be partly compensated by electron bombardment, but it was not possible, also after days of treatment, to heat the particle to high temperatures without loosing it from the trap. Further experiments are going on in order to understand and exclude this unwanted effect. It is planned to provide conditions such that the trapped SiO_2 particle reaches temperatures above 1000 K in order to (i) observe them via their blackbody radiation, (ii) to slowly evaporate them, and (iii) to reach in this way much smaller particles. Also reduction in an hydrogen environment is planned in order to get pure Si nanoparticles.

5. Conclusions

The wealth of applications, a few of which have been mentioned in this contribution, clearly shows that confinement of charged particles in oscillating fields is a very versatile tool. In gas phase activities presently going on in this laboratory [29], ion traps are combined with atomic sources including a beam of H-atoms from a cooled discharge and, in another experiment, a beam of C atoms or carbon molecules. The method of chemical probing has been used to distinguish between HCO^+ and HOC^+ isomers formed under interstellar conditions [30] and to probe the fine-structure states of atomic ions. Large progress has been made and is to be expected from the combination of low temperature traps with laser sources, especially also in the IR and FIR. So far such a combination has been applied successfully to study in great detail the reaction dynamics of the charge transfer $N_2^+ + h\nu + Ar \rightarrow AR^+ + N_2$ [31] and the rotational and fine-structure dependence of the endothermic reaction $C_2H_2^+ + H_2 \rightarrow C_2H_3^+ + H$ [32]. Some experiments, planned in combination with deuterated ions, are motivated from astrochemistry [16], e.g., the Herschel mission.

Concerning the nanoparticle traps, most of the current experiments are still far from what could be possible with today's techniques. The optical detection schemes can be significantly improved by using state of the art light sources including synchrotron radiation, ps- and fs-lasers and modern methods such as time correlated photon counting, CCD imaging, or nonlinear optical methods. For high-precision experiments the trap geometries and the electronics have to improved. Laser cooling can be used for getting sub-μm localization of the nanoparticles which is important for confocal detection or absorption experiments. Future characterization of the trapped particle will certainly include its optical properties, influenced by quantum confinement or the charge state, may be also excitation of innershell transitions. The study of ad- and desorption of gas, sputtering by fast particle bombardment and catalytic activities will be extended to low temperatures, the holy grail being the H–H recombination on interstellar grain equivalents. If, finally, atomic mass resolution is achieved which is possible for nm sized objects, chemical reactions can be followed by single jumps of the total mass, separated in time.

Acknowledgements

The development of rf devices involved the cooperation and assistance of many people both from the Freiburg team and from the researchers who helped in the past years to establish this laboratory in Chemnitz. Especially I would like to mention the contribution from PD. Dr. habil. Stephan Schlemmer. The projects mentioned in this report have been funded by the Deutsche Forschungsgemeinschaft, the VW foundation, and by the EC in a TMR network. Presently our activities are supported by the DFG-Forschergruppe Laboratory Astrophysics (FGLA).

References

1. Dawson, P. H., *Quadrupole Mass Spectrometry*, Elsevier Scientific Publishing, Amsterdam, 1976.
2. Gosh, P. K., *Ion Traps*, Claredon Press, Oxford, 1995.

3. Gerlich, D., *Adv. Chem. Phys.* **82** (1992), 1.
4. Teloy, E. and Gerlich, D., *Chem. Phys.* **4** (1974), 417.
5. Mark, S. and Gerlich, D., *Chem. Phys.* **209** (1996), 235.
6. Pullins, S., Dressler, R. A., Torrents, R. and Gerlich, D., *Z. Physik. Chemie* **214** (2000), 1279.
7. Zaifman, D., *Phys. Rev. A* **55** (1997), 1577.
8. Weibel, E. S. and Clarc, G. L., *J. Nucl. Energy C* **2** (1961), 112.
9. Wuerker, R. F., Shelton, H. and Langmuir, R. V., *J. Appl. Phys.* **30** (1959), 342.
10. Mark, S., Glenewinkel-Meyer, T. and Gerlich, D., *Int. Rev. Phys. Chem.* **15** (1996), 283.
11. Schlemmer, S., Luca, A., Glosik, J. and Gerlich, D., *J. Chem. Phys.* **116** (2002), 4508.
12. Gerlich, D., *J. Chem. Soc., Faraday Trans.* **89** (1993), 2199.
13. Luca, A., Schlemmer, S., Cermak, I. and Gerlich, D., *Rev. Sci. Instrum.* **72** (2001), 2900.
14. Paul, W., Lücke, B., Schlemmer, S. and Gerlich, D., *Int. J. Mass Spectrom. Ion Proc.* **150** (1995), 373.
15. Paul, W., Schlemmer, S., Lücke, B. and D. Gerlich, *Chem. Phys.* **209** (1996), 265.
16. Gerlich, D., Herbst, E. and Roueff, E., *Planetary and Space Science* **50** (2002), 1275.
17. Gerlich, D. and Schlemmer, S., *Planetary and Space Science* **50** (2002), 1287.
18. Böhringer, H., Glebe, W. and Arnold, F., *J. Phys. B* **16** (1983), 2619.
19. Gerlich, D. and Horning, S., *Chem. Rev.* **92** (1992), 1509.
20. Gerlich, D., In: I. Nenner (ed.), *Molecules and Grains in Space*, AIP Press, New York, 1994, p. 489.
21. Gerlich, D., *Physica Scripta T* **59** (1995), 256.
22. Owe Berg, T. G. and Gaukler, T. A., *Am. J. Phys.* **37** (1959), 1013.
23. Krämer, B., Hübner, O., Vortisch, H., Wöste, L., Leisner, T., Schwell, M., Rühl, E. and Baumgärtel, H., *J. Chem. Phys.* **111** (1999), 6521.
24. Schlemmer, S., Illemann, J., Wellert, S. and Gerlich, D., In: D. Dubin and D. Schneider (eds), *Trapped Charged Particles and Fundamental Physics*, AIP Press, New York, 1999, p. 80.
25. Schlemmer, S., Illemann, J., Wellert, S. and Gerlich, D., *J. Appl. Phys.* **90** (2001), 5410.
26. Schlemmer, S., Illemann, J., Wellert, S. and Gerlich, D., In: V. Ossenkopf, J. Stutzki, G. Winnewisser (eds), *The Physics and Chemistry of the Interstellar Medium*, GCA Verlag, Herdecke, 1999, p. 391.
27. Gerlich, D., Illemann, J. and Schlemmer, S., In: F. Combes and G. Pineau des Forets (eds), *Molecular Hydrogen in Space*, Cambridge University Press, Cambridge, 2000, p. 63.
28. Schlemmer, S., Wellert, S., Windisch, F., Grimm, M., Barth, S. and Gerlich, D., *Appl. Phys. A* (2003), accepted.
29. Gerlich, D., Luca, A. and Schlemmer, S., In: A. Giardini Guidoni (ed.), *XIX International Symposium on Molecular Beams, Book of Abstracts*, Universita di Roma, 2001, p. 9.
30. Smith, M. A., Schlemmer, S., v. Richthofen, J. and Gerlich, D., *Astrophys. J. Lett.* **578** (2002), 87.
31. Schlemmer, S., Kuhn, T., Lescop, E. and Gerlich, D., *Int. J. Mass Spectrom. Ion Proc.* **185** (1999), 589.
32. Schlemmer, S., Lescop, E., v. Richthofen, J. and Gerlich, D., *J. Chem. Phys.* **117** (2002), 2068.

Hyperfine Interactions **146/147:** 307–312, 2003.

Carbon Cluster Ions For a Study of the Accuracy of ISOLTRAP

A. KELLERBAUER[1,*], K. BLAUM[2], G. BOLLEN[3], F. HERFURTH[1], H.-J. KLUGE[2], M. KUCKEIN[4], E. SAUVAN[1], C. SCHEIDENBERGER[2] and L. SCHWEIKHARD[5]

[1]*CERN Division EP, 1211 Genève 23, Switzerland*
[2]*GSI, Planckstrasse 1, 64291 Darmstadt, Germany*
[3]*NSCL, Michigan State University, East Lansing, MI 48824, USA*
[4]*Inst. f. Physik, Technische Universität München, 85748 Garching, Germany*
[5]*Inst. f. Physik, Ernst-Moritz-Arndt-Universität Greifswald, 17487 Greifswald, Germany*

Abstract. Cyclotron frequency measurements of singly charged carbon clusters $^{12}C_n^+$ were carried out with the ISOLTRAP apparatus. The carbon cluster ions were produced externally by use of laser-induced desorption, fragmentation, and ionization of C_{60} fullerenes. They were injected into and stored in the Penning trap system. The observation of carbon clusters of different sizes has provided detailed insight into the final mass uncertainty achievable with ISOLTRAP and yielded a value of $u(m)/m = 8 \cdot 10^{-9}$. Since the unified atomic mass unit is defined as 1/12 the mass of the ^{12}C atom, ISOLTRAP can now be used to carry out absolute mass measurements.

Key words: carbon clusters, fullerenes, mass measurement, Penning trap.

1. Introduction and motivation

The ISOLTRAP experiment is a high-precision Penning trap mass spectrometer [1] for short-lived nuclides which is installed at the on-line isotope separator ISOLDE [2] at CERN. The masses of more than 200 radioactive nuclides have already been measured with ISOLTRAP.

The discovery and synthesis of C_{60} and C_{70} [3] has stimulated exciting scientific research and technical development. The fact that C_{60} fullerenes can now be produced in macroscopic amounts and with high purity has opened the way for the production and use of ionized clusters via laser-induced fragmentation and ionization of C_{60}. In this way, they can now be used to check the performance of the ISOLTRAP mass spectrometer. Apart from the negligible molecular binding energies, all carbon clusters that are composed solely of ^{12}C atoms have masses that are exact multiples of the unified atomic mass unit u.

Carbon clusters can also be used to perform absolute mass measurements with ISOLTRAP, i.e. mass measurements in which the reference ion has no mass un-

* Corresponding author; e-mail: a.kellerbauer@cern.ch.

certainty. In this case, ISOLTRAP's final result could directly be expressed as a mass in terms of the unified mass unit u. In addition, the masses of carbon clusters are evenly spaced throughout the entire chart of the nuclides, which minimizes the effect of any residual mass-dependent systematic errors of the apparatus.

2. Setup and technique

ISOLTRAP consists of a radiofrequency quadrupole (RFQ) ion guide beam preparation device [4] and a tandem Penning trap setup [1]. Figure 1 shows the overall setup of the ISOLTRAP apparatus with the newly installed carbon cluster ion source. The RFQ ion beam cooler and buncher serves the purpose of decelerating and cooling the 60-keV ISOLDE ion beam. The cooling trap is a cylindrical Penning trap which is used for the cooling and isobaric cleaning of the ion bunches.

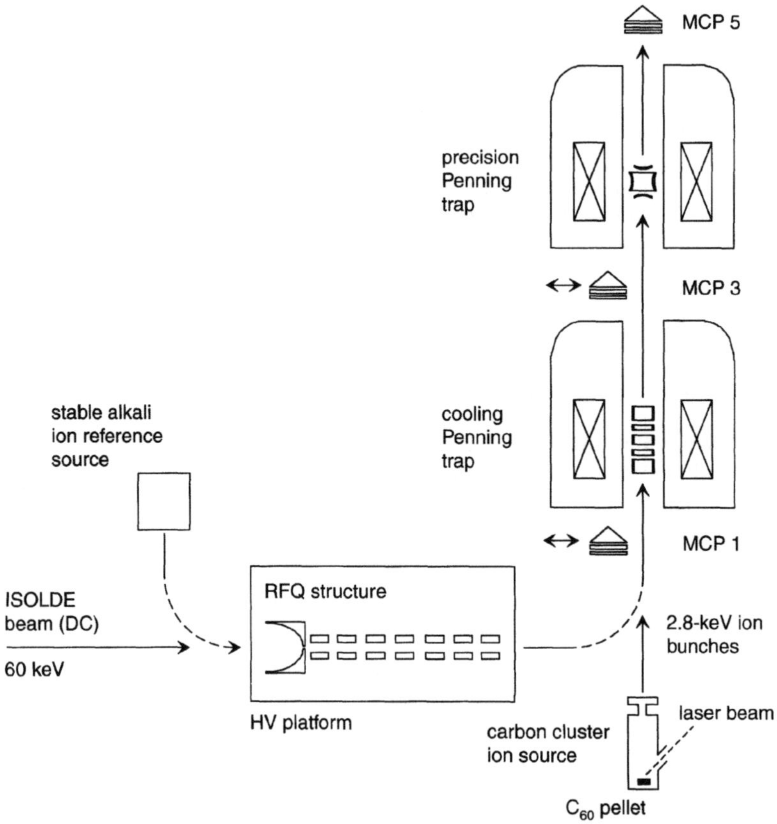

Figure 1. Schematic overview of the ISOLTRAP apparatus. Micro-channel plate (MCP) detectors are used to monitor the ion transfer as well as to record the time-of-flight resonance (MCP 5) for the determination of the cyclotron frequency.

The actual mass measurement is carried out using the precision trap which is a hyperbolical Penning trap. Both traps are situated inside the bores of superconducting magnets.

The mass measurement is based on a determination of the cyclotron frequency

$$\nu_c = \frac{1}{2\pi}\frac{qB}{m} \tag{1}$$

of an ion with unknown mass m in a magnetic field of magnitude B. The cyclotron frequency of the trapped ion is determined by ejecting the ion and measuring the time of flight (TOF) to the micro-channel plate (MCP) detector MCP 5 as a function of the frequency ν_{RF} of the applied azimuthal quadrupolar radiofrequency (RF) field. In resonance, the radial energy of the ion is increased. In the inhomogeneous-magnetic-field region between the center of the trap and the MCP detector, the magnetic moment associated with this radial energy then leads to an increase in axial energy, which results in a characteristic resonance curve [5]. The precision Penning trap is typically operated at a resolving power of about 10^6.

The magnitude of the magnetic field in Equation (1) is determined via a measurement of the cyclotron frequency $\nu_{c,ref}$ of a stable ion with well-known mass m_{ref}. The primary result of a mass measurement performed by ISOLTRAP is then expressed as the ratio of the cyclotron frequencies of the two ions:

$$r = \frac{\nu_{c,ref}}{\nu_c} = \frac{m}{m_{ref}}. \tag{2}$$

Since the magnetic field is subject to a long-term exponential decay as well as to short-term fluctuations, a reference measurement is carried out both before and after the measurement of the cyclotron frequency of the ion of interest. The combined uncertainty of the primary result is determined by the statistical uncertainty of the three measurements and the uncertainty due to systematic effects whose magnitude is unknown and which thus cannot be corrected for. If the final result is to be expressed in units of mass, the uncertainty of the knowledge of the mass of the reference ion contributes to the combined uncertainty as well.

The carbon cluster ions are produced by use of nanosecond laser desorption, fragmentation and photo-ionization of C_{60} at 532 nm. To this end, the frequency-doubled beam of a Q-switched Nd:YAG laser is focused on a C_{60} pellet (purity: $C_{60} > 99.4\%$). Further details on the carbon cluster ion source are given in [6, 7]. The charged fullerene fragments $^{12}C_n^+$ are electrostatically accelerated to 2.8 keV and transferred to the cooling Penning trap. From this point on, the measurement procedure is identical to that described above, which is used with ions from ISOLDE or from the reference ion source (Figure 1). A TOF spectrum recorded on detector MCP 1 at a laser pulse energy of about 10 mJ is shown in Figure 2(a). The main features are the high-mass, even-carbon-numbered fragments (fullerene fragments C_{34-58}^+), the low-mass fragments $C_{\leqslant 27}^+$, and the gap between C_{27}^+ and C_{34}^+ [6]. Figure 2(b) shows a typical TOF resonance of C_{10}^+ ions. The solid line is a fit of the expected line shape to the data points.

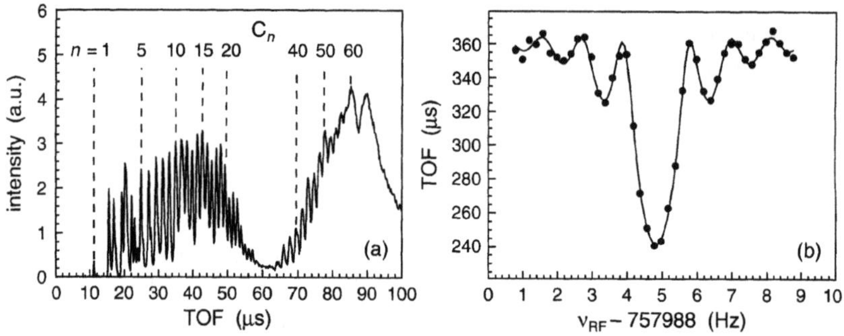

Figure 2. (a) TOF mass spectrum of carbon cluster ions at detector MCP 1 [7]; (b) TOF resonance curve of C_{10}^+. The solid line represents the theoretically expected line shape fitted to the data points. The error bars are smaller than the symbol size.

3. Measurement and results

In the case of a cross-reference mass measurement, a measurement which is carried out using carbon clusters both as the reference ion and as the ion of interest, the true value of the ratio of the cyclotron frequencies $\nu_{c,ref}/\nu_c$ is exactly known. The deviation of the measured ratio from this true value – beyond the statistical uncertainty of the measurement – is the combined error of the measurement apparatus and the procedure. If a large number of these measurements is conducted, they allow a detailed study of the influence of various experimental parameters on the error of the measurement. Taken together, they also allow a determination of the standard uncertainty distribution of the measurement result as well as the width of that distribution.

In order to characterize the deviation of the frequency ratio results from their expected true value, more than one hundred cross-reference measurements have been carried out. In three series of measurements, the singly charged ions of C_{10}, C_{12}, and C_{20} were in turn used as reference ions. All singly charged carbon clusters $^{12}C_n$ with $6 \leqslant n \leqslant 20$ were used as simulated ions of interest. As can also be seen from Figure 2(a), the production rates of the clusters with $n > 20$ were too low to allow a measurement of the cyclotron frequencies within a reasonable amount of time. For each TOF resonance, about 3000 ions were recorded. The duration of the quadrupolar excitation was varied between 900 ms and 3 s, corresponding to resolving powers between $6 \cdot 10^5$ and $3 \cdot 10^6$. Depending on the production rates and the excitation times, each resonance took between 15 min and 3 h. The relative statistical uncertainty $s(\nu_c)/\nu_c$ in the determination of the cyclotron frequency was a few times 10^{-8} for each single measurement.

Figure 3(a) shows the mean relative deviation $\varepsilon(r)/r$ of the measured cyclotron frequency ratios from the true values as a function of the difference in mass of the cluster reference ion and the ion of interest, $m - m_{ref}$. The fitted straight line suggests a linear mass-dependent systematic effect whose magnitude is $\varepsilon_m(r)/r = 1.6(4) \cdot 10^{-10} \cdot (m - m_{ref})/\text{u}$. Mass-dependent systematic effects can be due to

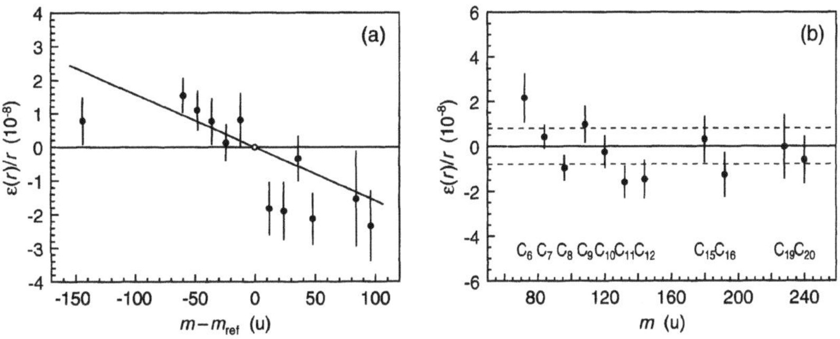

Figure 3. (a) Relative deviations of the measured cyclotron frequency ratios from the true values as a function of the difference in mass between the two ions. The straight line is a fit to the data points. The error bars represent the weighted mean statistical uncertainties of all single measurements that contribute to a data point. (b) Relative deviations of the weighted means of the measured frequency ratios from the true values as a function of the cluster ion mass, after correction for the mass-dependent effect. Again, the error bars represent the weighted mean uncertainties of all single measurements that contribute to a data point. The dashed lines indicate the residual systematic uncertainty that must be added to the uncertainties of the mean values in order to obtain a reduced χ^2 of one.

imperfections of the electrical quadrupole field or a misalignment of the precision trap's electrostatic trapping fields with respect to the magnetic-field axis. They are discussed in detail in [1]. The upper bound of the magnitude of these effects had previously been estimated to be $\varepsilon_{\mathrm{m}}(r)/r = 2 \cdot 10^{-9} \cdot (m - m_{\mathrm{ref}})/\mathrm{u}$ [8]. When the mass-dependent correction is applied, we find that the reduced χ^2 of the distribution of all individual measurements about their true values is $\chi^2/N = 0.96$.

Figure 3(b) shows the overall relative frequency ratio deviation for all measured carbon clusters as a function of the cluster mass after correction for the mass-dependent effect. As can be seen right away, these results agree well with the true values, represented by the horizontal line. A χ^2 test was then applied to these data. The reduced χ^2 of the distribution of the mean frequency ratios about their true values for all clusters is $\chi^2/N = 2.02$. This indicates the presence of a residual systematic effect whose magnitude can be determined from the condition $\chi^2/N \leqslant 1$ to be $u_{\mathrm{res}}(r)/r = 8 \cdot 10^{-9}$. Hence it could be shown that the limit of mass accuracy in ISOLTRAP measurements is about a factor of 10 smaller than previously estimated [1].

4. Conclusions and outlook

The present measurements have demonstrated that carbon clusters can be readily produced for injection into the ISOLTRAP mass spectrometer. They can be mass-selectively cooled in the cooling Penning trap and their cyclotron frequencies can be measured in the precision Penning trap.

Carbon clusters $^{12}C_n$ ($6 \leqslant n \leqslant 20$) were used to quantify systematic effects in the determination of masses with the ISOLTRAP apparatus. A mass-dependent systematic shift was identified which can be corrected for. If this correction is applied, the residual systematic uncertainty is found to be $u_{\text{res}}(r)/r = 8 \cdot 10^{-9}$, which also represents the present limit of mass accuracy of the ISOLTRAP system.

In the future, it will be possible to perform absolute mass measurements of radioactive nuclides using carbon clusters as reference ions. As an added benefit, the masses of these clusters are evenly spaced at intervals of 12 u throughout the entire chart of the nuclides such that the mass difference $m - m_{\text{ref}}$ will be at most 6 u. As a consequence, the relative mass-dependent shift will be smaller than $1 \cdot 10^{-9}$ and therefore negligible in future accurate high-precision mass determinations of short-lived nuclides.

References

1. Bollen, G. *et al.*, *Nucl. Instrum. Methods A* **368** (1996), 675.
2. Kugler, E. *et al.*, *Hyp. Interact.* **129** (2000), 23.
3. Kroto, H. W. *et al.*, *Nature* **318** (1985), 162.
4. Herfurth, F. *et al.*, *Nucl. Instrum. Methods A* **469** (2001), 254.
5. König, M. *et al.*, *Int. J. Mass Spectrom. Ion Proc.* **142** (1995), 95.
6. Scheidenberger, C. *et al.*, *Nucl. Instrum. Methods A* **701** (2002), 574c.
7. Blaum, K. *et al.*, *Eur. Phys. J. A* **15** (2002), 245.
8. Beck, D. *et al.*, *Nucl. Instrum. Methods B* **126** (1997), 374.

Hyperfine Interactions **146/147:** 313–317, 2003.

Atomic Spectroscopy and Collisions Using Slow Antiprotons – the ASACUSA Experiment at CERN-AD

E. WIDMANN
Department of Physics, University of Tokyo, Japan; e-mail: widmann@nucl.phys.s.u-tokyo.ac.jp

Abstract. The scope of the ASACUSA experiment is the study of atomic physics with low-energy antiprotons, namely high-precision spectroscopy of antiprotonic atoms, the study of antiprotonic atom formation processes, and stopping power and ionization cross section measurements in low-pressure gases. This talk gives an overview on the results achieved in the years 2000 and 2001, focussing mainly on the spectroscopy of antiprotonic helium.

Key words: antiprotonic helium, laser spectroscopy, hyperfine structure, ultra-low energy antiprotons, atomic collisions with ultra-slow antiprotons.

1. Introduction and overview of the ASACUSA program

The ASACUSA collaboration [1] is working at the Antiproton Decelerator (AD) facility of CERN in Geneva, Switzerland. The AD [2] is currently the only source of low-energy antiprotons in the world, delivering 200 ns long pulses of typically 4×10^7 antiprotons of energy 5.3 MeV (momentum 100 MeV/c). The distance between two pulses is about two minutes.

The physics program of the ASACUSA collaboration is divided into three phases corresponding to three different antiproton energy regimes. In *phase 1* the direct AD beam is used for laser and microwave spectroscopy of metastable antiprotonic helium (Section 2). For *phase 2*, a Radio Frequency Quadrupole Decelerator (RFQD) has been built by the CERN PS division and installed in the ASACUSA experimental area. The RFQD decelerates the AD beam to a variable energy of 10–100 keV with an efficiency of about 25%. With this beam first laser spectroscopy experiments of $\overline{p}He^+$ in ultra-low density helium gas have been performed in 2001. Already in 2000 the energy loss of antiprotons in various materials has been measured for the first time down to 1–2 keV energy. The results of the energy loss measurements are reported in a separate talk [3]. In *phase 3*, an antiproton capture trap is added after the RFQD to capture and cool antiprotons to liquid helium temperature. Then the antiprotons are again ejected from the trap to create a unique low-energy antiproton beam of 10 eV–several keV energy. This beam was named MUSASHI (Monoenergetic Ultra Slow Antiproton Source for High-precision Investigations) and will be used for the study of antiprotonic atom formation and ionization cross section measurements [1, 4]. MUSASHI is in its commissioning phase, and a first milestone was achieved in spring of 2001 when more than 3×10^5 antiprotons could be trapped in the large ASACUSA catching trap, and cooled to less than 100 eV [5]. The trap is now routinely capturing about 1 million antiprotons from one AD shot.

2. Spectroscopy of antiprotonic helium

The phase 1 of ASACUSA is exclusively devoted to the spectroscopy of antiprotonic helium, which has been performed previously by the PS205 collaboration at LEAR [6]. Here, highly excited states of the neutral three-body system $\bar{p}$–He^{2+}–$e^- \equiv \bar{p}He^+$ [7] (cf. Figure 1) are investigated by means of laser and microwave spectroscopy. These states are formed when antiprotons are stopped in helium (in the current experiments, helium gas of about 6 K and 0.1–2 bar). Antiprotons are captured by replacing one of the two ground-state electrons of helium and therefore

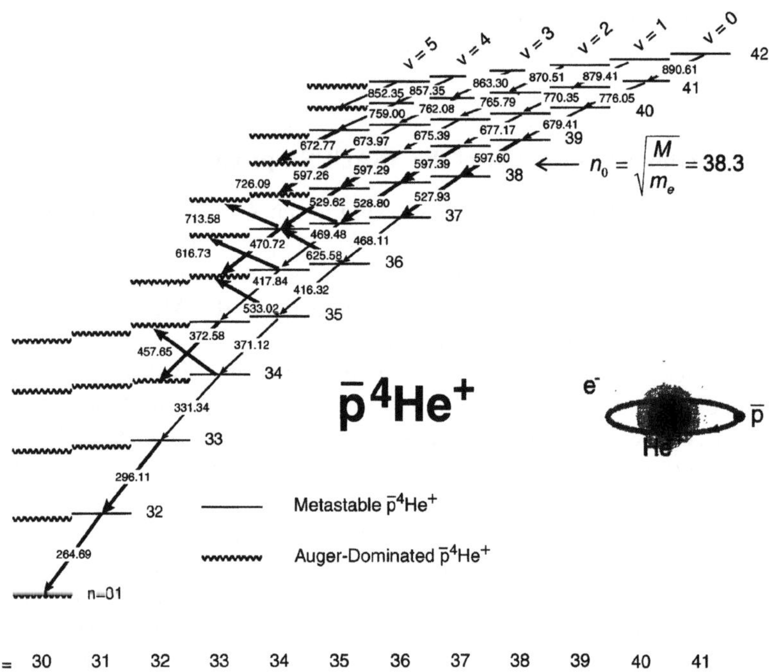

Figure 1. Right: Structure of antiprotonic helium. *Left*: energy level diagram of $\bar{p}^4He^+$. Metastable levels ($\tau \geqslant \mu s$) are denoted by solid lines, Auger-dominated short-lived ones ($\tau \leqslant 10$ ns) by wavy lines. Arrows symbolize radiative transitions following the propensity rule $\Delta v = 0$. Bold arrows correspond to experimentally observed transitions. They include so-called unfavoured ones with $\Delta v = 2$.

occupy states with principal quantum numbers around $n_0 = \sqrt{M^*/m}$ (=38.3 for $\bar{p}^4He^+$), where M^* is the reduced mass of the $\bar{p}$–He^+ system, and m the electron mass. Figure 1 shows the energy level diagram of $\bar{p}^4He^+$, which is divided in a metastable zone (solid lines) and a short-lived one (wavy lines). For the metastable levels the Auger transition rate is much smaller than the radiative transition rate, leading to lifetimes of $\tau \sim \mu s$, while for the short-lived ones the Auger rate is much higher than the radiative one, resulting in lifetimes of $\tau < 10$ ns. Antiprotons initially captured around n_0 undergo radiative transitions following cascades with $\Delta v \equiv \Delta(n - l - 1) = 0$ (n = principal and l = angular momentum quantum number) until they reach a short-lived state from which the $\bar{p}He^+$ subsequently ionizes via Auger transitions. The $\bar{p}He^+$ ion is then rapidly destroyed by Stark-mixing in the dense surrounding helium medium.

This picture was proven to be correct by a series of laser spectroscopy experiments at LEAR [6]. A laser pulse tuned to the transition at the end of a cascade will deexcite the antiprotons from the metastable to the short-lived state, thus forcing them to immediately annihilate. Using this signature, a total of 10 resonant transitions in $\bar{p}^4He^+$ and 3 in $\bar{p}^3He^+$ could be found at LEAR, thus confirming for the first time experimentally the long-held belief that exotic particles are initially captured around $n_0 = \sqrt{M^*/m}$.

2.1. PRECISION LASER SPECTROSCOPY AND CPT TEST OF ANTIPROTON CHARGE AND MASS

For the experiments at the AD a completely new experimental setup was designed and installed, adopted to the pulsed structure of the antiproton beam with very low repetition rate. Since middle of 2000, seven new transitions in $\bar{p}^4He^+$ and five in $\bar{p}^3He^+$ were found.

Our first goal was the improvement of the precision for the transition wavelengths in $\bar{p}He^+$ over the 0.5 ppm (5×10^{-7}) obtained at LEAR [8]. This was done by using an improved laser system and more sophisticated calibration of the wavelength measuring device against an iodine reference cell [9]. We extended the measurements to four transitions including two newly found ones in the UV region ($(n, l) = (35, 33) \rightarrow (34, 32)$ at 372 nm and $(33, 32) \rightarrow (32, 31)$ at 296 nm), where theoretical calculations are supposed to be most accurate. The experimental error was reduced by a factor of four to 1.3×10^{-7}, and all four transition wavelengths agreed with the most sophisticated three-body calculations by Korobov [10] and Kino *et al.* [11, 12] to better than 5×10^{-7} [9]. At this level of precision the calculations need to take into account relativistic corrections, the Lamb shift and QED corrections up to order α^4, so that this experiment constitutes a stringent test of the validity of three-body bound state QED calculations.

On the other hand, the agreement between experiment and theory can be used to perform a CPT test on the proton/antiproton charge and mass [13], since the theorists use in their calculations the numerically better known mass and charge of the proton. The energy levels itself are governed by the Rydberg constant $Ry \sim M_{\overline{p}} Q_{\overline{p}}^2$. Taking into account the fact that the antiproton cyclotron frequency $w_{\overline{p}} \sim Q_{\overline{p}}/M_{\overline{p}}$ is known to be equal to the proton one to better than 1 in 10^{10} [14], the agreement of theory and experiment for the transition energies in antiprotonic helium results in an upper limit for the equality of proton and antiproton charge and mass of 6×10^{-8} (90% confidence level) [9], a factor of 8 better than our own results at LEAR [8] and a factor of 300 better than previous measurements using X-rays of heavy antiprotonic atoms [15].

In 2001 we performed for the first time laser spectroscopy experiments with the RFQD beam, stopping ~60 keV antiprotons in helium gas of 30 K and 0.8 mbar. This corresponds to a factor 1000 lower density than before, and allows to essentially eliminate the systematic error coming from the extrapolation of the transition wavelength to zero density. We expect a further improvement of a factor 2...3 for the experimental accuracy after analyzing the new data.

2.2. HYPERFINE STRUCTURE OF ANTIPROTONIC HELIUM

The main highlight of the 2001 run was the first observation of the hyperfine splitting of the (37, 35) state using a 2-laser microwave triple resonance method. The uniqueness of $\overline{p}He^+$ comes from the fact that the antiproton in metastable states carries a very large angular momentum of $L_{\overline{p}} = 33 \ldots 39$, which leads through its interaction with the electron spin S_e to the dominant splitting (~12.9 GHz) which we call *hyperfine* (*HF*) splitting. The antiproton spin $S_{\overline{p}}$ leads to a further, about two orders of magnitude smaller splitting called *super-hyperfine* (*SHF*) structure, resulting in a quadruplet structure.

We developed a two-laser microwave triple resonance method [1, 16] which allows to measure transitions within the quadruplet that are associated with a flip of the *electron* spin. The preliminary analysis shows that the transition frequencies agree well with the calculated ones by Korobov and Bakalov [17, 18], but slightly less with published values of Yamanaka and Kino [19]. New results of Kino *et al.* presented at this conference [12], however, are very close to our experimental results. Both theoretical values now agree with our (preliminary) result on the level of the theoretical accuracy which is determined from omitting order-$\alpha^2 \approx 5 \times 10^{-5}$ corrections from the calculations. This again shows the impressive accuracy that the variational three body QED calculations are able to achieve.

The observed HF transitions are indirectly sensitive to the magnetic moment of the antiproton $\mu_{\overline{p}}$, and their determination to better than 10^{-6} could improve the value of $\mu_{\overline{p}}$ [20], which is experimentally known with an accuracy of only 3×10^{-3} [15, 21]. Transitions between the SHF levels are directly sensitive to $\mu_{\overline{p}}$ and therefore constitute a more promising way of performing another CPT test of an antiproton property. Such a measurement will, however, be rather difficult since it requires a quadruple resonance technique where in addition to the two laser pulses and the 13 GHz microwave radiation a 100–200 MHz radiation has to be applied.

References

1. ASACUSA collaboration, Atomic spectroscopy and collisions using slow antiprotons, CERN/SPSC 97-19, CERN/SPSC 2000-04 (1997/2000).
2. Maury, S. *et al.*, *Nucl. Phys. B, Proc. Suppl.* **56A** (1997), 349.
3. Moller, S., these proceedings.
4. Yamazaki, Y., In: J. Bollinger, R. Spencer and R. Davidson (eds), *Proceedings of Non-Neutral Plasma Physics III*, Princeton, New Jersey, 1999, American Institute of Physics, New York, 1999, p. 48.
5. Franzen, K.Y. *et al.*, In: F. Anderegg, C. F. Driscoll and L. Schweikhard (eds), *Proceedings of Non-Neutral Plasma Physics IV*, University of California, San Diego, 2001, American Institute of Physics, New York, 2002, p. 79.
6. Yamazaki, T., Morita, R,. N., Hayano, S., Widmann, E. and Eades, J., *Phys. Rep.* **366** (2002), 183.
7. Yamazaki, T. *et al.*, *Nature* **361** (1993), 238.
8. Torii, H.A. *et al.*, *Phys. Rev. A* **59** (1999), 223.
9. Hori, M. *et al.*, *Phys. Rev. Lett.* **87** (2001), 093401.
10. Korobov, V. I., *Nucl. Phys. A* **684** (2001), 663c; *Nucl. Phys. A* **689** (2001), 75c.
11. Kino, Y., Kamimura, M. and Kudo, H., *Hyp. Interact.* **119** (1999), 201.
12. Kino, Y., these proceedings.
13. Hughes, R. and Deutch, B. I., *Phys. Rev. Lett.* **69** (1992), 578.
14. Gabrielse, G. *et al.*, *Phys. Rev. Lett.* **82** (1999), 3198.
15. Groom, D. *et al.*, *The European Physical Journal C* **15** (2000), 1.
16. Widmann, E., *Hyp. Interact.* **119** (1999), 195.
17. Bakalov, D. and Korobov, V. I., *Phys. Rev. A* **57** (1998), 1662.
18. Korobov, V. and Bakalov, D., *J. Phys. B* **34** (2001), L519.
19. Yamanaka, N., Kino, Y., Kudo, H. and Kamimura, M., *Phys. Rev. A* **63** (2000), 012518.
20. Bakalov, D., *Nucl. Phys. A* **655** (1999), 301c.
21. Kreissl, A. *et al.*, *Z. Phys. C* **37** (1988), 557.

Hyperfine Interactions **146/147:** 319–323, 2003.

Practical Uses of Antiprotons

GERALD P. JACKSON
Pbar Oncology Inc., Suite 601, 396 West Fenton Lane, West Chicago, IL 60185, USA

Abstract. The production of commercial quantities of antiprotons has been a reality for many years now. The deceleration and trapping of antiprotons is a relatively new activity, but has been sufficiently proven to be translated into a business enterprise. Now that NASA has a portable Penning trap for transporting antiprotons, all the elements are in place to begin the commercial distribution of antiprotons. The list of potential customers for antiprotons is continuously growing, with detailed market analyses already performed on some medical and propulsion applications. In this paper these applications are reviewed, along with their appetite for antiprotons and the steps needed to bring them to market.

Key words: antimatter, propulsion, radioisotope, cancer, deceleration, ionization.

1. Deep space propulsion

There are a wide variety of interstellar phenomena which astrophysicists would like to observe directly by sending an unmanned probe outside of our solar system. The closest to us is the gravity lens point from our sun, which is thought to be near the orbit of Neptune. Imagine parking an observatory at this location, using the sun as the primary focusing lens and providing an angular resolution orders of magnitude higher than any other proposed observatory. In addition, this is also the focal point of neutrinos, making a neutrino telescope a real possibility. A bit further out is the termination shock at the heliopause, the boundary between the interstellar medium and the solar magnetic environment (heliosphere). One of the features expected to be observed is a hydrogen wall produced by the shock wave at this interface. Throughout this region is thought to be the Kuiper Belt, an inner cometary cloud which may be composed of approximately 100 million comets. A hundred times further away is thought to be the main Oort cometary cloud, containing approximately a trillion comets. Finally, it is thought that with enough antiprotons it is possible to send an unmanned probe to Alpha Centauri in the span of approximately 40 years. By deploying a solar wind parachute at the destination star, mankind would be able to explore that solar system for as long as the probe continues to operate.

There are a number of proposed propulsion system ideas presently under study at NASA [1]. As an example of their antiproton requirements a typical mission requires one microgram, which calls for a magnetic bottle which can hold 6E17 antiprotons. This represents 1500 years worth of antiproton production at the Fermi

National Accelerator Laboratory (Fermilab), the highest rate antiproton production factory in the world. Due to some recent proprietary proposals by the author, there is a prospect for building a new antiproton production facility which could produce the above quantity of antiprotons in a few years.

One of the key technologies required to fill any niches for commercial antimatter distribution is the invention of a portable Penning trap. Two such traps now exist. The later HiPAT trap was built for the NASA Marshal Space Flight Center (MSFC) by Synergistic Technologies, Inc. through a Small Business Innovative Research (SBIR) grant.

To this date the HiPAT trap has only been tested with protons. The protons are generated by ionizing residual hydrogen molecules in the trap volume with a 3 keV electron beam. These tests are taking place at the NASA Marshal Space Flight Center (MSFC) Propulsion Research Center in Huntsville, Alabama. The HiPAT trap was designed to store one trillion antiprotons in a 4 Tesla magnetic field bounded by a 20 kV potential well. Due to the fact that the solenoid is superconducting, the trap walls are acting as a cryopump, reducing the pressure in the volume to 1×10^{-10} Pa. At this pressure, a 400 day antiproton lifetime is anticipated.

2. PET isotope production

Positron emission tomography (PET) is the study and visualization of human physiology by electronic detection of short-lived positron emitting radioisotopes. These radioisotopes are mixed with an organic agent and introduced into a patient's body as a radio-pharmaceutical. PET is the only non-invasive technology that can routinely and quantitatively measure metabolic, biochemical, and functional activity in living tissue. PET scans are currently used in conjunction with Magnetic Resonance Imaging (MRI) and Computed Tomography (CT) scans to more accurately determine the location of tumors [2, 3].

The typical positron emitting isotopes used or desired for medical applications are C-11, N-13, O-15, and F-18. Their half-lives are respectively 20, 10, 2, and 110 minutes. Substantial technological research exists related to the diagnosis of diseases such as breast, brain, or liver cancer using very short-lived radioisotopes such as flourine-18 and oxygen-15. Unfortunately, the half-life of most of these radioisotopes is so short as to render them nearly inaccessible. The lack of readily available sources of these radioisotopes has severely inhibited visualization research for numerous diseases. The majority of all clinical PET scans are performed using radioisotope fluorine-18, which is currently only produced from cyclotrons. Other radioisotopes such a are not widely used because of their short half-life but may be better suited for specific kinds of diagnostics. The current treatments are primarily focused on cardiac (45%) and oncology (48%) diagnostics. The principle obstacle to expanding clinical PET scans is the accessibility of the radioisotopes.

In order to create these isotopes and inject them into the patient before the positron emission is depleted, it has been necessary to have a nearby accelerator for generating the radioisotopes. In the U.S. there are over 1700 institutions that have PET imaging capabilities, but only 40 isotope generation clinics exist. The current U.S. production capacity for sales of radioisotopes is $495 million, while the estimated demand is $2.6 billion. The use of PET has increased at an average annual rate of 33% since 1989 and is expected to increase similarly into the future. The problem is the high initial and operational costs of cyclotrons. Hospitals considering adding PET are looking at staggering costs of $5 million to $7 million for the building and cyclotron and about $1 million in annual operations expenses.

Using antiprotons as a source of inexpensive and portable radioisotope generation would revolutionize the PET industry. The concept is to transport antiprotons in portable Penning traps (magnetic bottles) and then deliver the antiprotons into a magnetized chamber containing the chemical which is to be converted into a PET isotope. By spilling antiprotons from the trap into the sample at a kinetic energy of approximately 1 MeV, PET isotopes can be produced in a small, portable, shielded enclosure. It would be possible to literally produce the radioisotopes at the patient bedside. This makes isotopes such as O-15, which holds so much promise for the diagnosis of brain tumors, viable for use in a hospital environment. The advantages of this concept include its portability, accessibility, and low cost generation of radioisotopes. Currently, patients must travel to select hospitals/scanning facilities that have a cyclotron or one close by. Antiproton generation of isotopes would give any hospital/scanning facility the ability to generate radioisotopes on site without the large capital expenditure required to build a cyclotron. The physician now has more time to administer the injection and perform the PET scan without concern for radioisotope expiration.

Calculations predict that less than 100 billion antiprotons are required to produce the 15 mCi source for a F-18 treatment. Similar numbers of antiprotons are required for production of other isotopes. In order to verify these calculations, reactions cross section measurements need to be made on carbon, oxygen, nitrogen, and fluorine.

3. Cancer therapy

Few words strike more fear into one's heart than "cancer". This disease strikes upward of a million people in the U.S. every year. Even though hundreds of billions of dollars have been spent on this disease, most oncologists will admit that a cure for cancer is not in sight – the industry fights an incremental war – there are no "magic bullets". Currently there are effective therapies for only 64% of patients with cancer. Radiation primarily kills cancer cells by attacking DNA, the cellular "software" that instructs the "hardware" (cells) how to survive and replicate. If the cell's DNA is sufficiently damaged, the cell dies and does not replicate. Conventional X-ray radiation therapy has two primary obstacles. First, it can only control tumors in two dimensions, which means that the skin and healthy tissue in front and behind the tumor are adversely affected. Second, X-ray therapy is not sufficiently lethal. It is ineffective at killing oxygen starved (hypoxic) and resting (non-replicating) cancerous cells.

The most popular particle therapies are proton, neutron, and heavy ion therapy. Because it has no electric charge, the neutron has the same two-dimensional illumination problem as X-rays, but the effectiveness of neutrons at killing hypoxic and resting cells is superior. The primary advantage of charged particle therapy is the ability to deposit energy (cell lethality) directly into the cancerous tumor in three dimensions. Regrettably, the killing power of protons is no greater than that of conventional X-ray radiation, requiring 30 to 40 treatments spaced over many weeks to try and catch each cell in the process of replication and to allow each cell to become better oxygenated as the tumor shrinks. The killing power of heavy ions is much greater than that of protons and can be precisely targeted to the tumor site. However, the collateral damage of heavy ions entering the body is greater than the damage of protons delivered to a tumor site. This ratio of killing power to collateral damage has been the central issue of radiation therapy for decades.

Antiprotons are similar to protons in their ability to penetrate healthy tissue with a minimum of damage and the ability to deposit most of the energy at the tumor. An experiment has already been performed on the energy deposition of slow antiprotons in tissue-like material [4]. Comparing with proton deposition, which is quite well understood, one finds a dramatic enhancement of deposited energy with antiprotons at the deposition (Bragg) peak. In this study polythene was used as the tissue-like material. At a depth of 0.5 g/cm^2 the proton and antiproton deposition rates (at 40 MeV kinetic energy) were normalized to unity. As deep as 1.5 g/cm^2, or down to 20 MeV, the proton and antiproton curves are equal. Between 1.5 and 2.0 g/cm^2 the deposition peak shows a dramatic difference. Whereas the proton deposition peak reaches a relative deposition rate of 5, the antiproton peak is as high as 9.

Antiprotons are very different from protons in that, after the antiproton slows and all of the kinetic energy is deposited, the antiproton will then annihilate on the nucleus of the nearest atom, thus depositing an extra and localized burst of cancer cell-killing energy beyond what could be delivered by a proton. The absorption of a pi-meson emitted in the annihilation causes the nucleus of the atom to recoil at an average kinetic energy of 30 MeV. At 30 MeV the newly created ion will travel about 20 to 30 microns before depositing all of its energy. Interestingly, a cell averages about 20 microns in width; therefore, this new ion will kill within a few cell diameters. Antiprotons now offer the option of delivering a dose of heavy ion therapy directly into a tumor, by creating the heavy ions within the tumor with minimal damage to intervening tissue. The actual lethality to cells due to heavy ion exposure has been measured for several types of ions and different types of cells [5]. A 500 R dose will yield a probability of cell survival of less than 0.01 (regardless of the replication state of the cell). In order to deposit 500 R in a 5 cc volume of tissue, about ten billion antiprotons are required. Therefore the radiation dose for antiproton treatment is smaller than a single proton treatment, and perhaps only a single treatment is required.

4. Antiproton availability

The present antiproton production rate at the Fermi National Accelerator Laboratory (Fermilab) is 100 billion antiprotons/hour. Given an initial consumption rate of 1% of this production rate, the flux of antiprotons for commercial applications would be approximately one billion antiprotons/hour.

There are a number of proposals to increases this antiproton production rate by a factor of ten. Due to limited resources, personnel, and time these upgrades have not been seriously pursued. There will surely be significant work on these upgrades in the next few years.

If one is not interested in generating an antiproton beam for the existing Fermilab scientific program, but lower energy antiprotons for the above applications or low energy particle physics, the traditional methods [6] for creating, capturing, and cooling antiprotons may no longer be optimum. In fact, recently some novel accelerator configurations have been proposed by the author which have the potential of increasing the antiproton stacking rate by two or three orders of magnitude. Such large increases would make it possible to fulfill even the demands of the more ambitious NASA interstellar missions, or provide PET diagnostic treatments which would meet much of the anticipated demand.

5. Conclusions

In this paper it has been shown that there are potential commercial applications for antiprotons. A great deal of work still needs to be done to verify the technical and economic validity of these applications, but enough antiprotons are now available from Fermilab to start commercial transactions. Though detailed plans are not yet sufficiently developed for presentation in peer reviewed articles, concepts have been proposed which have the potential for increasing the rate of antiproton production, up to a factor of thousand.

References

1. Schmidt, G. R. *et al.*, Internal NASA Note, 1999.
2. Quarles, R. P. *et al.*, *J. Cerebral Blood Flow and Metabolism* **13**(5) (1993), 733.
3. Nieweg, O. E. *et al.*, *Cancer* **71**(12) (1993), 3920.
4. Sullivan, A. H., *Phys. Med. Bio.* **30**(12) (1985), 1297.
5. Tilly, N. *et al.*, *Int. J. Radiat. Biol.* **75**(2) (1999), 233.
6. Church, M. D. and Marriner, J. P., *Ann. Rev. Nucl. Part. Sci.* 43 (1993), 253.

Hyperfine Interactions **146/147:** 325–329, 2003.

Energy Levels of Hydrogenlike Kaonic Atoms

J. P. SANTOS[1], F. P. PARENTE[2], S. BOUCARD[3] and P. INDELICATO[3]
[1]*Departamento de Física, Faculdade de Ciências e Tecnologia, Universidade Nova de Lisboa, Monte de Caparica, 2825-114 Caparica, Portugal, and Centro de Física Atómica da Universidade de Lisboa, Av. Prof. Gama Pinto 2, 1649-003 Lisboa, Portugal; e-mail: jps@cii.fc.ul.pt*
[2]*Departamento Física da Universidade de Lisboa and Centro de Física Atómica da Universidade de Lisboa, Av. Prof. Gama Pinto 2, 1649-003 Lisboa, Portugal*
[3]*Laboratoire Kastler-Brossel École Normale Supérieure et Université P. et M. Curie Unité Mixte de Recherche du CNRS n° C8552 Case 74, 4 place Jussieu, F-75252 Paris, CEDEX 05, France*

Abstract. Energies of the $[(n, \ell = n-1), 1 \leqslant n \leqslant 20]$ and the $[(n, \ell = n-2), 2 \leqslant n \leqslant 20]$ levels have been calculated for several hydrogenlike kaonic atoms throughout the periodic table, using the current world average kaon mass. Calculations were done in the framework of the Klein–Gordon equation, with finite nuclear size and all-order vacuum polarization corrections.

Key words: kaons, atoms, energy levels, circular transitions.

1. Introduction

An exotic atom can be formed when a particle with a negative charge and long enough lifetime slows down and stops in matter. The particle may replace an atomic electron, thus becoming bound in a high-n atomic Bohr orbit around the nucleus. The principal quantum number of this highly excited state is of the order of $n = \sqrt{m/m_e}$, where m and m_e are the masses of the particle and the electron, respectively [8]. The higher the overlap between the wave functions of the electron and the particle, the more probable the formation of exotic atoms [8].

Exotic atoms formed in this way are named after the exotic particle involved. When it is the negative kaon K^-, a spin-0 meson with a lifetime of 1.237×10^{-8} s, it is called a kaonic atom.

De-excitation of the kaonic atom will start via Auger processes, while there are electrons to be ejected, and then continue via radiative (E1) transitions, producing characteristic X-rays while cascading down its own sequence of atomic levels until some state of low principal quantum number.

The initial population of the atomic states is related to the available density of states, so for any given principal quantum number n higher orbital momenta will be favored to some extent because of their larger multiplicity. Stark mixing of states concurs to this effect. As the Auger transitions do not change the shape of the angular momentum distribution, the particle quickly reaches the $(\ell = n-1)$ orbits [8]. Once the radiative (E1) transitions begin to dominate, a $(n, \ell = n-1)$

state can only decay to $(n-1, \ell = n-2)$ level due to the E1 transition selection rules (Yrast decay). The so called parallel transition (e.g., $n, \ell = n-2 \rightarrow n, \ell = n-3$) are observed, but with an intensity of a few % of the Yrast line.

Finally, the particle in a state of low angular momentum will be absorbed by the nucleus through the kaon–nucleus strong interaction. This strong interaction causes a shift of the energy of the lowest atomic level from its purely electromagnetic value while the absorption reduces the lifetime of the state and the X-ray transitions to this final atomic level are broadened.

In this work we calculate with high accuracy the Quantum Electrodynamic (QED) contribution to energy either to allow extraction of strong energy shift from experiments or to provide accurate energies of highly circular transitions, where strong interaction is negligible, to be used as for calibrations of mass measurements [6, 11].

Recently, in the context of the DEAR experiment (strong interaction shift and width in kaonic hydrogen) at the Φ factory DAFNE in Frascati [7, 13] some transitions of kaonic atoms will be measured as a pre-test or in-line calibration, such as in the kaonic nitrogen, aluminum, titanium, neon, and silver.

Measurements in the exotic atoms permit the extraction of precise information about the orbiting particle [1], such as charge/mass, and magnetic moment, and the interaction of these particles with nuclei. In addition, properties of nuclei [2], as nuclear size, nuclear polarization, and neutron halo effects in heavy nuclei [12], have been studied.

Furthermore, as the X-ray transition energies are proportional to the reduced mass of the system, studies in the intermediate region of the atomic cascade were used to measure the masses of certain negative particles. In order to minimize the strong interactions for the mass determination it is mainly considered only transitions between circular orbits $(n, n-1) \rightarrow (n-1, n-2)$ far from the nucleus [10].

2. Calculation of the energy levels

2.1. THE KLEIN–GORDON EQUATION

To compute kaonic atoms bound states, as kaons are bosons, we solve the Klein–Gordon equation in a self-consistent way. This equation, which in the absence of a strong interaction may be written as

$$\left(\alpha^2\left(E - V_c(r)\right)^2 + \vec{\nabla}^2 - \mu^2 c^2\right)\psi(\mathbf{r}) = 0, \tag{1}$$

where μ is the kaon–nucleus reduced mass, E is the total energy in the atom and V_c is the sum of Coulomb potential, describing the interaction between the kaon and the finite charge distribution of the nucleus, and the vacuum-polarization Uëhling term $\alpha(\alpha Z)$. Atomic units $\hbar = 1$, $c = 1/\alpha$ are used. For spherically symmetric potentials, the bound state solutions of the KG equation (4) are of usual form $\psi_{n\ell m}(\mathbf{r}) = (u_{n\ell}(r)/r)Y_{\ell m}(\theta, \phi)$.

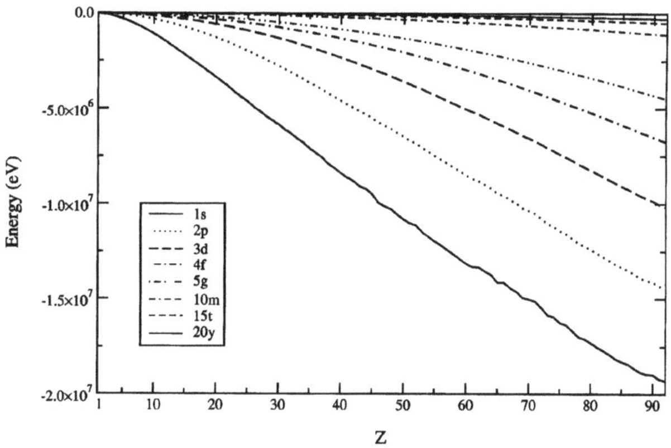

Figure 1. Energy levels with $\ell = n - 1$ of hydrogenlike kaonic atoms for $Z = 1$–92.

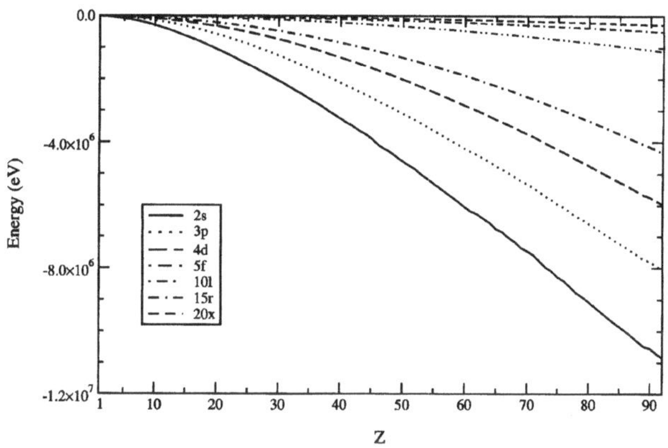

Figure 2. Energy levels with $\ell = n - 2$ of hydrogenlike kaonic atoms for $Z = 1$–92.

2.2. NUCLEAR STRUCTURE

For heavy elements, a change between a point-like and an extended nuclear charge distribution strongly modifies the wave function near the origin. One nuclear contribution is easily calculated by using a finite charge distribution in the differential equations from which the wave function are deduced. For atomic number larger than 45 we use a Fermi distribution with a thickness parameter $t = 2.3$ fm and a uniform spherical distribution otherwise.

2.3. QED EFFECTS

The effects of the vacuum polarization in the Uëhling approximation, which comes from changes in the bound-kaon wave function, can be relatively easily implemented in the framework of the resolution of the KGE using a numerical shooting method similar to the one used for the Dirac equation [5].

In practice one only needs to add the Uëhling potential to the nuclear Coulomb potential to get the contribution of the vacuum polarization to the wave function to all orders, which is equivalent to evaluating the contribution of all diagrams with one or several repeated vacuum polarization loops, leading to a contribution of the order of $[\alpha(\alpha Z)]^n$ with $n = 1, 2, 3, \ldots$ [3].

Other two vacuum polarization terms included in this work, namely the Källén and Sabry term and the Wichmann and Kroll term, were calculated by perturbation theory.

3. Results and conclusion

The theoretically orbital binding energy, E, is obtained as the sum of Klein–Gordon eigenvalue (including finite nuclear size, Uëhling potential and reduced mass) corrected for the relativistic recoil $-B^2/2M_A$, where B is the binding energy of the level. In Figures 1 and 2 we show the evolution of the orbital binding energy as function of Z for some circular levels and some parallel non-circular levels, respectively. In this work we have evaluated the energies of the $[(n, \ell = n - 1), 1 \leqslant n \leqslant 20]$ and the $[(n, \ell = n-2), 2 \leqslant n \leqslant 20]$ levels for several (hydrogenlike) kaonic atoms throughout the periodic table. The energies we obtain are in good agreement, but more accurate than earlier calculations (see, e.g., [4, 9, 10]), and also in good agreement with experiment.

Acknowledgements

This research was supported in part by JNICT (Portugal) under project Praxis/C/FIS/10030/98, and by the TMR Network Eurotraps Contract Number ERBFM-RXCT970144.

References

1. Batty, C. J., Eckhause, M., Gall, K. P., Guss, P. P., Hertzog, D. W., Kane, J. R., Kunselman, A. R., Miller, J. P., Brien, F. O., Phillips, W. C., Powers, R. J., Roberts, B. L., Sutton, R. B., Vulcan, W. F., Welsh, R. E., Whyley, R. J. and Winter, R. G., *Phys. Rev. C* **40** (1989), 2154.
2. Batty, C. J., Friedman, E., Gils, H. J. and Rebel, H., *Adv. Nucl. Phys.* **19** (1989), 1.
3. Boucard, S. and Indelicato, P., *Eur. Phys. J. D* **8**(1) (2000), 59.
4. Cheng, S. C., Asano, Y., Chen, M. Y., Dugan, G., Hu, E., Lidofsky, L., Patton, W., Wu, C. S., Hughes, V. and Lu, D. *Nud. Phys. A* **254** (1975), 381.
5. Desclaux, J. P., *Relativistic Multiconfiguration Dirac–Fock Package*, Vol. A., STEF, Cagliary, 1993.
6. Gotta, D., Anagnostopoulos, D. F., Augsburger, M., Borchert, G., Castelli, C., Chattelard, D., Egger, J. P., El-Khoury, P., Gorke, H., Hauser, P., Indelicato, P., Kirch, K., Lenz, S., Nelms, N., Rashid, K., Siems, T. and Simons, L. M., *Nud. Phys. A* **660** (1999), 283.
7. Guaraldo, C., Bianco, S., Bragadireanu, A. M., Fabbri, F. L., Iliescu, M., Ito, T. M., Lucherini, V., Petrascu, C., Bregant, M., Milotti, E., Vacchi, A., Zavattini, E., Augsburger, M., Chatellard, D., Knowles, P., Mulhauser, F., Schaller, L. A., Schellenberg, L., Schneuwly, H.,

Egger, J. P., Varidel, D., Breunlich, W. H., Cargnelli, M., Gartner, B., King, R., Lauss, B., Marton, J., Zmeskal, J., Ponta, T., Nakamura, S. N., Koike, T., Hayano, R. S., Hori, M., Ishikawa, T., Ishiwatari, T., Iwasaki, M., Akaishi, Y., Beer, G., Sanderson, A. C. and Seki, R., *Hyp. Interact.* **119** (1999), 253.

8. Horvath, D., *Nucl. Instrum. Meth. Phys. Phys. Res. B* **87** (1994), 273.
9. Iwasaki, M., Bartlett, K., Beer, G. A., Gill, D. R., Hayano, R. S., Ito, T. M., Lee, L., Mason, G., Nakamura, S. N., Olin, A., Outa, H., Salomon, M., Seki, R., Taniguchi, T., Terada, T. P., Trayling, G., Yamashita, Y. and Yen, S., *Nucl. Phys. A* **639** (1998), 501.
10. Kunselman, R., *Phys. Lett. B* **34** (1971), 485.
11. Lenz, S., Borchert, G., Gorke, H., Gotta, D., Siems, T., Anagnostopoulos, D. F., Augsburger, M., Chattelard, D., Egger, J. P., Belmiloud, D., El-Khoury, P., Indelicato, P., Daum, M., Hauser, P., Kirch, K. and Simons, L. M., *Phys. Lett. B* **416** (1998), 50.
12. Lubinski, P., Jastrzebski, J., Grochulska, A., Stolarz, A., Trzcinska, A., Kurcewicz, W., Hartmann, F. J., Schmid, W., Egidy, T. V., Skalski, J., Smolanczuk, R., Wycech, S., Hilscher, D., Polster, D. and Rossner, H., *Phys. Rev. Lett.* **73** (1994), 3199.
13. Maiani, L., *Nucl. Phys. A* **623** (1997), 16.

Hyperfine Interactions **146/147:** 331–336, 2003.

High-Precision Calculation of the Fine and Hyperfine Structure Splittings of Antiprotonic Helium-3,4 Atoms

Y. KINO[1], N. YAMANAKA[2], M. KAMIMURA[3] and H. KUDO[1]
[1]*Department of Chemistry, Tohoku University, Sendai 980-8578, Japan*
[2]*Department of Physics, University of Tokyo, Tokyo 113-0033, Japan*
[3]*Department of Physics, Kyushu University, Fukuoka 812-8581, Japan*

Abstract. Fine and hyperfine structure splittings of antiprotonic helium-3,4 atoms are calculated with the accurate three-body wave function obtained by the coupled rearrangement channel method. The result obtained is in good agreement with the previous calculation and the latest experimental results of the $(J, v) = (35, 1)$ state of the antiprotonic helium-4 atom.

Key words: antiprotonic helium atom, fine and hyperfine structure splittings.

1. Introduction

A few percent of antiprotons stopped in a helium target form antiprotonic helium atoms ($\bar{p}\mathrm{He}^+ = \bar{p} + \mathrm{He}^{2+} + e^-$) in metastable states with an extraordinary long lifetime ($\sim\mu$s). Using the metastability, transition wavelengths between the metastable states were measured with laser spectroscopy in CERN. The observed transition wavelengths [1] were in excellent agreement with the precise three-body calculations [2, 3]. This agreement gave the best limit of the antiproton mass with 60 ppb uncertainty. On the other hand, uncertainty of an antiproton magnetic moment is only 0.3%. A laser and microwave triple-resonance technique has been developed to reveal the hyperfine structure splittings caused by the antiproton spin [4]. The precise measurement with this technique is in progress at the Antiproton Decelerator (AD) in CERN [5]. In the present paper, we precisely calculate the fine and hyperfine structure splittings of $\bar{p}^{3,4}\mathrm{He}^+$ with the accurate wave function obtained by the coupled rearrangement channel (CRC) method [6].

2. Theory

The antiprotonic helium atom has atomic and molecular characters of three-body systems at the same time. In the CRC method, we introduce three rearrangement channels ($c = 1$–3) illustrated in Figure 1 to directly take the dual character of the antiprotonic helium atom and a correlation between the electron and the antiproton

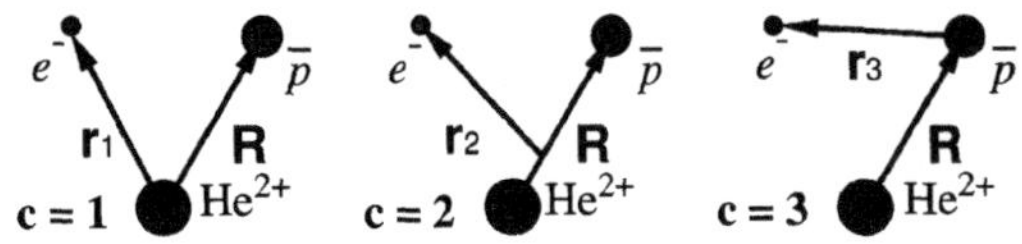

Figure 1. Three rearrangement channels.

[7] into account. Channel $c = 1$ is suited for describing the atomic picture where the helium nucleus attracts the electron and the antiproton like a helium atom. Channel $c = 2$ is for describing the diatomic molecular picture where the electron moves around two massive nuclei like a helium hydride diatomic molecule. Channel $c = 3$ is introduced to directory describe the correlation. The non-relativistic three-body wave function is given by a linear combination of three rearrangement channel wave functions,

$$\Psi_{vJM}^{\bar{p}\mathrm{He}^+} = \sum_{c=1}^{3} \Phi_{vJM}^{(c)}(\mathbf{r}_c, \mathbf{R}). \tag{1}$$

Each channel wave function is expanded in terms of Gaussian basis functions:

$$\Phi_{vJM}^{(c)}(\mathbf{r}_c, \mathbf{R}) = \sum_{n,N} \sum_{\ell_c, L_c} A_{v,n\ell_c N L_c} r_c^{\ell_c} R^{L_c} \mathrm{e}^{-(r_c/r_n)^2-(R/R_N)^2} \left[Y_{\ell_c}(\hat{\mathbf{r}}_c) \otimes Y_{L_c}(\hat{\mathbf{R}}) \right]_{JM}, \tag{2}$$

where v is the vibrational quantum number, J the total angular momentum, and M its z-component. Eigenenergies and coefficients A's are solved with the Rayleigh–Ritz variational principle.

The interactions which cause fine and hyperfine structures are derived from the Breit–Pauli Hamiltonian and are given by terms of α^2,

$$U_{ij}^{(SO)} = -\alpha^2 \left(\frac{z_i(z_j s_j - \mu_j)}{2m_j^2 s_j} \frac{(\mathbf{r}_{ij} \times \mathbf{p}_j) \cdot \mathbf{s}_j}{r_{ij}^3} + \frac{z_i \mu_j}{2m_i m_j s_j} \frac{(\mathbf{r}_{ij} \times \mathbf{p}_i) \cdot \mathbf{s}_j}{r_{ij}^3} \right), \tag{3}$$

$$U_{ij}^{(F)} = -\alpha^2 \frac{2\pi}{3} \frac{\mu_i}{m_i s_i} \frac{\mu_j}{m_j s_j} (\mathbf{s}_i \cdot \mathbf{s}_j) \delta(\mathbf{r}_{ij}), \tag{4}$$

$$U_{ij}^{(T)} = -\alpha^2 \frac{\mu_i}{m_i s_i} \frac{\mu_j}{m_j s_j} \left(\frac{\mathbf{s}_i \cdot \mathbf{s}_j}{r_{ij}^3} - 3 \frac{(\mathbf{r}_{ij} \cdot \mathbf{s}_i)(\mathbf{r}_{ij} \cdot \mathbf{s}_j)}{r_{ij}^5} \right), \tag{5}$$

where $\mathbf{p}_i$, s_i, m_i, z_i, μ_i and $\mathbf{r}_{ij}$ are the momentum of ith particle ($i = e^-$, $\bar{p}$ or $^3\mathrm{He}^{2+}$), its spin, its mass (in units of m_{e^-}), its electric charge, its magnetic dipole moment (in units of $e\hbar/2m_i c$) and the relative vector between ith particle and jth particle, respectively. For a spinless particle ($i = {}^4\mathrm{He}^{2+}$), the terms including s_i are dropped from Equations (3)–(5). The masses and magnetic moments used in this calculation are $m_{\bar{p}} = 1{,}836.152\,667\,5$, $m_{^3\mathrm{He}} = 5{,}495.885\,238$, $m_{^4\mathrm{He}} = 7{,}294.299\,508$, $\mu_{e^-} = -1.001\,159\,652\,1869$, $\mu_{\bar{p}} = -2.792\,847\,337$, and $\mu_{^3\mathrm{He}} = -6.368\,305\,996\,86$ [10]. The Rydberg constant is $3.289\,841\,960\,368 \times 10^6$ GHz/c.

Since the spin-dependent interactions are very weak, the splittings are calculated by the first order perturbation theory. The total wave function is written as

$$\Psi^{\text{tot}}_{v,JFM} = \sum_{FS} C^{vJF}_{KS}\Big[\big[[\Psi^{\bar{p}\text{He}^+}_{v,J} \otimes \chi_e]_K \otimes \chi_{\text{He}}\big]_S \otimes \chi_{\bar{p}}\Big]_{FM} \quad \text{for } \bar{p}^3\text{He}^+, \quad (6)$$

$$\Psi^{\text{tot}}_{v,JFM} = \sum_{K} C^{vJF}_{K}\Big[\big[\Psi^{\bar{p}\text{He}^+}_{v,J} \otimes \chi_e\big]_K \otimes \chi_{\bar{p}}\Big]_{FM} \quad \text{for } \bar{p}^4\text{He}^+, \quad (7)$$

where K and S are the intermediate angular momenta ($\mathbf{K} = \mathbf{J} + \mathbf{s}_e$, $\mathbf{S} = \mathbf{K} + \mathbf{s}_{\text{He}}$), F is the total angular momentum ($\mathbf{F} = \mathbf{S} + \mathbf{s}_{\bar{p}}$), and χ_i denotes a spin function of the corresponding particle. The energy splittings and the coefficients C's are determined by diagonalizing the spin-dependent interactions. Here, we define an effective Hamiltonian U_{eff} [11, 12],

$$(JKSFM\,|U_{\text{eff}}|\,JK'S'FM) = \sum_{i\neq j}\big\langle JKSF\big|U^{(SO)}_{ij} + U^{(F)}_{ij} + U^{(T)}_{ij}\big|JK'S'F\big\rangle, \quad (8)$$

with

$$|JKSFM\rangle = \Big|\Big[\big[[X_J \otimes \chi_e]_K \otimes \chi_{\text{He}}\big]_S \otimes \chi_{\bar{p}}\Big]_{FM}\Big\rangle, \quad (9)$$

$$|JKSFM\rangle = \Big|\Big[\big[[\Psi^{\bar{p}\text{He}^+}_{v,J} \otimes \chi_e]_K \otimes \chi_{\text{He}}\big]_S \otimes \chi_{\bar{p}}\Big]_{FM}\Big\rangle, \quad (10)$$

where X_J is the eigenfunction of $\mathbf{J}^2$ and J_z. The quantum number S is dropped when the helium isotope is ^{4}He. The effective Hamiltonian is given by

$$\begin{aligned} U_{\text{eff}} = {} & \sum_i \varepsilon^{SO,i}_{Jv}\,\mathbf{J}\cdot\mathbf{s}_i + \sum_{i<j}\varepsilon^{F,ij}_{Jv}\,\mathbf{s}_i\cdot\mathbf{s}_j \\ & + \sum_{i<j}\varepsilon^{T,ij}_{Jv}\big((\mathbf{J}\cdot\mathbf{s}_i)(\mathbf{J}\cdot\mathbf{s}_j) - J(J+1)\mathbf{s}_i\cdot\mathbf{s}_j\big). \end{aligned} \quad (11)$$

The coefficient $\varepsilon^{SO,e}_{Jv}$ is, for instance, calculated from

$$\varepsilon^{SO,e}_{Jv} = \frac{\langle v, JKSF||U^{(SO)}_{e\bar{p}} + U^{(SO)}_{e\text{He}}||v, JK'S'F\rangle}{(JKSFM||\mathbf{J}\cdot\mathbf{s}_e||JK'S'FM)}. \quad (12)$$

3. Results

In Tables I and II, the fine and hyperfine structure splittings are listed together with non-relativistic energies E_{Jv}. The notations are shown in Figure 2. The fine structure splittings $\text{FS}^{\pm(\pm)}$ are mainly caused by an electron spin flip. The hyperfine structure splittings $\text{HFS}^{\pm\pm}$ and $\text{HFS}^{\pm(\pm)}_{\bar{p}}$ are mainly caused by a helium-3 spin flip and an antiproton spin flip, respectively. We should note that the energy level of the (J, v) state of $\bar{p}^3\text{He}^+$ is roughly equal to that of the $(J+1, v)$ state of $\bar{p}^4\text{He}^+$ because of the difference between reduced masses of $\bar{p}$–He^{2+} system. The results

Table I. Fine and hyperfine structure splittings (GHz) of $\bar{p}^3He^+$. The notations are shown in Figure 2

(J,v)	FS^{++}	FS^{+-}	FS^{-+}	FS^{--}	$-E_{Jv}$
(33,0)	17.1992223	17.1361438	12.9658836	12.9028051	3.082114099
(33,1)	16.3390934	16.3298475	11.6736443	11.6643983	2.983373114
(33,2)	15.4750575	15.5082896	10.3893051	10.4225373	2.897192280
(33,3)	14.6362182	14.7005830	9.1513279	9.2156927	2.821963014
(33,4)	13.8461775	13.9311184	7.9913887	8.0763296	2.756217715
(34,0)	17.0515754	17.0378292	12.5054102	12.4916640	2.970628286
(34,1)	16.1104406	16.1428481	11.1251569	11.1575644	2.884912610
(34,2)	15.1917147	15.2580506	9.7884848	9.8548207	2.810261077
(34,3)	14.3252489	14.4140676	8.5338379	8.6226566	2.745174137
(34,4)	13.5356134	13.6370049	7.3917103	7.4931019	2.688292937
	HFS^{-+}	HFS^{--}	HFS^{++}	HFS^{+-}	
(33,0)	2.2594689	2.2692908	1.9738699	1.9640479	
(33,1)	2.5790923	2.5865723	2.0863568	2.0788769	
(33,2)	2.9112503	2.9159836	2.1745021	2.1697687	
(33,3)	3.2523500	3.2538850	2.2325403	2.2310053	
(33,4)	3.5979607	3.5958878	2.2568281	2.2589010	
(34,0)	2.4880596	2.4950143	2.0581056	2.0511510	
(34,1)	2.8240176	2.8284421	2.1612661	2.1568415	
(34,2)	3.1695678	3.1709908	2.2336621	2.2322392	
(34,3)	3.5204055	3.5183754	2.2710055	2.2730356	
(34,4)	3.8716176	3.8658120	2.2722855	2.2780910	
	$HFS_{\bar{p}}^{-+}$	$HFS_{\bar{p}}^{--}$	$HFS_{\bar{p}}^{+-}$	$HFS_{\bar{p}}^{++}$	
(33,0)	0.2405790	0.2504009	0.1873224	0.1775005	
(33,1)	0.1863749	0.1938549	0.1846090	0.1771290	
(33,2)	0.1418221	0.1465554	0.1797876	0.1750542	
(33,3)	0.1065938	0.1081288	0.1724936	0.1709586	
(33,4)	0.0799551	0.0778821	0.1628231	0.1648960	
(34,0)	0.1749809	0.1819355	0.1681893	0.1612347	
(34,1)	0.1298511	0.1342757	0.1666832	0.1622586	
(34,2)	0.0943484	0.0957714	0.1621072	0.1606842	
(34,3)	0.0677512	0.0657212	0.1545400	0.1565700	
(34,4)	0.0488889	0.0430833	0.1444748	0.1502804	

Table II. Fine and hyperfine structure splittings (GHz) of $\bar{p}^4He^+$. The notations are shown in Figure 2

(J,v)	FS^+	FS^-	$HFS_{\bar{p}}^-$	$HFS_{\bar{p}}^+$	$-E_{Jv}$
(34,0)	14.39323435	14.32401299	0.24400540	0.17478403	3.093466899
(34,1)	13.30923588	13.29472467	0.18965264	0.17514143	2.996335438
(34,2)	12.20556668	12.23508465	0.14409877	0.17361673	2.911180930
(34,3)	11.11113642	11.17459215	0.10673490	0.17019064	2.836524586
(34,4)	10.05138135	10.13926592	0.07697642	0.16486099	2.771011547
(35,0)	14.07225938	14.05273870	0.17897332	0.15945265	2.984020953
(35,1)	12.89607407	12.92396407	0.13299325	0.16088325	2.899282176
(35,2)	11.72542403	11.78986764	0.09545313	0.15989673	2.825146801
(35,3)	10.59049540	10.68120416	0.06578814	0.15649690	2.760233321
(35,4)	9.51776810	9.62557117	0.04314628	0.15094935	2.703283195
(35,1) [13]	12.8963462	12.9242428	0.1329887	0.1608853	2.89928217806256(2)

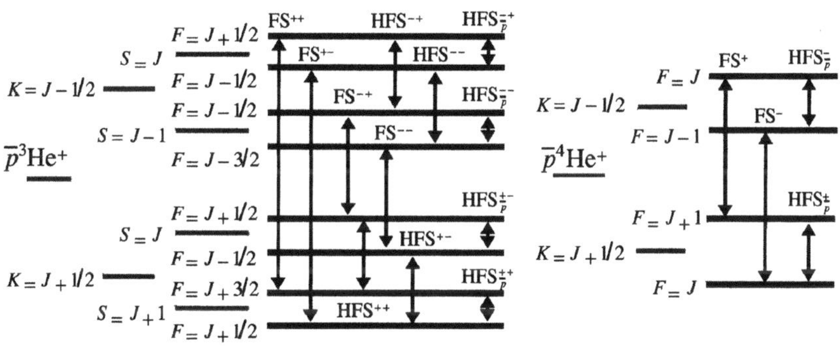

Figure 2. Fine and hyperfine structures of $\bar{p}^3He^+$ and $\bar{p}^4He^+$.

of $\bar{p}^4He^+$ are in good agreement with the previous calculation [11, 13] and are consistent with the preliminary experimental result of the $FS^{\pm}$ of the (35,1) state [5]. The antiproton spin flip is approximately independent of the helium-3 spin, because the $HFS_{\bar{p}}^{-\pm}$ and $HFS_{\bar{p}}^{+\pm}$ are close to the $HFS_{\bar{p}}^-$ and $HFS_{\bar{p}}^+$, respectively. The helium-3 spin enlarges the difference between the $FS^{\pm+}$ and $FS^{\pm-}$ of the (34,1) state which is similar to the (35,1) state of $\bar{p}^4He^+$.

Acknowledgements

We are grateful to P. Froelich, V. I. Korobov, D. Bakalov, E. Widmann, M. Hori and R. S. Hayano for helpful discussions. Y. K. wishes to acknowledge financial support by Inoue Foundation for Science. The computations were carried out on the FACOM/VPP700E at RIKEN and FACOM/VPP5000 at JAERI.

References

1. Hori, M. *et al.*, *Phys. Rev. Lett.* **87** (2001), 093401.
2. Korobov, V. I., Bakalov, D. and Monkhorst, H. J., *Phys. Rev. A* **59** (1999), 919.
3. Kino, Y., Kamimura, M. and Kudo, H., *Hyp. Interact.* **119** (1999), 201.
4. Widmann, E., *Hyp. Interact.* **119** (1999), 195.
5. Widmann, E., Talk at 3rd Euroconference on Atomic Physics at Accelerators (APAC2001).
6. Kamimura, M., *Phys. Rev. A* **38** (1988), 621.
7. Kino, Y., Kamimura, M. and Kudo, H., *Nucl. Phys. A* **631** (1998), 649.
8. Kino, Y. *et al.*, *Hyp. Interact.* **138** (2001), 179.
9. Mohr, P. J. and Taylor, B. N., *Rev. Mod. Phys.* **72** (2000), 351; CODATA recommended values of the fundamental physical constants, http://physics.nist.gov/cuu/.
10. Firestone, R. B. and Baglin, C. M. (eds), *Table of Isotopes*, 8th edn, 1998 update, John Wiley and Sons, Inc., New York, 1998.
11. Bakalov, D. and Korobov, V. I., *Phys. Rev. A* **57** (1998), 1662.
12. Yamanaka, N., Kino, Y., Kudo, H. and Kamimura, M., *Phys. Rev. A* **63** (2001), 012518.
13. Korobov, V. I. and Bakalov, D., *J. Phys. B* **34** (2001), L519.

Hyperfine Interactions **146/147:** 337–341, 2003.

Time-Dependent Coupled-Channel Calculations of the Annihilation Rate in Positron–Hydrogen Collisions

NOBUHIRO YAMANAKA[1] and YASUSHI KINO[2]

[1]*Department of Physics, University of Tokyo, Tokyo 113-0033, Japan; e-mail: yam@nucl.phys.s.u-tokyo.ac.jp*

[2]*Department of Chemistry, Tohoku University, Sendai 980-8578, Japan; e-mail: kino@mail.cc.tohoku.ac.jp*

Abstract. We calculate the positron annihilation rate in positron–hydrogen collisions for positron energies of 250 meV–6.5 eV using a time-dependent coupled channel method. A semiempirical model recently proposed in Phys. Rev. Lett. 79 (1997), 2241 predicts that the annihilation rate prominently enhances through virtual positronium formation as the positron energy approaches the positronium formation threshold (6.8 eV) from 4 eV. We illustrate time evolution of wave functions to show that such a prominent enhancement of the annihilation rate is not found and contribution of virtual positronium formation is not important below 6.5 eV.

Key words: positron annihilation, time-dependent coupled channel method.

1. Introduction

A semi-empirical model of electron–positron annihilation in positron–atom (or positron–molecule) collisions has been recently proposed by Laricchia and Wilkin [1]. According to this model, the annihilation mechanism is classified into direct annihilation and annihilation following positronium (Ps) formation. The former dominates the annihilation for very low-positron energy, while the latter above the Ps formation threshold. In the vicinity of the threshold, the annihilation rate is predicted to be prominently enhanced through virtual Ps formation as the positron energy approaches the threshold. However, Reeth and Humberston have been reported that this enhancement was not clearly found in a variational calculation for positron–hydrogen collisions [2].

In the present paper, we study electron–positron annihilation in positron–hydrogen collisions with a time-dependent coupled channel (TDCC) method [3, 4]. In this method, coupled-channel equations for scattering wave functions are solved time-dependently. Hence, a physical picture of dynamics can be clearly obtained in a full-quantal framework. By numerically describing the radial part of wave functions, closed channels are automatically incorporated. We calculate the annihilation rate for positron energies of 250 meV–6.5 eV. The result is compared with previous results. We illustrate time evolution of wave functions to show that contribution of virtual Ps formation is not important.

In the present paper, atomic units (a.u.), $m = \hbar = e = 1$, are used unless otherwise stated.

2. Annihilation rate and numerical method

The rate $\lambda_{2\gamma}$ of electron–positron annihilation into two gamma rays with an energy of 511 keV is written [5] as

$$\lambda_{2\gamma} = \pi r_0^2 c N Z_{\text{eff}}, \tag{1}$$

where $r_0 = e^2/(mc^2)$ is the classical radius of the electron, c the speed of light, N the number density of target atoms, and Z_{eff} the effective electron number per a target atom and calculated with the TDCC wave function $\boldsymbol{\Psi}$ as

$$Z_{\text{eff}} = A \int \mathrm{d}t \int \mathrm{d}\boldsymbol{R}\mathrm{d}\boldsymbol{r} |\boldsymbol{\Psi}(\boldsymbol{R}, \boldsymbol{r}, t)|^2 \delta(\boldsymbol{R} - \boldsymbol{r}), \tag{2}$$

where $\boldsymbol{R}$ and $\boldsymbol{r}$ are the coordinates of the positron and the electron in the target hydrogen. The normalization factor is taken to be $A = v$, where v is the speed of the positron, to satisfy $Z_{\text{eff}} = 1$ in the Born approximation in which the wave function is not distorted through the collision.

The TDCC wave function is given in a form

$$\boldsymbol{\Psi}(\boldsymbol{R}, \boldsymbol{r}, t) = \frac{\sqrt{\pi}}{kRr} \sum_{JLl} i^{L-1} \sqrt{2L+1} \psi_{Ll}^{JM_J}(R, r, t) \mathcal{Y}_{Ll}^{JM_J}(\hat{\boldsymbol{R}}, \hat{\boldsymbol{r}}), \tag{3}$$

where $\mathcal{Y}_{Ll}^{JM_J}$ is the angular momentum eigen function, k the wave number of the positron, J the total orbital angular momentum, and L and l the orbital angular momenta associated with $\hat{\boldsymbol{R}}$ and $\hat{\boldsymbol{r}}$. The radial part of the wave function at $t_0 = 0$ is given by the product of the ground-state wave function ϕ_{1s} of the hydrogen and an incoming wave packet g_{kL} of the positron,

$$\psi Y_{Ll}^{JM_J}(R, r, t_0) = g_{kL}(R) \phi_{1s}(r) \delta_{LJ} \delta_{l0}. \tag{4}$$

The wave packet is given by

$$g_{kL}(R) = \frac{1}{(\sigma^2 \pi)^{1/4}} \exp\left[-\frac{(R - R_0)^2}{2\sigma^2} \right] h_L^-(kR), \tag{5}$$

where σ and R_0 are the width and initial position of the wave packet and h_L^- is an asymptotic form of the spherical Hankel function.

In TDCC calculations, the radial function (4) is numerically described on nonuniform grid points of the two-dimensional radial space (R, r) used in [4]. The extent of the space is $R_{\max} = r_{\max} = 6\sigma$. In Equation (3), the upper limit of l is taken to be $l_{\max} = 15$ for $J = 0$–3. The width and initial position of the wave packet is taken to be $6\pi/k$ and $R_0 = R_{\max}/2$; the energy uncertainty is $\Delta E/E = 2\Delta k/k \simeq 10\%$. The wave function at t is obtained by solving the TDCC Equation (6) in [4].

3. Results and discussion

Figure 1 shows the effective electron number Z_{eff}. The present result has sufficiently converged with respect to the partial waves J within a few percents. The present result is in good agreement with previous results of a variational calculation by Reeth and Humberston [2] and a close-coupling calculation by Ryzhikh and Mitroy [6] within several percents. The S-wave ($J = 0$) contribution decreases as the positron energy increases; it is dominant for low energy. This decrease can be explained in terms of the hydrogen having less time to adjust to distortion of the positron field as the energy increases [2]. The higher partial-wave contributions increase as the energy increases. The semi-empirical model proposed by Laricchia and Wilkin [1] predicts that Z_{eff} is prominently enhanced as the energy approaches the threshold (6.8 eV) from 4 eV. However, such an enhancement is not found below 6.5 eV in the present result. The previous calculations show that more prominent enhancement of Z_{eff} at extremely vicinity of the threshold. This enhancement has been confirmed to lead to spurious divergence due to break down of a perturbative treatment of positron annihilation [6].

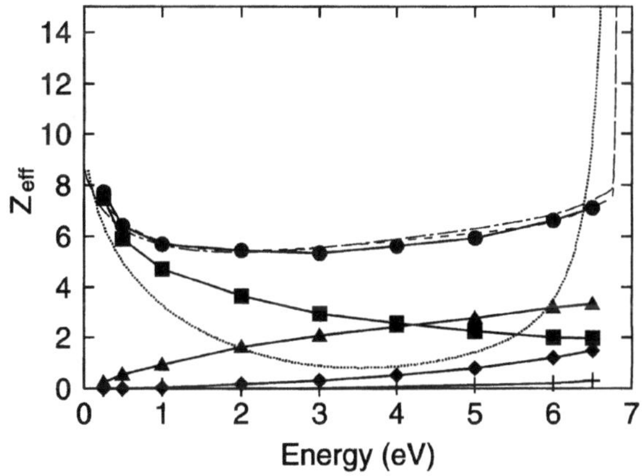

Figure 1. The effective electron number Z_{eff}. The solid lines with symbols represent the present result; the closed circle represents total of partial wave contributions of $J = 0$–3 and the closed square, the closed triangle, the closed diamond, and the plus are each partial wave contribution of $J = 0$–3. The dotted line represents the semi-empirical calculation by Laricchia and Wilkin [1]; the broken line, a variational calculation by Reeth and Humberston [2]; the dash-dotted line, a close-coupling calculation by Ryzhikh and J. Mitroy [6].

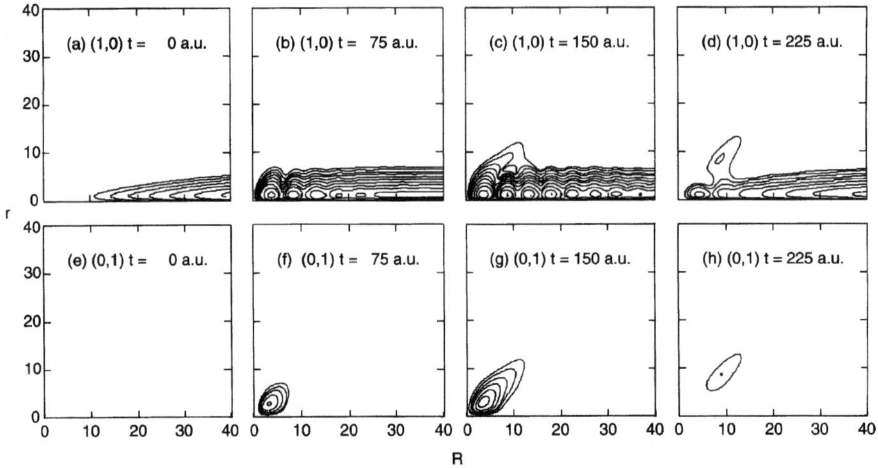

Figure 2. Time evolution of wave functions for $J = 1$ at the positron energy of 6.5 eV: (a)–(d) for the $(L, l) = (1, 0)$ channel and (e)–(h) for (0, 1) channel [7].

We show time evolution of the wave function for $J = 1$ in Figure 2 to examine contribution of virtual Ps formation [7]. We take the positron energy of 6.5 eV lower than the threshold, considering the wave packet has an energy uncertainty of about 0.7 eV around the threshold. In the figures, the wave density is drawn with contours in a logarithmic scale. Figure 2(a) shows the initial condition (4) in the $(L, l) = (1, 0)$ channel; the shape of the contours is determined by the width of the wave packet in the R direction and the mean radius of the ground-state hydrogen in the r direction. The incoming wave propagates to the left along the R axis as the time increases. The wave reflects at $r = 0$ and becomes an outgoing wave that propagates to the right. In Figure 2(b), interference of the incoming and outgoing waves is found as several nodes of the contours. In Figure 2(c), most components, which represent elastic scattering, propagate along the R axis. The wave is slightly distorted and a mound temporally stays around the origin. A very small component, which represents virtual Ps formation, extends to the right-diagonal. This component gradually disappears in Figure 2(d).

On the other hand, in the (0, 1) channel, the wave density is zero at t_0 by the initial condition (Figure 2(e)). As the time increases, a wave emerges in the vicinity of the origin (Figure 2(f)). It transfers from the (1, 0) channel through the dipole terms of the interaction. The wave emerged grows to a large mound and localizes around the origin. In Figure 2(g), a virtual Ps component extents to the right diagonal. However, similarly to the (1, 0) channel, it has very low amplitude and gradually disappears in Figure 2(h). Thus, the annihilation is dominated by the direct annihilation mechanism and contribution of virtual Ps formation is not important.

4. Conclusion

We have calculated the positron annihilation rate in positron–hydrogen collisions with the TDCC method. The present result is in agreement with previous results. We have investigated annihilation mechanism in the vicinity of the Ps formation threshold to demonstrate that contribution of virtual Ps-formation is not important below 6.5 eV.

References

1. Laricchia, G. and Wilkin, C., *Phys. Rev. Lett.* **79** (1997), 2241.
2. Van. Reeth, P. and Humberston, J. W., *J. Phys. B* **31** (1998), L231.
3. Yamanaka, N. and Kino, Y., *Phys. Rev. A* **64** (2001), 042715.
4. Yamanaka, N. and Kino, Y., *Phys. Rev. A* **65** (2002), 062709.
5. Charlton, M. and Humberston, J. W., *Positron Physics*, Cambridge University Press, 2001.
6. Ryzhikh, G. G. and Mitroy, J., *J. Phys. B* **33** (2000), 2229.
7. Corresponding animations are presented at http://nucl.phys.s.u-tokyo.ac.jp/~yam/tdcc/.

Hyperfine Interactions **146/147:** 343–347, 2003.

Measurement of the Strong Interaction Shift and Width of the Ground State of Pionic Hydrogen

B. MANIL[1,*], D. F. ANAGNOSTOPOULOS[2], S. BIRI[3], G. L. BORCHERT[4], W. BREUNLICH[5], J. P. EGGER[6], D. GOTTA[4], A. GRUBER[5], M. HENNEBACH[4], P. INDELICATO[4], T. JENSEN[7,8], Y. W. LIU[9], V. MARKUSHIN[9], N. NELMS[10], L. M. SIMONS[9] and H. ZMESKAL[5]

[1]*Laboratoire Kastler-Brossel, Unité Mixte de Recherche du CNRS n° C8552, Université Pierre et Marie Curie, 4 place Jussieu (Case 74), F-75252 Paris CEDEX 05, France; e-mail: manil@spectro.jussieu.fr*
[2]*Department of Material Science, University of Ioannina, Ioannina, Greece*
[3]*ATOMKI, Hungary Academie of Science, Debrecen, Hungary*
[4]*Institut für Kernphysik, Forschungszentrum Jülich, Jülich, Germany*
[5]*IMEP, Österreichisch Akademie der Wissenschaften, Wien, Austria*
[6]*Institut de Physique de l'Université de Neuchâtel, Neuchâtel, Switzerland*
[7]*University of Aarhus, Aarhus, Denmark*
[8]*ETH Zürich, Switzerland*
[9]*Paul Scherrer Institut, Villigen PSI, Switzerland*
[10]*Department of Physics and Astronomy, University of Leicester, England*

Abstract. An experiment to measure the strong interaction width (Γ_{1s}) and shift (ε_{1s}) of the ground state of pionic hydrogen (π^-p), in order to determine the isospin separated scattering lengths (a^+ and a^-) with a relative accuracy of better than 10^{-2}, has been undertaken at the Paul Scherrer Institute (PSI). This measurement is based on high-precision X-ray spectroscopy of the $np \to 1s$ transitions pionic hydrogen (π^-p). A preliminary analysis of the data taken in December 2000 is presented.

Key words: pionic atoms, strong interaction, X-ray spectroscopy.

1. Introduction and theoretical background

Bound pion–nucleon systems permit to test strong-interaction properties at low energies. Many experiments with these bound systems have been performed during the last twenty years [1–3], which allowed to obtain some accurate results [4–6]. Particularly, the pionic hydrogen atom, which is the simplest of these bound systems, offer a lot of experimental advantages in order to obtain an accurate determination of the hadronic scattering lengths a^h [7, 8]. These scaterring lengths provide an important test of chiral perturbation theory (χPT).

Pionic atom energies, hovewer, are dominated by the electromagnetic interaction of its constituents. The effect of the strong interaction is only significant if the wave functions of the pion and the proton have a large overlap. In the ground state of the pionic hydrogen atom, the strong interaction induces a broadening $\varepsilon_{1s} \approx 1$ eV and a shift $\Gamma_{1s} \approx 7$ eV for an electromagnetic binding energy $E_{1s} = 3238$ eV. These measured quantities are connected to the hadronic scattering lengths a^h, which describe respectively the $\pi^-\mathrm{p} \to \pi^-\mathrm{p}$ and $\pi^-\mathrm{p} \to \pi^0 n$ process, by Deser-type formulae [9, 10]

$$\frac{\varepsilon_{1s}}{E_{1s}} = -\frac{4}{r_B} a^h_{\pi^-\mathrm{p}\to\pi^-\mathrm{p}}(1+\delta_\varepsilon), \tag{1}$$

$$\frac{\Gamma_{1s}}{E_{1s}} = \frac{8\mathcal{Q}_0}{r_B}\left(1+\frac{1}{\mathcal{P}}\right)\left(a^h_{\pi^-\mathrm{p}\to\pi^0 n}(1+\delta_\Gamma)\right)^2, \tag{2}$$

where r_B is the Bohr radius of the pionic hydrogen atom ($r_B = 222.56$ fm), $\mathcal{Q}_0 = 0.142\ \mathrm{fm}^{-1}$ is a kinematic factor and $\mathcal{P} = (1.156 \pm 0.009)$ is the Panofsky ratio [11]. The electromagnetic corrections δ_ε and δ_Γ have recently been calculated with a model potential with an accuracy of about 0.5% [4]. The relations of the hadronic scattering lengths a^h with the isospin separated scattering lengths a^+ (isoscalar) and a^- (isovector) are given by

$$a^h_{\pi^-\mathrm{p}\to\pi^-\mathrm{p}} = (a^+ + a^-) \quad \text{and} \quad a^h_{\pi^-\mathrm{p}\to\pi^- n} = \sqrt{2}(a^-). \tag{3}$$

The shift and the width of the ground state in pionic hydrogen and deuterium atoms have been determined in a recent series of experiments of the *ETHZ-Neuchâtel-PSI* collaboration by measuring the $3p$–$1s$ transition at 2886 eV with a reflection-type crystal spectrometer [4]. The results improved the value for the strong interaction shift ε_{1s} by almost two order of magnitude compared to earlier works. In addition, the first accurate measurement of the width Γ_{1s} was obtained. But, the error in the width is still almost an order of magnitude bigger than the one in the shift and is strongly influenced by the very poor knowledge of accelerating mechanisms (contributions of Doppler broadening) during the de-excitation of the pionic atoms [12]. This excludes all determinations of the isospin separated scattering lengths with a %-level relative accuracy. To remedy that, a new experiment has been planned at PSI to measure accurately the ground state width (and shift) from the pionic hydrogen np–$1s$ transitions at several pressures between 3 and 15 bar and in liquid hydrogen [7]. This measurement of the width can be extracted from a simultaneous fit of all transitions, which lets the different Doppler broadenings vary freely. Using the fact that the strong interaction width is the same for all transitions, a common value can be obtained with an relative accuracy of about 2.5%.

2. Experimental set-up and data analysis

For this experiment, we work with a reflection Bragg crystal spectrometer using the following technical developments:

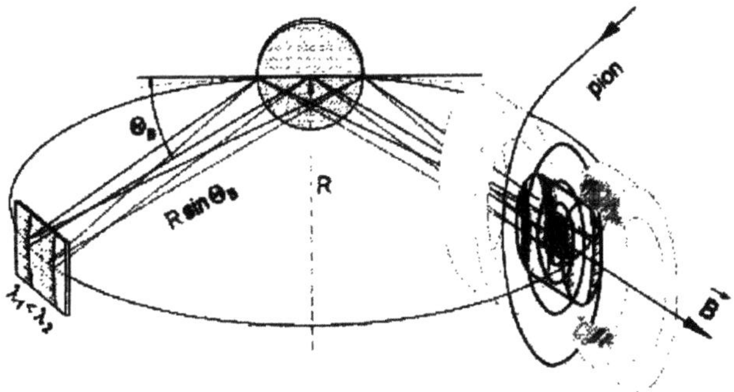

Figure 1. Principle of the experimental set-up. The apparatus is composed of three main components: a cyclotron trap, a Bragg crystal spectrometer, and a CCD cooling detector. Moreover, it is installed inside a concrete shielding used to suppress the high neutron induced background.

- A superconducting split coil magnet (cyclotron trap) increased the stop densities by two orders of magnitude for pions and allowed to work with an external cryogenic gaseous target [13].
- A Si(111) spherically crystal with a size of the order of 100 cm^2 and a bending radius of ~3 m.
- A CCD cooling detector for the detection of low energy X-rays with excellent, spatial and energy resolution [14]. The spatial resolution is decisive to reduce the background events recorded by the CCD. They are removed by using the large discrepancy between the topology of energy deposition by low-energy X-rays or by high-energy background events. These X-rays deposit their energy in one or two adjacent pixels, while background events deposit theirs in large clusters of pixels. The complex data processing that is required to extract the meaningful information from a very large number of pixel hits, is completely explained in [8, 15, 16].

The principle of this experimental set-up, already used for an improved pion mass measurement and pionic deuterium studies, is presented on the Figure 1 and is also described in detail in [8, 15].

3. Preliminary results

The first step of this experiment, we have obtain two beam times, in December 2000 (3 weeks) and in April–May 2001 (6 weeks) in order to optimize the beam injection, the cyclotron trap, and the gas targets, and to realize the preliminary data taking.

During the beam time of December 2000, we have also measured simultaneously, on the same spectrum, the pionic hydrogen and pionic oxygen X-ray transitions to provide an internal calibration, reduce the effect of temporal drift, and provide a reasonably accurate evaluation of the spectrometer response function (see Figure 2). This preliminary spectrum is not sufficient to obtain a precise measure-

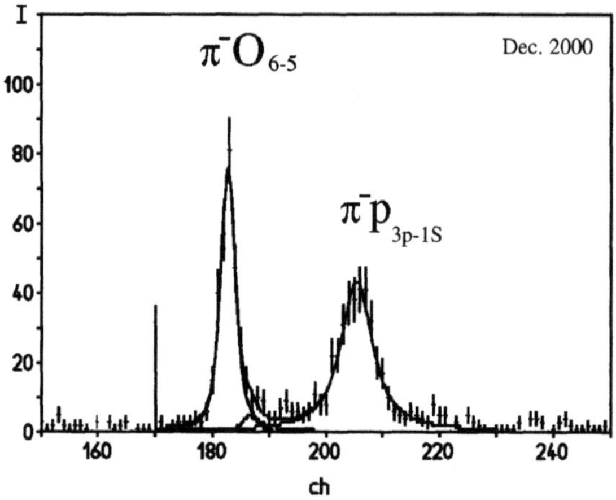

Figure 2. The $3p \to 1s$ transition of the pionic hydrogen calibrated by the $6h \to 5g$ transition of the pionic oxygen, which is close in energy and not influenced by the strong interaction. The vertical line, on the left of the picture, shows the position of the $\pi^{-}p(3p \to 1s)$ X-ray line without the effects of the strong interaction.

ment of the width. But we can determine, from this one, the shift with an accuracy comparable to the most recent experimental result [17]. For the second beam time, which has been accomplished only a couple of months ago, the data analysis has just started. Then, we do not discuss theses measurements in the present paper.

4. Conclusion and perspectives

This preliminary result is very encouraging. But, the still necessary increase in accuracy requires an additional effort. We must understand perfectly the different process of the atomic cascade. A simultaneous spectroscopy of pionic and muonic hydrogen atoms is planned as the X-rays transitions providing from the muonic atoms do not show any strong interaction broadening, but exhibit Doppler broadening similar to pionic atoms [18]. Moreover, we must also obtain an optimal determination the intrinsic response function of our spectrometer. This will be provided in the future by the use of electronic H-like ion X-ray lines produced by an Electron-Cyclotron-Resonance Ion Trap, built from the cyclotron trap and optimized for X-ray production.

References

1. Forster, A. *et al.*, *Phys. Rev. C* **28** (1983), 2374.
2. Bovet, D. *et al.*, *Phys. Lett. B* **153** (1985), 231.
3. Beer, W. *et al.*, *Phys. Lett. B* **261** (1996), 16.
4. Sigg, D. *et al.*, *Nucl. Phys. A* **609** (1996), 269.
5. Chatellard, D. *et al.*, *Nucl. Phys. A* **625** (1997), 855.

6. Hauser, P. *et al.*, *Phys. Rev. C* **58** (1998), 1869.
7. Anagnostopoulos, D. *et al.*, PSI proposal R-98-01.1, 1998.
8. Manil, B., Ph.D. thesis, University of Paris, 2001.
9. Deser, S., Goldberger, M. L., Baumann, K. and Thirring, W., *Phys. Rev.* **96** (1954), 774.
10. Rasche, G. and Woolcock, W. S., *Nucl. Phys. A* **381** (1982), 405.
11. Spuller, J. *et al.*, *Phys. Lett. B* **67** (1977), 479.
12. Leon, M. and Bethe, H. A., *Phys. Rev.* **127** (1962), 636.
13. Simons, L. M., *Hyp. Interact.* **81** (1993), 253.
14. Nelms, N. *et al.*, *Nucl. Inst. Meth. A*, to be published.
15. Borchert, G. L., Manil, B. *et al.*, *Hyp. Interact.* **132** (2001), 195.
16. Sigg, D., *Nucl. Instr. Meth. Phys. Res. A* **345** (1994), 107.
17. Schröder, H.-Ch., *Phys. Lett. B* **469** (1999), 25.
18. Markushin, V. E. and Jensen, T. S., *Hyp. Interact.*, to be published.

Author Index to Volumes 146/1–4 and 147/1–4 (2003)

MIX
Papier aus verantwortungsvollen Quellen
Paper from responsible sources
FSC® C105338
FSC
www.fsc.org

If you have any concerns about our products,
you can contact us on
ProductSafety@springernature.com

In case Publisher is established outside the EU,
the EU authorized representative is:
Springer Nature Customer Service Center GmbH
Europaplatz 3, 69115 Heidelberg, Germany

Printed by Libri Plureos GmbH
in Hamburg, Germany